Gross Electrical
Feed Drives for
Machine Tools

Electrical Feed Drives for Machine Tools

Edited by Hans Gross

Siemens Aktiengesellschaft
John Wiley & Sons Limited

Authors: Institut für Steuerungstechnik und Fertigungs-
Gottfried Stute einrichtungen an der Universität Stuttgart
Karl-Heinz Böbel (the Institute of Control Theory for
Jürgen Hesselbach Machine Tools and Manufacturing Equipment.
Ulf Hodel at University of Stuttgart)
Peter Stof

Translated by Daniela Vogelgesang

CIP-Kurztitelaufnahme der Deutschen Bibliothek
Electrical feed drives for machine tools / ed. by Hans Gross. [Authors: Gottfried Stute... Transl. by Daniela Vogelgesang]. – Berlin; München: Siemens-Aktiengesellschaft; Chichester: Wiley, 1983.
    Dt. Ausg. u.d.T.: Elektrische Vorschubantriebe
    für Werkzeugmaschinen
    ISBN 3-8009-1377-1 (Siemens-Aktienges.)
    ISBN 0-471-26273-0 (Wiley)
NE: Gross, Hans [Hrsg.]

British Library Cataloguing in Publication Data:
Electrical feed-drives for machine tools.
1. Machine-tools–Electric equipment
I. Gross, Hans
621.9′02   TJ1180
ISBN 0-471-26273-0

Library of Congress Cataloging in Publication Data:
Elektrische Vorschubantriebe für Werkzeugmaschinen.
    English.
    Electrical feed-drives for machine tools.
    Translation of: Elektrische Vorschubantriebe für Werkzeugmaschinen.
    Includes bibliographies and index.
    1. Machine-tools–Electric driving. I. Gross, Hans. II. Title.
TK4059.M32E4413 1983 621.9′02 83-9410   ISBN 0-471-26273-0 (Wiley)

Title of German original edition:
Elektrische Vorschubantriebe für Werkzeugmaschinen
Siemens Aktiengesellschaft 1981
ISBN 3-8009-1338-0

© 1983 by Siemens Aktiengesellschaft, Berlin and München; John Wiley & Sons Ltd., Chichester

All Rights Reserved. No part of this publication may be reproduced, stored in a retrieval system, or transmitted, in any form or by any means electronic, photocopying, recording or otherwise, without the prior permission of Siemens Aktiengesellschaft and John Wiley & Sons Ltd.

Printed in the Federal Republic of Germany

Preface

Due to the steadily increasing manufacturing and personnel costs, increasing productivity has recently become a stringent necessity. Electrical drives have considerably contributed to solving this problem in the area of the machine tool industry. They represent a cost-efficient engineering technique, which is economical even for use on feed units that were previously moved manually or with a central drive. Also, today's simple to operate, freely programmable numerical controls require independently controlled drives with a wide speed range.

Siemens has significantly contributed to this transition from mechanical and hydraulic systems to electrical ones, and has brought on the market highly dynamic converter-fed servo drives.

Due to their simple, robust construction, good dynamic properties, as well as their high durability, these drives are opened to a wide field of applications. The feed units of numerically controlled machine tools and tracing machines are positioned along programmed paths or with position control loops. The increasing automation puts ever stricter demands on these position control loops. The velocity and the accuracy of the feed movements can be significantly increased with the electrical drive.

For the further development of this engineering technique, Siemens AG sponsored special research studies centered on position control and feed drives at the University of Stuttgart, Institute of Control Theory for Machine Tools and Manufacturing Equipment. The theoretical technical basics for position control optimization have been worked out at this Institute, and procedures for practical measurements on applied position control loops have been developed. The long practical experience that Siemens has in this area was an important part of the project. The present volume summarizes the results of the scientific work at the University, and the practical experience of start-ups, design, development, and construction of these drives.

This book gives the basics necessary for the daily tasks and the procedures to be applied to make the right interpretations and decisions, for the staff in design and development offices, for sales and field engineers, and for service technicians.

Students of mechanical, electrical, and control engineering who would like to get aquainted with the area of feed drive technique, find here a comprehensive presentation of the different fields.

Erlangen/Stuttgart, April 1983

(Siegfried Waller)
Manager of the
Manufacturing Industries Division
of the Power Engineering and
Automation Group of Siemens AG

(Prof. Dr.-Ing. Gottfried Stute)
Manager of the Institute
of Control Theory for Machine Tools
and Manufacturing Equipment
at the University of Stuttgart

Contents

1	**Feed Drive and Position Control Loop**	13
1.1	Basics of Control Theory	13
1.1.1	Open and Closed Loop Control	13
1.1.2	Stationary Systems Behavior	14
1.1.2.1	Characteristic Curves	14
1.1.2.2	Linearization	15
1.1.2.3	Normalized Values	16
1.1.2.4	Referenced Values	18
1.1.3	Dynamic Systems Behavior	18
1.1.3.1	Differential Equations	18
1.1.3.2	Transient Response	22
1.1.3.3	Frequency Response Curve	26
1.1.3.4	Bode Diagram	29
1.1.3.5	Linkage of Transfer Elements	32
1.1.3.6	Frequency Response Curve Symbol for Control Loops	34
1.1.3.7	Experimental Determination of the Frequency Response Curve	35
1.2	Basics of the Position Control Loop	36
1.2.1	Construction and Functioning of the Position Control Loop	36
1.2.2	Localization	38
1.2.2.1	Measuring Value Detection	38
1.2.2.2	Measuring Procedures	39
1.2.2.3	Location of Measurement	39
1.2.3	Characteristic Values and Properties of Position Control Loops	40
1.2.3.1	Dynamic Systems Behavior of Single Control Loop Elements	41
1.2.3.2	Explanation of Terms	44
1.2.3.3	Behavior of Linear Position Control Loops	44
1.2.3.4	Non-linearities in the Position Control Loop	50
1.3	Position Loop Gain	52
1.3.1	Position Control Loop Model	52
1.3.2	Settling Time and Overshoot Width at Positioning	53
1.3.3	Ramp Distance for Thread Cutting	56
1.3.4	Contour Deviations on Corners	61
1.3.5	Contour Deviations on Circles	63
1.3.6	Disturbance Transient Response	66
1.3.7	Requirements for the Position Loop Gain	67
1.4	Bibliography	69
2	**DC Motors for Feed Drives**	70
2.1	Requirements for the Feed Drive	70
2.1.1	Requirements for Stationary Operation	70
2.1.2	Requirements for Non-stationary Operation	70

2.2	DC Shunt Motor	71
2.2.1	Construction Forms	71
2.2.1.1	Permanent Magnet-excited DC Servo Motors of the Series 1HU	71
2.2.1.2	Electrically Excited DC Servo Motors of the Series 1GS	74
2.2.2	Drive Behavior	76
2.2.2.1	Differential Equation and Block Diagram	76
2.2.2.2	Stationary Behavior of the Uncontrolled Drive	80
2.2.2.3	Dynamic Behavior of the Uncontrolled Drive	82
2.2.2.4	Dynamic Behavior of the Speed-controlled Drive	84
2.2.2.5	Calculating the Drive Nominal Angular Frequency	86
2.2.2.6	External Inertia Effect	92
2.2.2.7	Dead Time Effect	94
2.3	Selection of DC Servo Motors	96
2.3.1	Methods of Calculation	96
2.3.2	Stationary Load	97
2.3.2.1	Friction and Losses	98
2.3.2.2	Machining Force	100
2.3.3	Dynamic Load	102
2.3.4	Heating Behavior	111
2.3.4.1	Periodic Load Changes	111
2.3.4.2	Overload Behavior	112
2.3.4.3	Thermal Time Constants	116
2.3.5	Additional Optimization Criteria	118
2.3.5.1	Energy Content of Moved Masses	118
2.3.5.2	Attainable Acceleration	119
2.3.5.3	Drive Comparisons	121
2.3.6	Calculation Scheme	123
2.3.7	Summary of Calculation Formulas	132
2.4	Bibliography	133
3	**Current Converters for DC Servo Motors**	**134**
3.1	Notions and General Overview	134
3.1.1	Requirements	134
3.1.2	Dead Time	136
3.1.3	Form Factor	137
3.1.4	Efficiency	139
3.1.5	Costs	140
3.2	Transistor DC Choppers	141
3.2.1	Applications	141
3.2.2	Principle of Operation	141
3.2.2.1	Driving, Clockwise, I. Quadrant	142
3.2.2.2	Driving, Counterclockwise, III. Quadrant	144
3.2.2.3	Braking, Clockwise, IV. Quadrant	144
3.2.2.4	Braking, Counterclockwise, II. Quadrant	145
3.2.3	DC Bus	145
3.3	Line Synchronized Thyristor Controllers	146

3.3.1	Generalities	146
3.3.2	Rectifier and Inverter Operation	147
3.3.3	Circulating Curent-free Connection	149
3.3.4	Circulating Current-conducting Connection	150
3.3.5	6-Pulse Circulating Current-conducting Cross Connection	151
3.3.6	3-Pulse Circulating Current-conducting Anti-parallel Connection	153
3.3.7	6-Pulse Circulating Current-free Anti-parallel Connection	157
3.3.8	2-Pulse Circulating Current-conducting Anti-parallel Connection	159
3.4	Controllers	163
3.4.1	Structure of the Control Loop	163
3.4.2	Overcurrent Limitation	164
3.4.3	Subordinated Current Control	165
3.4.4	Speed Control with Current Limit	166
3.4.5	Status Monitor	168
3.4.6	Adaptive Speed Control Influence	169
3.5	Selection Criteria	172
3.5.1	Characteristic Values of Converter Circuits	172
3.5.2	Applications for Converter Circuits	172
3.6	Bibliography	175
4	**Design Versions of Mechanical Transmission Elements**	**177**
4.1	Generalities	177
4.2	Requirements for the Mechanical Transmission System	178
4.2.1	Nominal Angular Frequency	180
4.2.2	Stiffness	184
4.2.3	Damping Gradient	185
4.2.4	Non-linearities	189
4.2.4.1	Reversing Errors	189
4.2.4.2	Friction	190
4.2.4.3	Effects of Reversing Errors and Frictional Reversing Errors	194
4.2.4.3.1	Indirect Position Control Measurement	195
4.2.4.3.2	Direct Position Control Measurement	198
4.2.5	Moment of Inertia	202
4.2.5.1	Cylindrical Bodies	202
4.2.5.2	Linearly Moved Masses	204
4.2.5.3	Gear Ratio	205
4.3	Calculating the Slide Elements of the Drive	205
4.3.1	Feed Screw Drives	205
4.3.1.1	Construction and Requirements	205
4.3.1.2	Feed Screw	208
4.3.1.2.1	Stiffness of the Feed Screw	209
4.3.1.2.2	Moment of Inertia of the Feed Screw	213
4.3.1.2.3	Efficiency	215
4.3.1.2.4	Feed Screw Lead	216
4.3.1.2.5	Lead Accuracy	216
4.3.1.2.6	Critical Speed	217

4.3.1.2.7	Buckling Strength	218
4.3.1.3	Ball Screw Nut	220
4.3.1.4	Feed Screw Bearings	223
4.3.1.4.1	Stiffness and Preload	224
4.3.1.4.2	Effect of the Bearing Mounting Elements	226
4.3.1.4.3	Friction and Temperature Behavior	227
4.3.2	Rack and Pinion Drives	229
4.3.2.1	Range of Applications, Influencing Variables	229
4.3.2.2	Stiffness	230
4.3.2.2.1	Torsional Stiffness of Shafts	230
4.3.2.2.2	Bending Strength of Shafts	231
4.3.2.2.3	Radial Stiffness of Shaft Bearings	231
4.3.2.2.4	Bending Strength of the Gear Wheel, Pinion, and Rack Teeth	233
4.3.2.2.5	Example	233
4.3.2.3	Preload at the Pinion	236
4.3.3	Machine Slide and Guides	238
4.3.3.1	Machine Slide	238
4.3.3.2	Table Guides	239
4.3.3.2.1	Requirements	239
4.3.3.2.2	Friction Guides	240
4.3.3.2.3	Roller Guides	241
4.3.3.2.4	Hydrostatic Guides	243
4.3.3.2.5	Aerostatic Guides	244
4.3.3.3	Damping	245
4.4	Gears	247
4.4.1	Overview of Applications	247
4.4.2	Requirements	251
4.4.2.1	Moment of Inertia	251
4.4.2.2	Torsion and Bending	253
4.4.2.3	Reversing Errors	254
4.4.3	Toothed Belt Gear	255
4.4.4	Advantage Limits of a Gear Ratio	259
4.4.4.1	Motors of Equal Frame Size	259
4.4.4.2	Motors of Different Frame Size	261
4.4.4.3	Disturbance Transient Response	264
4.4.4.4	Comparison of two Motors	264
4.5	Couplings	266
4.6	Bibliography	268
5	**Command Value Modification**	271
5.1	Purpose of the Command Value Modification	271
5.2	Smoothing the Command Values	271
5.3	Effects of Command Value Smoothing	275
5.3.1	Stress and Contour Deviation Characteristics	275
5.3.2	Contour Movements	276
5.3.3	Diagrams for Contour Deviation and Stress Characteristics	279

5.4	Determining the Command Value for Acceleration	284
5.5	Conclusions	287
5.6	Bibliography	287
6	**Measurements on Feed Drives**	**288**
6.1	Substantiation, Aim	288
6.2	Measuring Procedures	289
6.2.1	Measurements on the Mechanical Transmission Elements	289
6.2.1.1	Reversing Error (Back Lash)	290
6.2.1.2	Total Stiffness, Nominal Angular Frequency	295
6.2.1.3	Friction Characteristics	296
6.2.2	Measurements on the Speed Control Loop	298
6.2.2.1	Step Response, Transient Response	299
6.2.2.2	Current Limit	301
6.2.2.3	Motor Heat	301
6.2.2.4	Nominal Angular Frequency of the Feed Drive	302
6.2.3	Measurements on the Position Control Loop	304
6.2.3.1	Contour Deviations	304
6.2.3.2	Measuring the Position Loop Gain	306
6.2.3.3	Measurements for the Smallest Position Increments	308
6.2.3.4	Positioning Response	309
6.3	Measuring Values of Machines in Use	310
6.3.1	Evaluation of Measured Frequency Response Curves	310
6.3.2	Measurements on a Timing Belt Gear	311
6.3.3	Effect of the Moment of Inertia	314
6.3.4	Motors in Short Version	314
6.3.5	Effects of Optimization of the Speed Control	316
6.3.6	Frequency Response Curve of a Shell Magnet Motor	316
6.3.7	Influence of the Converter Circuit	318
6.3.8	Frequency Response Curve of an Electrically Excited DC Servo Motor	320
6.3.9	Conclusions	321
7	**Technical Data**	**322**
7.1	DC Servo Motors	322
7.1.1	General Data	322
7.1.2	Design Series 1 HU	325
7.1.3	Design Series 1GS 3	331
7.2	Current Converter Units	334
7.2.1	Transistor DC Choppers	334
7.2.2	Line-synchronized Thyristor Controllers	336
8	**Technical Appendix**	**339**
8.1	Symbols and Units Used	339
8.2	SI Units	350
8.3	Conversion Tables	353
8.4	Equations	356
	Index	358

Introduction

In order to understand the characteristics of feed drives in position control loops on machine tools, a basic knowledge of control theory is necessary. The control technical procedures useful in explaining these characteristics in context are presented in the first chapter of this book. They represent a rapid refresher course for the expert and give the interested reader fast access to the subject. A comprehensive bibliography is also listed.

In the further chapters, sophisticated mathematical computations are purposefully omitted in favor of the methods and presentations commonly used in practice. These are based on the *frequency response curve* with the nominal angular frequency of a transfer element. It is presented with the *Bode diagram*, which can be easily determined with simple measurements by means of a frequency generator and a recorder or a storage oscilloscope. The advantage of the frequency response curve method is that algebraic calculation procedures can still be used for complex transfer control systems without mastering infinitesimal calculation; only knowledge of time differentials and integrals is necessary.

The inclusion of the newest standards (DIN) allows on one hand, the establishment of a consistent terminology, and on the other hand, comparisons between different drive concepts. The basis for the units is the SI-system. Relationships are shown by dimensional equation, supplemented in some cases with adjusted dimensional equations. The conversion tables presented in the technical appendix allow the engineer to make simple conversion from the formerly used units into SI-units.

1 Feed Drive and Position Control Loop

1.1 Basics of Control Theory

1.1.1 Open and Closed Loop Control

Open loop control is to be understood as the effect of information on an energy or material flow.

Here, *informations* mean instructions or statements subject to storage, transmission, and processing.

The *signals* are the carriers of informations. The information is presented as the value or the value sequence of the signal parameter (DIN 44300), e.g. of the amplitude.

The *closed loop control* differs from the open loop control in that the control variable x is measured, and the adjustment of the energy flow is dependent upon the control deviation between command value w and control parameter x. The principle presentation of the control loop is shown in fig. 1.1. The most important terms can be derived from it (DIN 19221 and DIN 19226).

The closed action loop is characteristic of the closed loop control. Sections of the control loop, like the *control device* and the *control system,* respectively their subdivisions, are designated as *control loop elements,* and are represented by rectangular blocks. The representation of connections between blocks results in the *block diagram.* The *input values u,* respectively the input signals, act upon the single elements of the control loop. The signals derived from them for further processing, also known as *output variables v,* are determined by the dynamic systems behavior of the single elements (see DIN 19229).

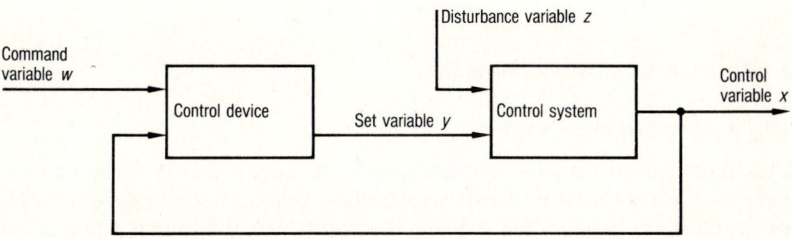

Fig. 1.1 Control loop elements

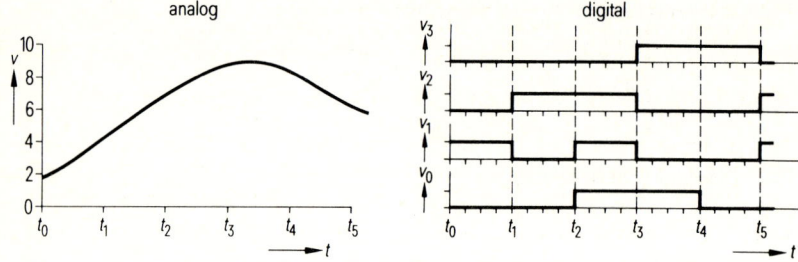

Fig. 1.2   Presentation of analog and digital signals

There is a distinction to be made here between static and dynamic systems behavior. The *static behavior* can be described, e.g. through characteristic curves. This marks the behavior in a stationary state, after all the transient responses have subsided. The *dynamic behavior* designates the time response of an element, and refers primarily to responses.

The flow of signals, respectively variables, is shown by means of lines with directional arrows. These lines represent the connections between individual blocks in a block diagram. The signals themselves can be either analog or digital. Examples are shown in fig. 1.2.

Analog signals can have steady or non-steady courses. An analog acting element or system is characterized by the fact that it reconstructs an analog output signal out of an analog input signal. No linear dependency is necessary between the two signals. Examples of analog systems are potentiometers with linear, logarithmic, cosine-, or sinusoidal characteristic curves.

Digital signals consist of characters, which can only assume discrete states. Examples of such characters are numbers. The *binary signal* belongs to this category; it is a signal with only two states. A digital acting system assigns digital output signals to corresponding digital input signals.

1.1.2   Stationary Systems Behavior

1.1.2.1   Characteristic Curves

The transfer elements of a system can be sorted according to their behavior in the stationary state. Fig. 1.3 shows stationary characteristic curves of transfer elements. If the relationship between the input and the output value of an element is a steady function, the total element is designated as steady; if the relationship is not steady, the total element is considered unsteady.

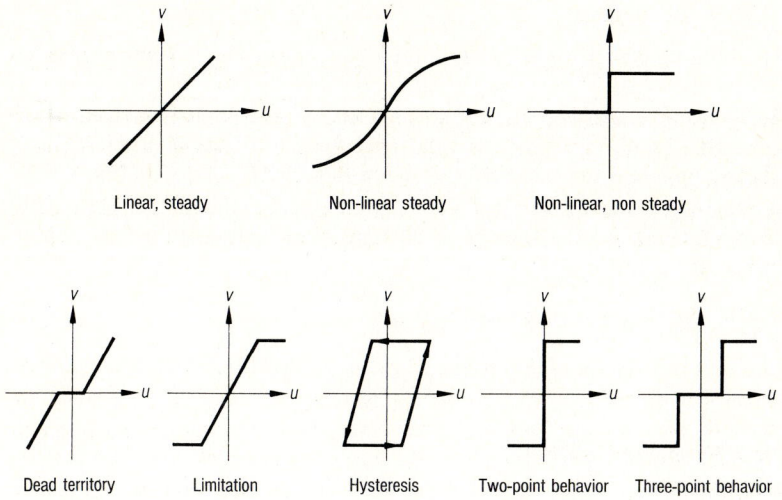

Fig. 1.3 Stationary characteristic curves of transfer elements

A special case of the steady characteristic curve is the linear characteristic curve. For systems consisting of transfer elements with linear characteristic curves, there is an extensive theory which describes the transient response completely. The mathematical treatment of systems containing transfer elements with non-linear characteristic curves is very difficult, and is presented mostly with approximating procedures.

1.1.2.2 Linearization

The theory of linear systems can also be used in connection with non-linear systems with steady characteristic curve. The prerequisite for this is that the proposed work-range should be small, i.e. only small deviations from a fixed operating point would be allowed. The steady but non-linear characteristic curve will be replaced by the tangent, at the operating point; hereby the operating point-dependent gain as an operating point-dependent relation value between the input and the output value is derived. Fig. 1.4 shows as an example the determining of the operating point-dependent gain for a DC generator, and presents it graphically, together with the time behavior, in a block diagram.

The starting point for the determination of the gain is the excitation characteristic curve $U_G = U_G(U_E)$ for the DC generator (fig. 1.4a).

15

The individual steps are:

1. Establishing the *operating point* on the characteristic curve, with coordinates U_{0G} and U_{0E}.
2. Introducing a new coordinates system whose zero point is located in the operating point, and which is valid only for the deviation in question. In the example, it is the coordinate system with axes ΔU_E and ΔU_G (fig. 1.4b).
3. Linearizing in the operating point, i.e. approximating the characteristic curve through the tangent. The slope of the tangent indicates the operating point-dependent gain

$$K = \frac{\Delta U_G}{\Delta U_E} = \frac{U_{0G}}{U_{HE}} \tag{1.1}$$

U_{HE} is an auxiliary variable determined by the intersection point of the tangent with the abscissa.

1.1.2.3 Normalized Values

To normalize, means to refer a value to a certain characteristical value of the same magnitude, in order to render that value dimensionless and thus obtain manageable numerical values. This procedure significantly simplifies the mathematical treatment of control technical problems. The generator voltage U_G, for instance, can be derived dependently on the speed n, at constant excitation, as $U_G = c_E \cdot n$ where c_E is the excitation constant, depending on the type of generator.

If the reference value selected for the speed n is the nominal speed n_0, and with $U_{0G} = c_E \cdot n_0$ for the normalized generator voltage dependent on the normalized speed, we will get

$$\frac{U_G}{U_{0G}} = \frac{n}{n_0} \tag{1.2}$$

The machine-specific constant c_E is thus eliminated.

The normalization can also be used for systems which are linearized. One example of this might be the DC generator with its corresponding non-linear characteristic curve, presented in fig. 1.4. In order to normalize the operating point-dependent gain, the auxiliary coordinates ΔU_E and ΔU_G as well as the original coordinates U_G and U_E must be normalized. The reference values are the rated voltages U_{0E} and U_{0G} of the operating point. Thus, one obtains the normalized auxiliary coordinate system with the axes $\frac{\Delta U_E}{U_{0E}}$ and $\frac{\Delta U_G}{U_{0G}}$. From the original coordinate system one obtains the normalized coordinate system with the normalized coordinate axes $\frac{U_E}{U_{0E}}$ and $\frac{U_G}{U_{0G}}$ (fig. 1.4c). The normalized gain at the operating point, derived, is:

$$K_N = \frac{\Delta U_G / U_{0G}}{\Delta U_E / U_{0E}} = K \frac{U_{0E}}{U_{0G}} \tag{1.3}$$

Physical equivalent diagram

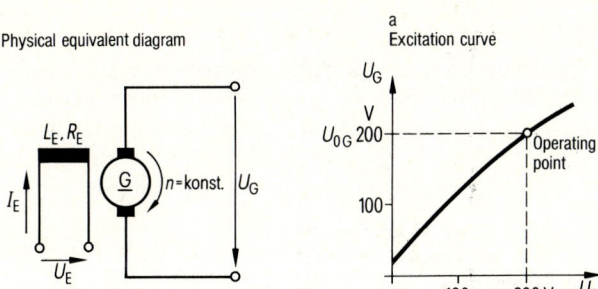

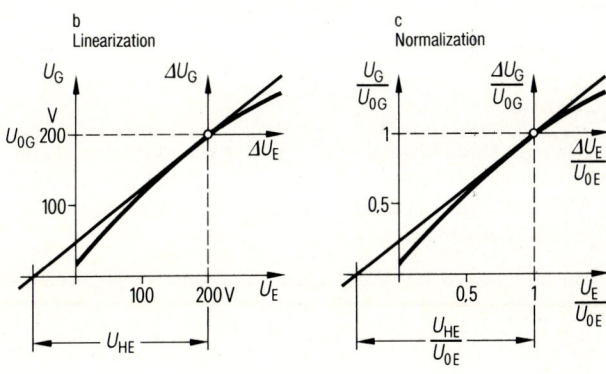

Block diagram

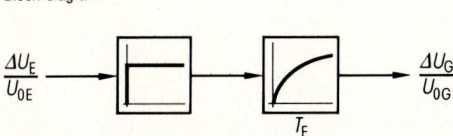

Fig. 1.4
Linearization and normalization in reference to the DC generator example

1.1.2.4 Referenced Values

Referenced values are quotients of values of different types belonging to tangible matter or bodies (DIN 5490). Examples are the referenced corner deviation $\dfrac{60 \cdot e_{max\,E}}{v_B/\omega_{0\,A}}$ and the referenced position loop gain $K_v/\omega_{0\,A}$ (see fig. 1.32). The referenced values are used in diagrams, to reduce the number of parameters.

1.1.3 Dynamic Systems Behavior

The time behavior of a system shows the way the output signal follows a variable input signal over time. The time behavior can be deduced by different methods:

▷ as described by differential equations
▷ as described by the transient response
▷ as described by the frequency response curve equations, respectively their graphic presentation, e.g. in the Bode diagram or Nyquist plots.

1.1.3.1 Differential Equations

Differential equations are the mathematical models of physical transfer elements and systems. With given input values and boundries, the corresponding output values can be calculated with their help, by solving the differential equation or the system of differential equations involved.

The first column of fig. 1.5 shows the differential equations of the so-called basic transfer elements.

Along with the differential equations, this figure also shows the further description possibilities given by the transient response (see section 1.1.3.2) and the frequency response curve $F(j\omega)$ (see section 1.1.3.3).

Examples for the basic transfer elements described in fig. 1.5.

Electrical Proportional Element

Ideal amplifier with resistor circuit (see fig. 1.6).

The output voltage is

$$U_2 = -\frac{R_2}{R_1} U_1 = -K \cdot U_1 \qquad (1.4)$$

with the gain

$$K = \frac{R_2}{R_1} \qquad (1.5)$$

Type of behavior	Differential equation		Transient response for the step function as an input value		Freq. response curve equation $F(j\omega)$	Type of element
	System equation		Solution equation	Graph		
Proportional	$v = K \cdot u$		$v = K \cdot u_s$	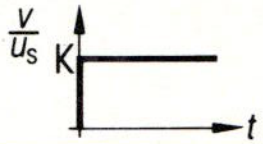	K	P-element
Proportional with 1st order delay	$T\dfrac{dv}{dt} + v = u$		$v = u_s(1 - e^{-\tfrac{t}{T}})$	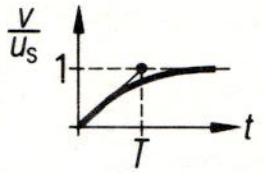	$\dfrac{1}{1 + j\omega T}$	PT_1-element
Proportional with 2nd order delay	$T^2\dfrac{d^2 v}{dt^2} + 2DT\dfrac{dv}{dt} + v = u$ $D < 1$		$v = u_s\left\{1 - e^{-D\tfrac{t}{T}}\left[\cos\left(\dfrac{t}{T}\sqrt{1 - D^2}\right)\right.\right.$ $\left.\left.+ \dfrac{D}{\sqrt{1 - D^2}}\sin\left(\dfrac{t}{T}\sqrt{1 - D^2}\right)\right]\right\}$	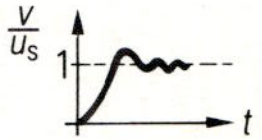	$\dfrac{1}{1 + j\omega 2DT + (j\omega)^2 T^2}$	PT_2-element
Integral	$T\dfrac{dv}{dt} = u$		$v = \dfrac{t}{T} \cdot u_s$	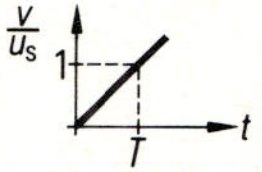	$\dfrac{1}{j\omega T}$	I-element
Differential	$v = T\dfrac{du}{dt}$		$v = T\dfrac{du_s}{dt}$	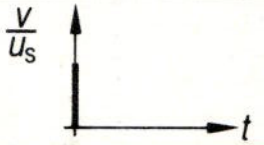	$j\omega T$	D-element
Dead time	$v = u(t - T_T)$		$v = u_s(t - T_T)$	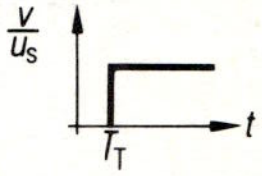	$e^{-j\omega T_T}$	Dead time element

v = output value u = input value u_s = input step

Fig. 1.5   Summary of the basic transfer elements

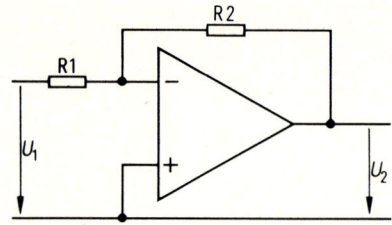

Fig. 1.6
Ideal amplifier with resistor network in input and feedback

At a stepwise change in the input voltage U_1, the output voltage U_2 also executes a stepwise change.

Electrical Proportional Element with 1st Order Delay

The DC generator of fig. 1.4, with an operating point-dependent gain $K=\dfrac{U_{0G}}{U_{HE}}$ and excitation time constant $T_E=\dfrac{L_E}{R_E}$. Following applies:

$$\Delta U_G = K \cdot \Delta U_E (1 - e^{-\frac{t}{T_E}}) \tag{1.6}$$

At a step change in the excitation voltage U_E by ΔU_E, the output voltage follows according to an exponential function to the value $U_G + \Delta U_G$ (see fig. 1.10).

Mechanical Proportional Element with 2nd Order Delay

Spring-mass system on a feed drive, according to fig. 1.7.

For an impulse-type deflection due to force F, the slide position shows a transient response.

As shown in section 1.1.3.4 for the 2nd order delay, and in section 4.2 for the mechanical transmission elements, the transient response of this transfer element is characterized by

the nominal angular frequency $\omega_0 = \dfrac{1}{T} = \sqrt{\dfrac{k}{m}}$

and the damping gradient $D = \dfrac{1}{2} c_v \sqrt{\dfrac{1}{k \cdot m}}$

k spring constant
m mass
c_v damping coefficient

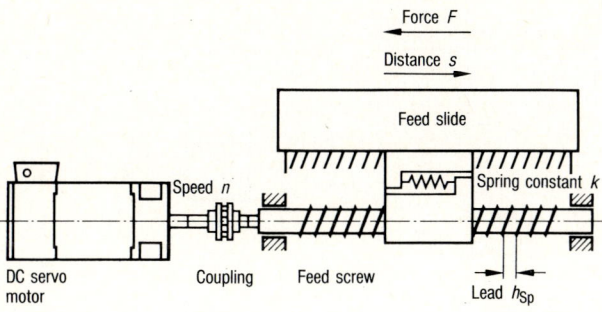

Fig. 1.7   Feed unit

Mechanical Integral Element

The transformation of a feed screw speed n, into the slide movement s, on a feed unit (see fig. 1.7).

The output value for the slide movement s is

$$s = h_{Sp} \cdot \int n \cdot dt \tag{1.7}$$

At constant input value, feed screw speed n, the distance is

$$s = s_0 + h_{Sp} \cdot n \cdot t \tag{1.7.1}$$

i.e. the slide movement increases linearly from the start value s_0, over time t.

Electrical Differential Element

Approximation: the charging current i_C of a capacitor with capacitance C dependent on the voltage U_C.

$$i_C = C \cdot \frac{dU_C}{dt} \tag{1.8}$$

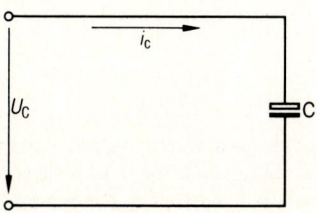

Fig. 1.8
Ideal capacitor on DC voltage

21

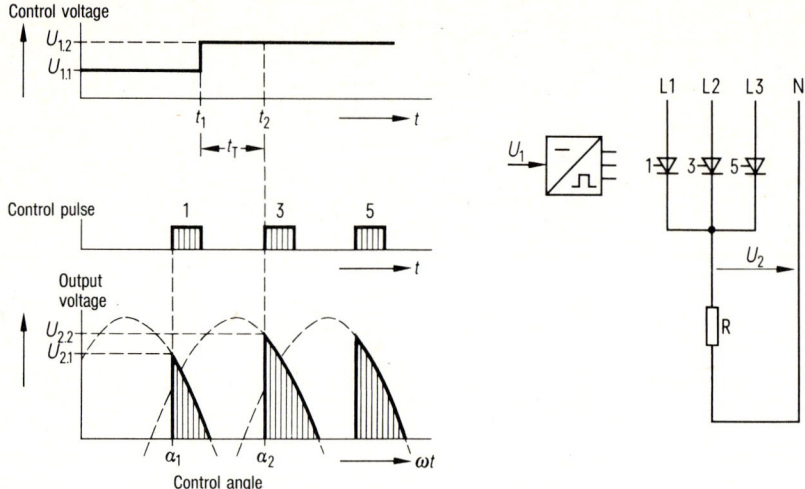

Fig. 1.9 Thyristor control of a 3-pulse current converter circuit

At a stepwise voltage change, charging current i_C for an ideal capacitor is theoretically infinite. In practice, it results in an impulse of limited amplitude and exponentially declining edge.

Electrical Dead Time Element

Approximation: a current converter with line synchronized gate control.

A control pulse on the thyristor can result in current flow only when the voltage is positive relative to the direction of conduction. For instance, according to fig. 1.9, if a control voltage change from $U_{1.1}$ to $U_{1.2}$ occurs at time t_1, the output voltage U_2 will change only at time t_2. The control angle shift from α_1 to α_2 becomes effective only with the control of the following thyristor 3. The dead time is $t_T = t_2 - t_1$. (For dead times of current converter circuits, see table 2.3 on page 95.) The following proportionally applies for the course of the output voltage over time:

$$U_2(t) \sim U_1(t + t_T) \tag{1.9}$$

1.1.3.2 Transient Response

The *transient response* indicates the evolution over time of the output value of a system, in response to random over time changes in the input value. It is a specific solution to the differential equation, used to describe the system.

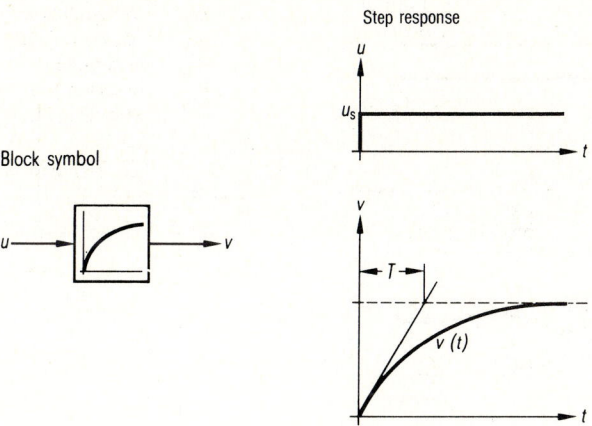

Fig. 1.10
Description of the time behavior of a 1st. order delay element (PT_1-element), through the step response

A characteristical input function is, for instance, the *step function*. With this function applied to a system, the output will show the step response, respectively the transient response.

The *step response* is the development over time in the output signal, as a response to a stepwise change in the input signal. If the change in the output signal is referred to the step amplitude of the input signal, the result is the *transient response function* (referenced step response, DIN 19226). It is used very often for the graphic description of transient responses of transfer elements in block symbols.

Figure 1.10 shows the step response as well as the transient response in the block symbol for a 1st. order delay element.

Transient responses are especially apparent when the transfer elements, respectively systems, are switched on or off. One way of determining the time behavior of a system whose differential equation is unknown, is by evaluating its transient response. For this, the input signal must be varied step-by-step, and the output signal must be recorded.

With this method, for instance, one can check the setting of the control parameters of a control loop; the relevant values are according to fig. 1.11 (also see DIN 19226/DIN 19229):

The *setting tolerance;* represents the difference between the largest and smallest agreed-upon allowable deviation of the control variable, and the limit value.

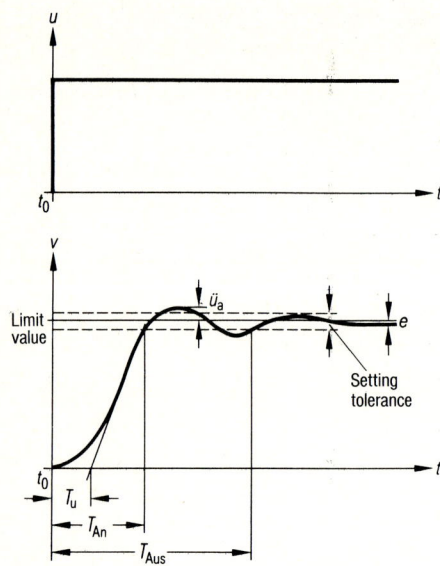

Fig. 1.11
Characteristic values of the step response of a transfer element, respectively system

The *overshoot width* $ü_a$ of the control variable; is the maximal deviation of the step response from the limit value, after initially exceeding one of the limits of the response tolerance.

The *response time* T_{An}; represents the time elapsed between time t_0, and the time when the step response reaches the limit of the setting tolerance for the first time.

The *settling time* T_{Aus}; represents the time elapsed between time t_0, and the time at which the control parameter reaches the limit of the setting tolerance for the last time before settling within its limits.

The *delay time* T_u; represents the time elapsed from the beginning of the input signal step t_0, up to the intersection of the first inflectional tangent with the time axis.

The *end deviation e*; is the remaining deviation of the control variable from the limit value, in the stationary state.

As an example of transient response function, the switch-on of a DC shunt motor is shown.

According to fig. 1.12, at time $t=0$, the DC voltage U_A is put to the terminals of the externally excited DC shunt motor. The presumptions are:

1. The electrical time constant is negligeably small, relative to the mechanical one.
2. The load torque is zero, so that an "idle start" can be assumed.

The factor to be determined is the course of motor speed over time $n_M = f(t)$, or normalized with the maximal motor speed $\dfrac{n_M}{n_{\max M}} = f(t)$.

The notions used are:

M_{StM} short-circuit torque of the motor
M_M motor torque
J_M motor inertia
T_{mechM} mechanical time constant of the motor
n_M motor speed
$n_{\max M}$ motor maximal speed
ω_M motor angular velocity

The resulting torque characteristic curve and the equations for the feed drive are presented in section 2.2.2. They also apply to DC shunt motors alone (replacing indices A through M). The presumptions under which the solution to the differential equation (2.14) is given, are that the load torque is zero and the initial condition at $t=0$, is $n_M/n_{\max M} = 0$.

$$\frac{n_M}{n_{\max M}} = 1 - e^{-\frac{t}{T_{mech\,M}}} \tag{1.10}$$

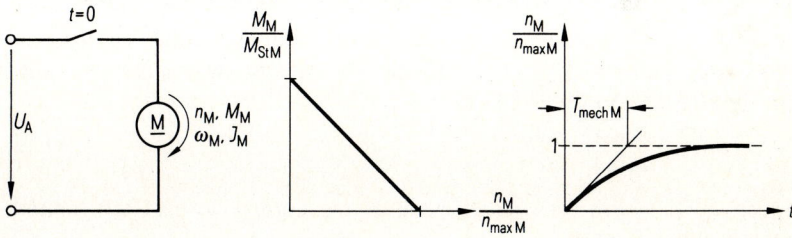

Fig. 1.12   Switch-on of a DC shunt motor

The speed course follows a natural exponential function with time constant $T_{\text{mech M}}$, which can be calculated from the motor data. It conforms to the mathematical considerations of equation (2.13), respectively (2.13.1):

$$T_{\text{mech M}} = J_M \frac{\omega_{\text{max M}}}{M_{\text{St M}}} = 2\pi \frac{J_M \cdot n_{\text{max M}}}{M_{\text{St M}}} \tag{1.11}$$

The normalized motor speed reaches after $t = 3\, T_{\text{mech M}}$, the value 0.95.

1.1.3.3 Frequency Response Curve

The time behavior of linear transfer elements or systems can also be described by using the *frequency response curve F*, for all frequencies between 0 and ∞. It also characterizes unequivocally the behavior of the transient response. In order to determine the frequency response curve, the system must be excited with a sinusoidal input value. The output signal is a sinusoidal oscillation whose amplitude and phases differ from the values of the input signal, as a function of the frequency.

For the input variable

$$u(t) = \hat{u} \cdot \sin \omega t \tag{1.12}$$

For the output variable

$$v(t) = \hat{v} \cdot \sin(\omega t + \varphi) \tag{1.13}$$

This means that relative to the input value of amplitude $\hat{u}$, the output value of amplitude $\hat{v}$ has the phase shift φ.

According to EULER [1.1], the relations (1.12) and (1.13) can also be expressed in complex terms. In these terms, the input and output values are not regarded as time variants $u(t)$ respectively $v(t)$ any longer, but are treated as frequency variant values $u(j\omega)$, respectively $v(j\omega)$. j indicates the imaginary part of complex figures $j = \sqrt{-1} \cdot j\omega$ is designated as the imaginary angular frequency.

The relationship between the frequency variant input and output variables can be determined from the complex frequency response curve equation, called for short frequency response curve:

$$F(j\omega) = \frac{v(j\omega)}{u(j\omega)} = \frac{\hat{v}(\omega)}{\hat{u}(\omega)} \cdot e^{j\varphi(\omega)} \tag{1.14}$$

Equation (1.14) can also be presented as the normal form of the complex number and it becomes:

$$F(j\omega) = X(\omega) + j \cdot Y(\omega) \tag{1.15}$$

where $X(\omega)$ is the real, and $Y(\omega)$ is the imaginary part.

The frequency response curve absolute value thus becomes:

$$|F(j\omega)| = \sqrt{X^2(\omega) + Y^2(\omega)} = \frac{\hat{v}(\omega)}{\hat{u}(\omega)} \tag{1.16}$$

The phase shift between the input and the output signals is calculated to:

$$\varphi(\omega) = \arctan \frac{Y(\omega)}{X(\omega)} \tag{1.17}$$

The frequency response curve equation of a linear system is determined by substituting the oscillation equations (1.12) and (1.13) in the differential equation refering to the transfer element.

If the differential equation is known, it can be very easily formally derived. If it is assumed that the starting conditions of the process are at zero, we replace in the differential equation the differentiation to the n'th time:

$\dfrac{d^n v(t)}{dt^n}$ is substituted by expression $(j\omega)^n \cdot v(j\omega)$, and

$\dfrac{d^n u(t)}{dt^n}$ is substituted by expression $(j\omega)^n \cdot u(j\omega)$, $\tag{1.18}$

and then we create the ratio $v(j\omega)/u(j\omega)$. This relationship between the *time range* and the *frequency range* is presented in tabular form in (1.19):

| Time range | Frequency range | |
differential equation	Frequency response curve equation	Transfer function
$\dfrac{d..}{dt}$	$\mathrel{\hat=} j\omega$	$\mathrel{\hat=} p = s$
$\int ... dt$	$\mathrel{\hat=} \dfrac{1}{j\omega}$	$\mathrel{\hat=} \dfrac{1}{p} = \dfrac{1}{s}$

(1.19)

Besides the frequency response curve, it also gives the transfer *function* as defined by DIN 19229. In this context, the transfer behavior is regarded as being dependent on the *complex angular frequency* p, also designated s in the literature. The presentation and computation of the transient response function in the time range is omitted here. Solutions, e.g. through the Laplace transformation and expansion into partial fractions, can be found in the specialty literature [1.2], [1.3], [1.4]. A very clear presentation of the Laplace transform method for the practical engineer is given in [1.5].

The description of the time behavior of a transfer element or system with the frequency response curve equation simplifies the mathematical treatment, and also offers the possibility to present the transfer behavior graphically, with the help of the frequency characteristic curves or the Nyquist curves.

As an example of how to derive the frequency response curve equation, the proportional element with 1st order delay is considered here:

The differential equation for this element in the time range, is as in fig. 1.5

$$T\frac{dv(t)}{dt}+v(t)=u(t) \tag{1.20}$$

If, according to rules (1.18) and (1.19), $\frac{d}{dt}$ is replaced by $j\omega$, and the time variant values $v(t)$ and $u(t)$ are replaced by the frequency variant values $v(j\omega)$ and $u(j\omega)$, in the frequency range we obtain:

$$v(j\omega)\cdot T\cdot j\omega+v(j\omega)=u(j\omega) \tag{1.21}$$

From this, according to equation (1.14), we derive the frequency response curve to:

$$F(j\omega)=\frac{v(j\omega)}{u(j\omega)}=\frac{1}{1+j\omega T} \tag{1.22}$$

The absolute value of the frequency response curve is determined by converting the equation (1.22) into the normal of the complex number, and by substituting the real and imaginary parts in equation (1.16):

$$|F(j\omega)|=\frac{\hat{v}(\omega)}{\hat{u}(\omega)}=\frac{1}{\sqrt{1+(\omega T)^2}} \tag{1.23}$$

The phase angle φ is calculated with equation (1.17):

$$\varphi(\omega)=-\arctan(\omega T) \tag{1.24}$$

If the inverse value of the time constant T is equaled to the so-called nominal angular frequency ω_0,

$$\omega_0=\frac{1}{T} \tag{1.25}$$

then the frequency response curve of the first order delay element will be:

$$F(j\omega)=\frac{1}{1+j\omega\dfrac{1}{\omega_0}} \tag{1.22.1}$$

The absolute value of the frequency response curve becomes:

$$|F(j\omega)|=\frac{1}{\sqrt{1+\left(\dfrac{\omega}{\omega_0}\right)^2}} \tag{1.23.1}$$

and the phase angle is:

$$\varphi(\omega)=-\arctan\frac{\omega}{\omega_0} \tag{1.24.1}$$

(These equations are represented in fig. 1.13.)

1.1.3.4 Bode Diagram

The Bode diagram presents separately, the absolute value and the phase of the frequency response curve as a function of frequency. They are referred to as the *amplitude curve* and the *phase curve*. The amplitude response curve is plotted in a double-logarithmic coordinate net, while the phase curve is presented in a single logarithmic net. The abscissa will show logarithmically either the frequency f, respectively the angular frequency $\omega = 2\pi f$, or the normalized values of f or ω. The ordinate shows for the amplitude response curve, the absolute value of the frequency response curve, respectively the amplitude ratio $\hat{v}(\omega)/\hat{u}(\omega)$ logarithmically, and for the phase response curve it shows linearly the phase angle $\varphi(\omega)$.

As examples of frequency response representation, we have the Bode diagram of the delay elements. Fig. 1.13 shows the Bode diagram for a 1st order delay element.

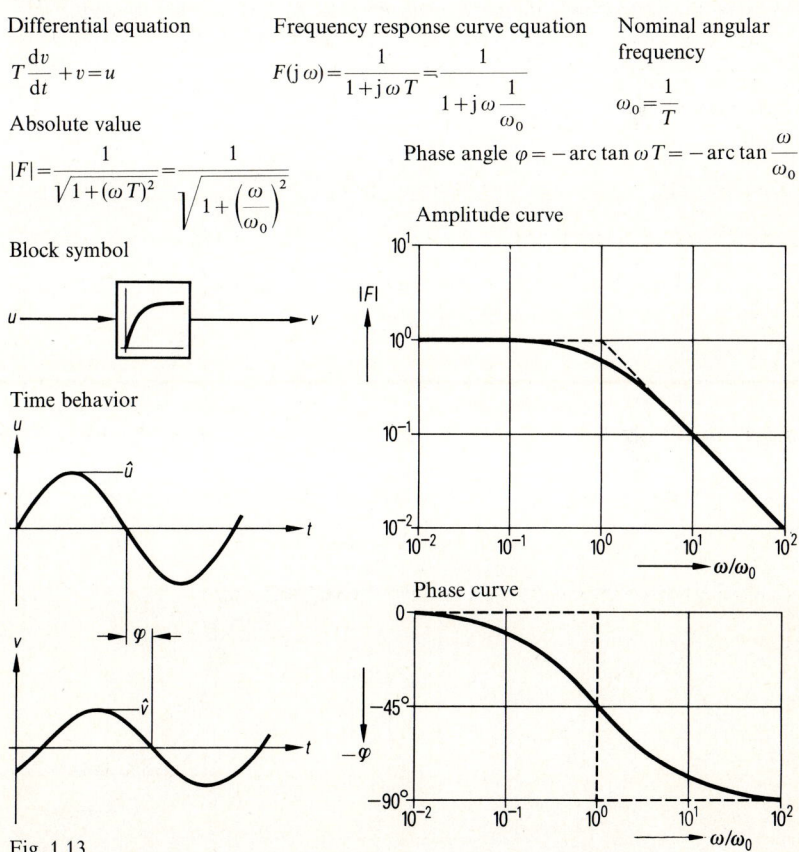

Differential equation
$$T\frac{dv}{dt} + v = u$$

Absolute value
$$|F| = \frac{1}{\sqrt{1+(\omega T)^2}} = \frac{1}{\sqrt{1+\left(\dfrac{\omega}{\omega_0}\right)^2}}$$

Frequency response curve equation
$$F(j\omega) = \frac{1}{1+j\omega T} = \frac{1}{1+j\omega\dfrac{1}{\omega_0}}$$

Nominal angular frequency
$$\omega_0 = \frac{1}{T}$$

Phase angle $\varphi = -\arctan \omega T = -\arctan \dfrac{\omega}{\omega_0}$

Fig. 1.13
Frequency response curve (Bode diagram) for a 1st order delay element

Examples of 1st order delay elements are:

▷ Serial connection of resistor and inductivity (e.g. the armature circuit of a DC motor or the excitation winding of a generator).
The input variable is the voltage; the output variable is the current.

▷ Resistor with capacitance switched to ground (e.g. the smoothing element of the reference or actual value input of a controller).
The input variable is the voltage; the output variable is voltage at the capacitor.

▷ Damped spring without mass, as physical model of a buffer storage for hydraulic systems.
The input variable is the pressure difference in the input or output line; the output variable is the pressure in the storage.

Fig. 1.14 shows the Bode diagram example of a 2nd order delay element.

Differential equation $T^2 \dfrac{d^2 v}{dt^2} + 2DT \dfrac{dv}{dt} + v = u$ Nominal angular frequency $\omega_0 = \dfrac{1}{T}$

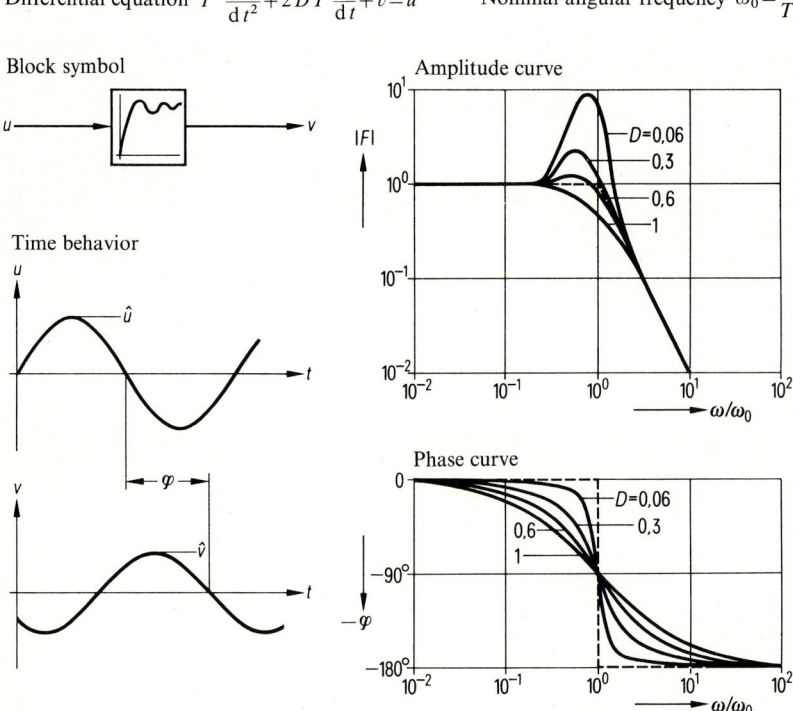

Fig. 1.14 Bode diagram for a 2nd order delay element

The differential equation for 2nd order delay elements is:

$$T^2 \frac{d^2 v}{dt^2} + 2DT \frac{dv}{dt} + v = u \tag{1.26}$$

If, here too, the characteristic nominal angular frequency is used,

$$\omega_0 = \frac{1}{T} \tag{1.27}$$

the result becomes:

$$\frac{1}{\omega_0^2} \frac{d^2 v}{dt^2} + 2 \frac{D}{\omega_0} \frac{dv}{dt} + v = u \tag{1.26.1}$$

Second order delay elements are also called oscillation elements; they exhibit an oscillating behavior, characterized by the damping gradient D. Damping gradient D is a measure of the transfer element's ability to oscillate. It holds:

$D=0$ The system is in continuous oscillation.
$D<1$ The system executes transient response functions with single or multiple overshoots around the end value.
$D \geqq 1$ Transient response functions occur without overshoot.

The frequency response curve equation for the frequency response curve of a 2nd order delay element can be determined from the relationships (1.18) and (1.19):

$$F(j\omega) = \frac{1}{1 + j\omega \frac{2D}{\omega_0} + (j\omega)^2 \cdot \frac{1}{\omega_0^2}} \tag{1.28}$$

The amplitude curve is derived with equation (1.16) as,

$$|F(j\omega)| = \frac{1}{\sqrt{\left[1 - \left(\frac{\omega}{\omega_0}\right)^2\right]^2 + \left(2D \frac{\omega}{\omega_0}\right)^2}} \tag{1.29}$$

and with equation (1.17) the phase curve can be calculated to:

$$\varphi(\omega) = -\arctan \frac{2D \frac{\omega}{\omega_0}}{1 - \left(\frac{\omega}{\omega_0}\right)^2} \tag{1.30}$$

Examples of 2nd order delay elements are:

▷ Electrical oscillating circuits consisting of serially or in parallel connected capacitance, inductivity, and resistor.
The input variable is the voltage; the output variable is the current.
▷ DC shunt motor with constant excitation.
The input variable is the voltage; the output variable is the speed.
▷ Spring-mass oscillator with damping (e.g. transmission elements on machine tools). The input variable is, for instance, the motor speed; the output variable is the table (slide) velocity.
▷ Speed control loops for position control loops on machine tools.
The input variable is the speed reference value; the output variable is the actual speed value.

From the amplitude response curve of figs. 1.13 and 1.14, it can be derived that for low frequency oscillations, delay elements behave like proportional elements. By comparison, oscillations of higher frequencies are heavily damped.

For first order delay elements, the amplitude drop approaches a straight line with a slope of -1 asymptotically, through the point $\omega/\omega_0=1$, $|F|=1$. The phase shift ranges from $0°$, for lower frequencies oscillations, to maximum $-90°$, for oscillations at higher frequencies. At nominal angular frequency $\omega_0=\frac{1}{T}$, the phase shift is $-45°$.

For second order delay elements, the amplitude drop asymptotically approaches a straight line with a slope of -2, through the point $\omega/\omega_0=1$, $|F|=1$. The phase shift ranges from $0°$, for lower frequencies, to a maximum of $-180°$ for higher frequencies. For nominal angular frequency ω_0, the phase shift is $-90°$. Depending on the damping gradient D value, the amplitude response curve gets more or less distorted upwards, in the vicinity of $\omega=\omega_0$. As demonstrated in section 6.3.1, the damping gradient can be determined from the absolute value of the frequency response curve, for the point $\omega=\omega_0$.

1.1.3.5 Linkage of Transfer Elements

A system consisting of interconnected transfer elements can be handled easier mathematically, with the help of a frequency response curve description than with a differential equation. A total frequency response curve can be assembled from the individual frequency response curves; the basic rules necessary for this are summarized in fig. 1.15.

This demonstrates that for a *serial linkage* of transfer elements, the resulting frequency response curve is obtained by multiplying the individual frequency curves. It is:

$$F_{res}=F_a \cdot F_b \tag{1.31}$$

For the *linkage in parallel* of transfer elements, the resultant frequency response curve is given by the sum of the individual frequency response curves.

$$F_{res} = F_a + F_b \tag{1.32}$$

For a *feedback circuit* (with either positive or negative feedback), the resulting frequency response curve is determined most conveniently by surveying the signal flow, as follows:

$$v = u \cdot F_a \genfrac{(}{)}{0pt}{}{+}{-} v \cdot F_b \cdot F_a$$

(+ for positive feedback, and − for negative feedback)

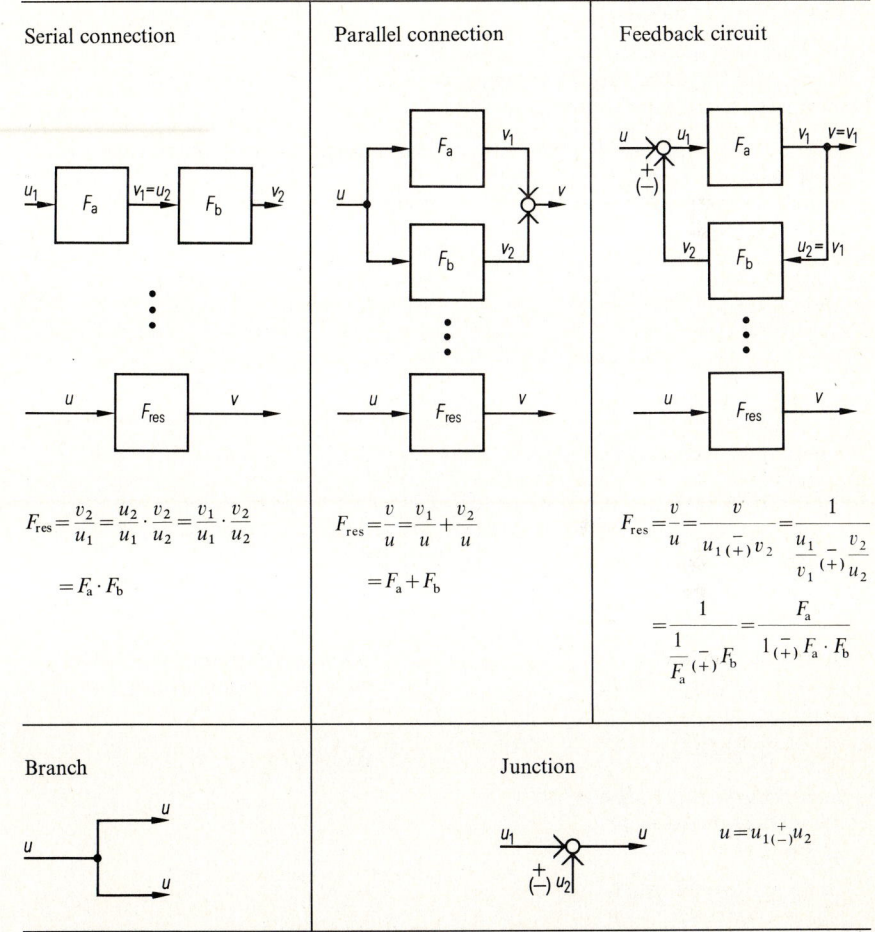

Fig. 1.15
Basic rules for the linkage of single transfer elements into a system

From this follows:

$$F_{res} = \frac{v}{u} = \frac{F_a}{1 \, (\overline{+}) \, F_b \cdot F_a} = \frac{1}{\frac{1}{F_a} \, (\overline{+}) \, F_b} \tag{1.33}$$

(+ for positive feedback, and − for negative feedback)

1.1.3.6 Frequency Response Curve Symbol for Control Loops

In order to better characterize and evaluate control loops, the following distinctions should be made:

Frequency response curve of an open control loop F_o
Command frequency response curve F_w
Disturbance frequency response curve F_z

For a single loop control loop, the conditional equations of these frequency response curves are presented in fig. 1.16.

Block diagram

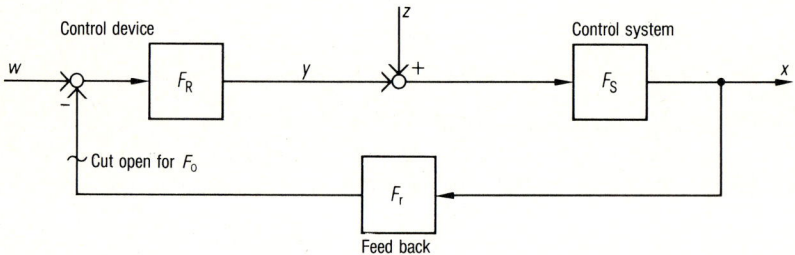

Frequency response curve
of an open control loop F_o

$F_o = (-1) \cdot F_R \cdot F_S \cdot F_r$

Command frequency response
curve of the control loop, F_w
($z = 0$)

$F_w = \frac{x}{w} = \frac{1}{F_r} \cdot \frac{(-F_o)}{(1-F_o)} = \frac{F_R \cdot F_S}{1-F_o}$

Disturbance frequency response
curve of the control loop F_z
($w = 0$)

$F_z = \frac{x}{z} = \frac{F_S}{1-F_o}$

Fig. 1.16 Frequency response curve of a single-loop control loop

The *frequency response curve of the open control loop* F_o is often used to evaluate or describe and designate the characteristics of the closed control loop. For this purpose, the control loop is "cut open" at any position with no reaction within the action loop.

The "cut open" control loop consists of the linkage in series of the individual transfer elements. The sign change of the value introduced, for comparison, into the control loop, must be taken into consideration. With the single frequency response curves given in fig. 1.16, one can determine the frequency response curve of the open control loop:

$$F_o = -F_R \cdot F_S \cdot F_r \qquad (1.34)$$

The *command frequency response curve* F_w describes the signal-transient response behavior of the closed control loop, between the command variable w and the control variable x in the frequency range. This command frequency curve is the frequency response curve normally shown for transfer elements. For the control loop shown in fig. 1.6, according to a signal flow survey with $z = 0$,

$$F_w = \frac{F_R \cdot F_S}{1 - F_o} \qquad (1.35)$$

If $F_r = 1$, the command frequency response curve assumes the following relationship:

$$F_w = \frac{1}{1 - \dfrac{1}{F_o}} \qquad (1.35.1)$$

The *disturbance frequency response curve* F_z of a control loop describes the time behavior which interferes in the control loop between control value x, and disturbance variable z. The accompanying command value stays constant. For the control loop presented in fig. 1.16,

$$F_z = \frac{F_S}{1 - F_o} \qquad (1.36)$$

The control process refering to an arbitrary course of the command, respectively disturbance variables, can be calculated within the frequency range with the help of the command and disturbance frequency response curves.

1.1.3.7 Experimental Determination of the Frequency Response Curve

The dynamic behavior and the characteristic data of an unknown control loop element or system can also be determined by measuring the frequency response curve.

For this purpose, a variable-frequency, sinusoidal input voltage of constant amplitude is applied at the input of the system to be tested, and the output signal is measured. A direct recording of the Bode diagram is possible by means of suitable computing circuits in conjunction with a 2-coordinates recorder.

The Bode diagram can also be derived point-by-point through the evaluation of oscillograms. The approximate order (1st, 2nd, or higher) of the system can be determined from the course of the amplitude response curve and the phase response curve. Frequency response curves so measured are presented and explained in chapter 6.

1.2 Basics of the Position Control Loop

1.2.1 Construction and Functioning of the Position Control Loop

The machine tool machining of a part, in order to create a desired form, requires a relative motion between tool and part. The individual feed units must therefore be adjustable in their position. Today, this is mainly accomplished through position control loops with speed controlled DC servo motors. Position adjustments with controls alone, e.g. using step motors, are limited to small drive power ranges, and are not significant any more. Therefore, the following discussion concerns only the position control loop position adjustments.

The requirement on the position control loops of numerically controlled machine tools as well as tracer mills is that they adjust the corresponding feed units dependently on the stored geometry and velocity data with as little delay and distortion as possible, from the position command values given by the control devices. These data are in their totality the position command values.

The position control loop, respectively position control, is characterized by the loop structure of the system. It creates a closed signal action flow between the position command value as an input signal of the position control loop, and the actual position value.

The closed action loop of the position control means that the actual position value must be recorded with the proper measuring device, and is fed back for comparison with the position command value. In addition, the position command value's progress over time can be altered with the use of a suitable command value modifier (see chapter 5). The position control deviation controls the actuator of the feed unit – the feed drive.

Based on the action flow, the position control can therefore be designated as a process in which the controlled variable – the position, respectively state of a mechanical transport unit (feed slide, table or rotary table) – can be continuously recorded, automatically compared to the command value, and depending on the result of the comparison, it can be modified to approach the command value. Fig. 1.17 shows the signal flow chart of a position control loop. Like any other control system, the position control loop consists of a control device and a control system.

The control device consists of:

▷ Position control, including comparator and control amplifier
▷ Position measuring system

The control system consists of:

▷ Electrical drive with the DC servo motor
▷ The mechanical transmission elements of the machine tool

Together, control device and control system will be further designated as feed drive.

The individual function units of the position control have the following tasks: The *position control* compares the position command value x_s (command variable w) to the actual position value x_i^*, which is delivered by the position measuring system. It creates out of the control difference Δx, the command value for the position velocity v_s of the feed unit, the correcting variable y.

x_i actual position value of feed unit (control variable x)
x_i^* actual position value signal of measuring system
x_s position command value (command variable w)
Δx position control deviation
v_i actual velocity value
v_s velocity command value (correcting variable y)
F_v feed force (disturbance variable z)

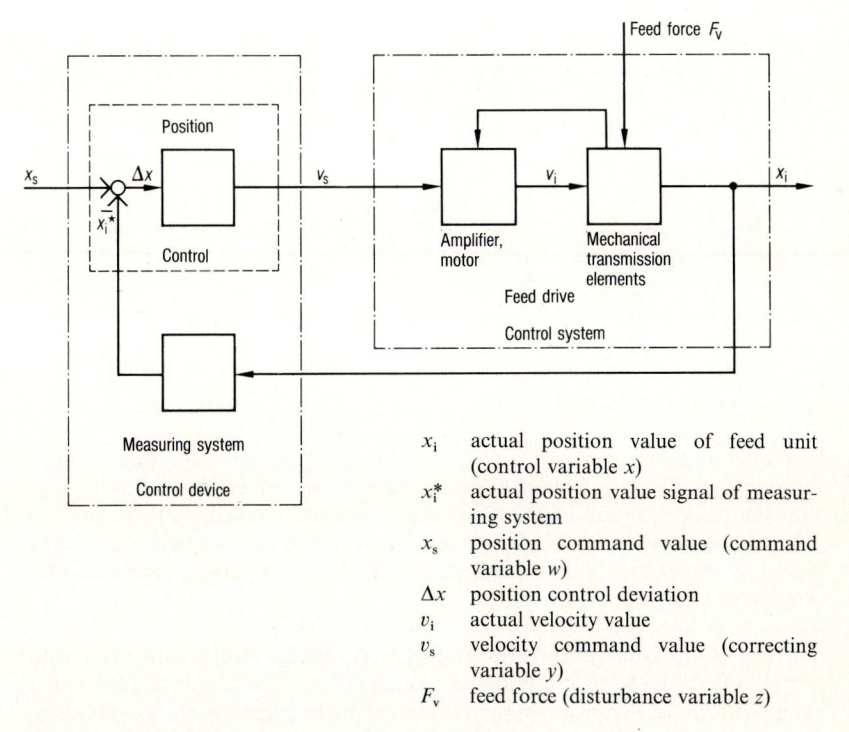

Fig. 1.17   Signal flow chart of a position control loop

The input variables are the position command value x_s, and the actual position value x_i^*; the output variable is the command value of the position velocity v_s.

The *position measuring system* creates out of the actual position value x_i of the coupled machine part, the measuring value of the position x_i^*.

The *feed drive* transforms the velocity command value of a translatoric velocity unit into the corresponding actual distance x_i; for rotary velocity units (rotary table), into corresponding angle of rotation (control variable x). The feed drive consists of a speed controlled motor and its coupled mechanical transmission elements, which reach to the contact point of the cutting forces on the part through coupling, respectively gear and feed screw. The feed force F_v acts as the disturbance variable z which results from the external load; it is fed into the position control loop through the mechanical transmission elements. The input variable is the velocity command value v_s, and the output variable is the actual position value x_i, appearing on the position control velocity unit.

1.2.2 Localization

1.2.2.1 Measuring Value Detection

As mentioned under 1.1.1, informations between single elements of the control loop can be transmitted through analog or digital signals. The distinction between analog and digital position measuring system is made according to the type of measuring value detection used.

For *analog* working measuring systems, the geometrical value, distance or angle, is proportionally transformed into another corresponding physical value, e.g. into the phase position or the amplitude of an electrical signal more suitable for position value processing. The advantage is the clear cut relation of the actual position value to the measuring signal; the disadvantage is in the accompanying hardware complexity.

For *digital* working systems, the analog position distance is subdivided into a multitude of equal length position increments Δs. The size of Δs determines the resolution of the positioning system. The advantages are the simple hardware required and the safe measuring value transmission.

1.2.2.2 Measuring Procedures

Besides the type of measuring value detection, the kind of measuring procedure is also important for the position measuring system. There are absolute, incremental, and cyclic absolute measuring procedures.

Absolute measuring procedures apply to both analog and digital working position measuring systems. Each measuring value is assigned an unequivocal signal value. The advantage is a machine fixed zero point which can be the reference point for the measurement of these absolute measuring values. The high costs of practical application are, however, disadvantageous. Examples of analog-absolute measuring procedures are potentiometers or cascaded resolver gears. Digital-absolute measuring procedures are used for position measurement with a digital-absolute coder.

Incremental position measuring procedures apply only for digital measuring value detection. In this case there is no fixed zero point, but any point can be arbitrarily declared the zero point by setting the position counting device accordingly. A machine-fixed zero point can also be established, with a zero marker in the measuring device. The advantage lies in the low hardware requirements for measuring value detection and transmission, as well as in the simplicity of the zero point selection and shift. The disadvantage is, that after the supply voltages are shut off, the measuring device cannot reconstruct the momentary position any more. For the position values to remain stored, appropriate storages must be provided. Examples are the digital incremental decoder or the digital linear scale.

Cyclic-absolute measuring procedures apply to both analog and digital working position measuring systems. A small, absolute detected measuring range (e.g. 2 mm) is cyclically repeated, and the number of repetitions is counted incrementally. This combines the advantage of high resolution with the one offered by the simpler storage of larger distances.

Examples of the application of cyclic-absolute measuring procedure with analog measuring value detection are the resolver and the INDUCTOSYN.

1.2.2.3 Location of Measurement

Position control can be direct or indirect, depending on the location of the position value detection on the machine tool. Figure 1.18 shows these two possibilities in the block diagram of the position control, using as example the feed screw drive.

For *indirect position control,* the position of the table is recorded indirectly from the angular position of the screw or feed motor. The reversing errors behind the measuring location are active outside the position control loop. They appear as permanent position deviations, and thus affect the positioning accuracy of parts to be machined (see 4.2.4.3.1). This residual position deviation can be partially eliminated by compensating with the position command value.

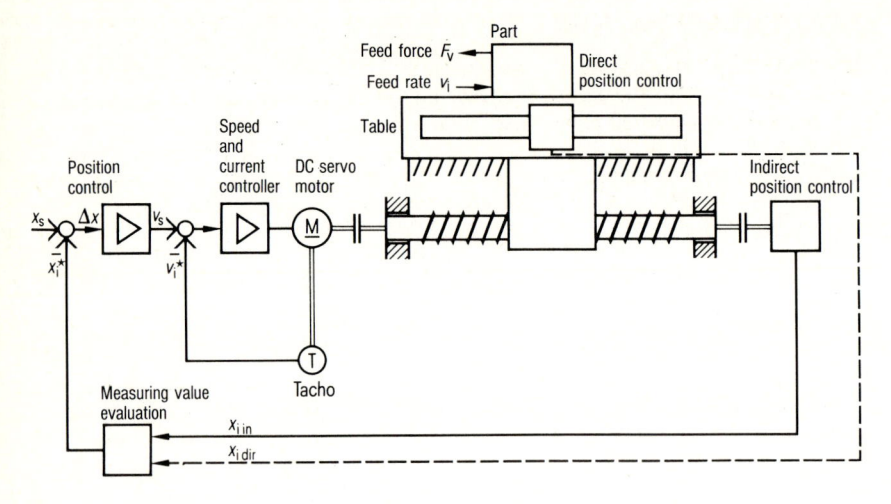

Fig. 1.18   Position control loop with direct and indirect position control

$x_{i\,in}$ Indirectly measured actual position value
$x_{i\,dir}$ Directly measured actual position value

A reversing error within the position control loop (e.g. at the couplings of motor to feed screw, respectively measuring system to feed screw) has the same consequences on the adjustment of position and speed control loops as the ones described below for the direct position control.

For *direct position control,* the measuring system is mounted directly between the fixed and the moving machine parts. The reversing errors therefore act only within the position control loop; permanent position deviations do not occur. However, for the design and the optimization of the position control loops, the reversing error ought to be small; this requires appropriate constructive dimensioning of the mechanical transmission elements. Reversing errors within the position control loop result in oscillations around the position controlled or, at certain damping gradients, in an inaccurate position approach (see section 4.2.4.3.2). The direct measuring location should be as close as possible to the machining location (Abbee's principle). Tilt errors on feed slides (tables) result in part errors.

1.2.3 Characteristic Values and Properties of Position Control Loops

Knowledge of the signal flow between the individual control loop elements is not sufficient for the characterization of the position control loop. The transient response behavior of individual elements and of the total system must also be known, and the static and dynamic behaviors are decisive factors in the machining result.

1.2.3.1 Dynamic Systems Behavior of Single Control Loop Elements

Usually, position controllers and measuring devices have a proportional behavior, i.e. they react without delay to input variable changes. Because of this, the measuring devices can be overlooked in as far as the dynamics are concerned. The proportional behavior of the control device is thus considered only as the proportionality factor of the control. This amplification, the so-called position loop gain K_v – also called K_v-factor – is defined as the ratio of the command velocity v_s to the position control deviation Δx:

$$K_v = \frac{v_s}{\Delta x} \tag{1.37}$$

Based on this relationship, for a straight line motion with constant velocity, the resulting position control deviation, so-called following error, is:

$$\Delta x = \frac{v_s}{K_v} \tag{1.37.1}$$

The position control deviation indicates the amount by which the actual position value lags behind the position command value. The following error causes no contour deviations for linear movements, when the position loop gains in all the position control loops participating to the contour generation are identical. If the position loop gains are uneven, for linear motion a parallel offset of the actual contour to the command contour results, and for circular or parabolic interpolation of command contours a distortion of the actual contour is created.

The unit for the position loop gain K_v, is the s^{-1}. Other common units are:

$$\frac{mm/min}{\mu m} \quad \text{or} \quad \frac{m/min}{mm} \quad \text{and} \quad \frac{mm/min}{0{,}1\ mm}$$

The relationship between the individual units is presented again in table 1.1.

The frequency response curve of the position control, including the position measuring device is given by:

$$F_{LR}(j\omega) = \frac{v_s(j\omega)}{\Delta x(j\omega)} = K_v \tag{1.38}$$

The feed drive, as actuator of the position control loop, is equipped with a speed control loop; its speed regulator has proportional-integral behavior (PI). By adjusting the controller appropriately, the drive with the coupled mechanical transmission elements can be given the behavior of a 1st or 2nd order delay element. The mean dead time for thyristor-fed drives, and the reactions of

Table 1.1 Conversion factors for the units of the position loop gain K_v

Unit	s^{-1}	$\dfrac{mm/min}{\mu m}$ or $\dfrac{m/min}{mm}$	$\dfrac{mm/min}{0.1\ mm}$
s^{-1}	1	$6 \cdot 10^{-2}$	6
$\dfrac{mm/min}{\mu m}$ or $\dfrac{m/min}{mm}$	16.67	1	100
$\dfrac{mm/min}{0.1\ mm}$	0.167	$1 \cdot 10^{-2}$	1

the mechanical transmission elements can be approximated with a dead time element switched in series with the delay element.

The frequency response curve will be (from fig. 1.5):

$$F_T(j\omega) = e^{-j\omega T_T} \qquad (1.39)$$

The frequency response curve $F_A(j\omega)$ of a drive with the behavior of a 1st order delay element, according to equation (1.22.1) is, without dead time:

$$F_A(j\omega) = \frac{v_i(j\omega)}{v_s(j\omega)} = \frac{1}{1 + j\omega \dfrac{1}{\omega_{0A}}} \qquad (1.40)$$

and when considering the dead time element:

$$F_A(j\omega) = \frac{e^{-j\omega T_T}}{1 + j\omega \dfrac{1}{\omega_{0A}}} \qquad (1.40.1)$$

With the behavior of a 2nd order delay element, the frequency response curve of the drive, according to equation (1.28), without dead time, becomes:

$$F_A(j\omega) = \frac{v_i(j\omega)}{v_s(j\omega)} = \frac{1}{1 + j\omega \dfrac{2D_A}{\omega_{0A}} + (j\omega)^2 \dfrac{1}{\omega_{0A}^2}} \qquad (1.41)$$

and with dead time element, it is:

$$F_A(j\omega) = \frac{e^{-j\omega T_T}}{1 + j\omega \frac{2D_A}{\omega_{0A}} + (j\omega)^2 \frac{1}{\omega_{0A}^2}} \tag{1.41.1}$$

Where:

v_i actual velocity value $\qquad D_A$ damping gradient of the drive
v_s velocity command value $\qquad \omega_{0A}$ nominal angular frequency of the drive

A common characteristic of all position control loops on machine tools is the integral behavior of the control system. This integral behavior describes the physical relationship between the velocity v_i of a feed slide or table, and the actual position value x_i to be controlled.

Generally, it holds that:

$$v_i = \frac{dx_i}{dt} \text{ that is } x_i = \int v_i \cdot dt \tag{1.42}$$

x_i is distance for translational motion; the angle to be moved, for rotational motion.

For frequency response curve presentations, according to rules (1.18) and (1.19) for this integral element, it holds that:

$$F_I(j\omega) = \frac{x_i(j\omega)}{v_i(j\omega)} = \frac{1}{j\omega} \tag{1.43}$$

From this, we obtain for the linear position control loop, the block diagram 1.19.

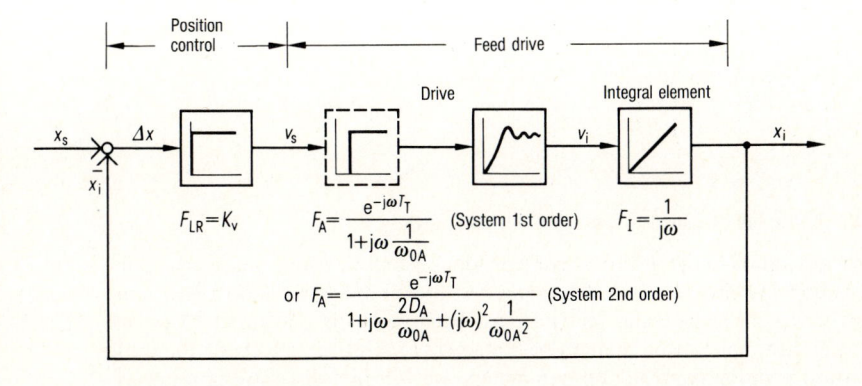

Fig. 1.19 Simplified block diagram of a linear position control loop

1.2.3.2 Explanation of Terms

The most important position control loop parameters, according to the previously presented frequency response curve equations (1.38), (1.39), (1.40) respectively (1.41) and (1.43), are:

    the position loop gain K_v
    the nominal angular frequency of the drive system ω_{0A}
    the damping gradient of the drive system D_A
    the dead time of the drive system T_T.

The *position loop gain* K_v is the ratio of the command velocity of the position controlled travel unit (feed slide) to the position deviation in the settled state with no external load. It is most significant for the generation of contours with multiple simultaneously working feed axes.

The *nominal angular frequency* ω_{0A} of a drive system offers information about its signal transfer behavior. A high nominal angular frequency in reference to the frequency spectrum of the input signals, means a low degree of distortion of these signals through the system.

The *damping gradient* D_A of the drive system of 2nd order characterizes the ability to avoid continuous oscillations. The oscillations appearing during transient responses are counteracted by a corresponding damping gradient D_A, so that the settling processes subside more or less rapidly. The smaller the damping gradient, the closer the system is to the stability limit, i.e. the oscillating tendency and thus the settling time for a settling response becomes larger.

The nominal angular frequency ω_{0A} and the damping gradient D_A of electrical feed drives are determined by the mechanical time constant $T_{\text{mech}A}$, the electrical time constant T_{elA}, and the setting of the control parameters.

The *dead time* T_T as an approximate value, offers information about the effects of the delays in the converter and the mechanical transmission elements on the signal transfer behavior of the drive. The converter portion is determined from the pulse number p_{SR} and the line or the pulse frequency f_N of its supply voltage (see 2.2.2.7).

1.2.3.3 Behavior of Linear Position Control Loops

The behavior of linear position control loops can be treated mathematically as a closed system; this is cumbersome for position control loops with non-linearities as they appear in real systems. Because of this, in order to get an overview and make some qualitative statements about the effects of the most important position control loop parameters, we will first disregard non-linearities.

Such a simplified position control loop is shown in fig. 1.19, in a block diagram. The position controller will be assigned the behavior of a proportional element, and the drive is assigned the behavior of a delay element with dead time element switched in series. The feed screw/slide system, respectively the rack and pinion system, are assigned the integral element.

One possibility for describing the behavior of the linear position control loop consists in the determination of the frequency response curve according to section 1.1.3.6.

The command frequency response curve of the position control loop of a drive with the behavior of a 1st order delay element without dead time (F_A is substituted according to equation (1.40)) is:

$$F_{wL}(j\omega) = \frac{1}{1 + j\omega \frac{1}{K_v} + (j\omega)^2 \frac{1}{K_v \cdot \omega_{0A}}} \quad (1.44)$$

The equation of this command frequency response curve applies to one of a 2nd order delay element. Thus, according to equation (1.28), we can define:

Nominal angular frequency of the position control loop

$$\omega_{0L} = \sqrt{K_v \cdot \omega_{0A}} \quad (1.45)$$

Damping gradient of the position control loop

$$D_L = \frac{1}{2}\sqrt{\frac{\omega_{0A}}{K_v}} \quad (1.46)$$

If the drive exhibits the behavior of a 2nd order delay element without dead time, the command frequency response curve becomes:

$$F_{wL}(j\omega) = \frac{1}{1 + j\omega \frac{1}{K_v} + (j\omega)^2 \frac{2D_A}{K_v \cdot \omega_{0A}} + (j\omega)^3 \frac{1}{K_v \cdot \omega_{0A}^2}} \quad (1.47)$$

(F_A is substituted according to equation (1.28).

For systems with higher than 2nd order transfer function, the damping gradient and nominal angular frequency cannot be defined any more in the sense of 2nd order elements. In this case, the following notions, according to fig. 1.20, are characteristic of the behavior:

▷ cut-off angular frequency of the amplitude response curve ω_E
▷ limit angular frequency ω_G
▷ 90° angular frequency of the phase response curve ω_{90}
   (90° angular frequency ω_{90} and cut-off angular frequency ω_E for a 2nd order delay element are the same as the nominal angular frequency ω_0)
▷ resonance angular frequency ω_R ; multiple resonance positions are possible.

Knowledge of the cut-off or limit angular frequency allows the behavior of the control loop to be divided into two parts. In the range below these characteristic values, sinusoidal signal portions are transferred by the control loop with

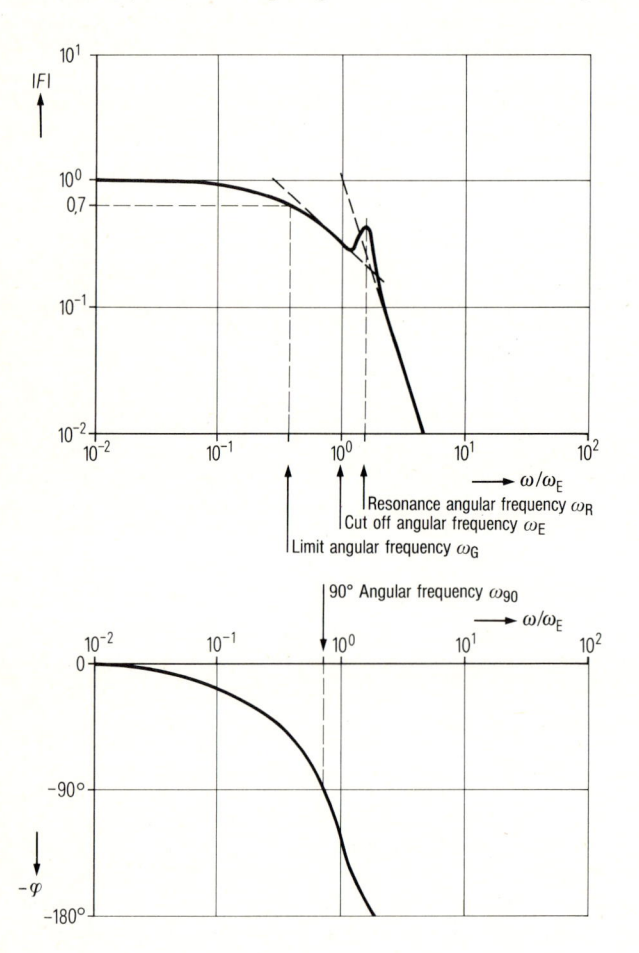

Fig. 1.20
Characteristic values of a system with a 3rd order transfer function

only slight distortions; at angular frequencies above these values, the resulting distortions are very heavy. This means that the higher the cut-off or limit angular frequency is, the higher the accuracy with which the control loop can follow the command value changes.

For the control loop, this means that at high values of the stated angular frequencies, positioning processes can be executed without overshoots at higher velocities.

For a position control loop with a 3rd order transfer function, the cut-off angular frequency is:

$$\omega_{EL} = \sqrt[3]{K_v \cdot \omega_{0A}^2} = \omega_{0A} \sqrt[3]{\frac{K_v}{\omega_{0A}}} \qquad (1.48)$$

The relationship found in section 1.3.7, equation (1.52), between the position loop gain K_v and the nominal angular frequency ω_{0A} of the drive

$$0{,}2\,\omega_{0A} \leq K_v \leq 0{,}3\,\omega_{0A}$$

is used to derive the cut-off angular frequency of the position control loop:

$$\omega_{EL} = (0{,}6\ldots 0{,}7)\,\omega_{0A} \qquad (1.49)$$

The damping gradient within the position control loop must be large enough, so that no oscillations could possibly occur. D_L must be larger than 1.

A further possibility for the description, respectively evaluation of the behavior of position control loops, results from the positioning processes. The characteristic values important in this case are the following error Δx (see section 1.2.3.1) and the overshoot width $\ddot{u}_a$. For position control loops on machine tools, $\ddot{u}_a$ is the largest command-actual value deviation of the velocity unit, which occurs at the transition from motion with constant velocity ($v = $ const.) to standstill ($v = 0$), during position approach.

Since the overshoot width depends not only on the characteristics of the position control loop but also on the type and manner in which the position command value is generated, on one hand we can establish for the control loop a velocity limit at which the overshoot width is just inside its resolution, and on the other hand, we can maintain the overshoot within low limits with a command value modifier (see chapter 5), so that there are no visible effects on the part.

The effect of the position control loop parameters on the positioning processes can be demonstrated with the help of a computer simulation. The results are shown as examples of a position control loop whose drive exhibits the behavior of a 1st order delay element without dead time element. Figure 1.21 presents the travel-time functions of the position command value and the actual position values, for a positioning process executed under the influence of the position loop gain K_v and of the nominal angular frequency ω_{0A} of the drive.

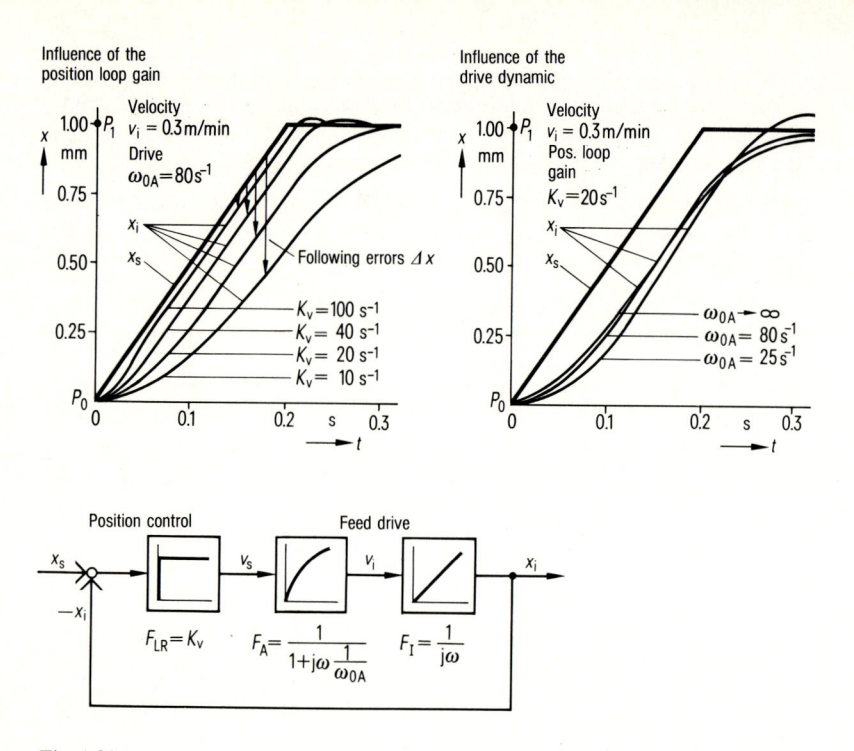

Fig. 1.21
Positioning process of a position control loop (drive acting as a 1st order delay element without dead time element)

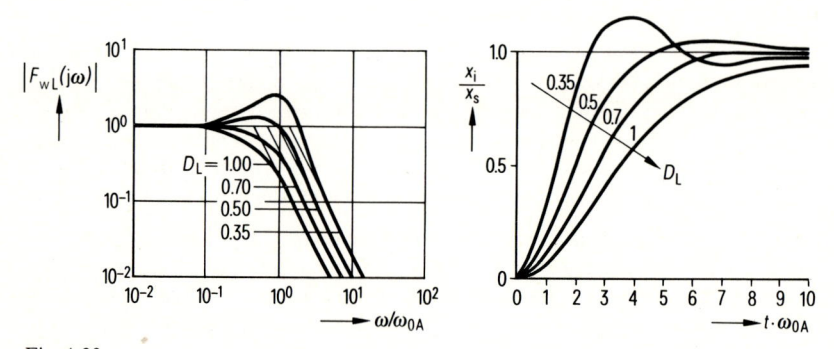

Fig. 1.22
Amplitude response curves and transfer functions of a position control loop acting as a 2nd order delay element

The diagram shown on the left side of fig. 1.21 presents the time difference between position command value and actual position value at a given nominal angular frequency ω_{0A} and for different position loop gains K_v. The difference between command and actual position values is given as the following error Δx. One recognizes that for a given nominal angular frequency of $\omega_{0A} = 80 \text{ s}^{-1}$, at a position loop gain of 20 s^{-1}, the aperiodic limit case is reached. Higher position loop gains will lead to overshoots of the actual position over the command position. Lower position loop gains result in following errors that are too large, thus in errors at directional changes. Permanent position deviations do not occur, due to the integral element of the control process.

The diagram shown on the right side of fig. 1.21 presents in the same manner, how a nominal angular frequency ω_{0A} that is too small, can lead to overshoot during the positioning process at a given position loop gain.

The amplitude response curves of the command frequency F_{wL} response corresponding to this example are shown in fig. 1.22. The parameter is the damping gradient of the position control loop,

$$D_L = \frac{1}{2} \sqrt{\frac{\omega_{0A}}{K_v}}$$

according to equation (1.46) of the 2nd order delay element with behavior equivalent to the one of the control loop.

The left side of the figure shows the amplitude response curve of the command frequency response curve plotted over the angular frequency ω, referenced to the nominal angular frequency ω_{0A} of the drive. This makes a general valid statement concerning position control loops at different drive nominal angular frequencies possible. It can be recognized that with decreases in the damping gradient D_L and with increases in the position loop gain K_v, the nominal angular frequency of the position control loop $\omega_{0L} = \sqrt{K_v \cdot \omega_{0A}}$ shifts to larger values. However, the resonance ratio increase of the corresponding amplitude response curve, also increases simultaneously. On the other hand, at the same position loop gain K_v, a larger damping gradient D_L and a higher nominal angular frequency ω_{0L} can be reached for the position control loop with a higher drive nominal angular frequency ω_{0A}.

The right side of the figure shows the transfer functions of the position control loop. Decreases in the damping gradient D_L of the control loop lead to actual position overshoots of the command value.

We can derive from these presentations, that in order to maintain a desirable position loop gain K_v, the drive nominal frequency ω_{0A} of the drive system must not get below a certain minimal value, which would allow it to overshoot positioning. It also must be recognized that for a fixed position loop gain K_v, further increases in the nominal angular frequency of the drive ω_{0A} will not result in improvements of the transient response. From this, we can deduce

49

that an appropriate position control loop behavior is characterized by a compromise between a high nominal angular frequency ω_{0L}, and a sufficient damping gradient D_L.

1.2.3.4 Non-linearities in the Position Control Loop

The statements concerning the behavior of position control loops made thus far, have taken into consideration only linear transfer elements. The disadvantage is, that the non-linearities known to be found in every real system are not included in these statements. Such non-linearities are, for example, the insensitivity range, the reversing error, limitations, etc. They considerably influence the behavior of the position control.

Figure 1.23 shows the block diagram of a position control loop containing the most important non-linearities [1.6].

The block diagram is simplified in regard to the type and number of non-linearities, in order to maintain the general applicability. For the sake of clarity, the individual function units, according to the hardware construction of the position control loop, are divided into a linear and a non-linear block, where the linear block has been assigned its corresponding time behavior. Despite this considerable simplification, tests on analog computers have shown that the behavior of a position control loop can be reconstructed reasonably well with the aid of this block diagram. In addition, the parameters given can be measured with relative ease on the actual equipment.

The behavior of the mechanical parts of the feed drive system can be approximated with the spring mass system, as a 2nd order mass oscillator (compare

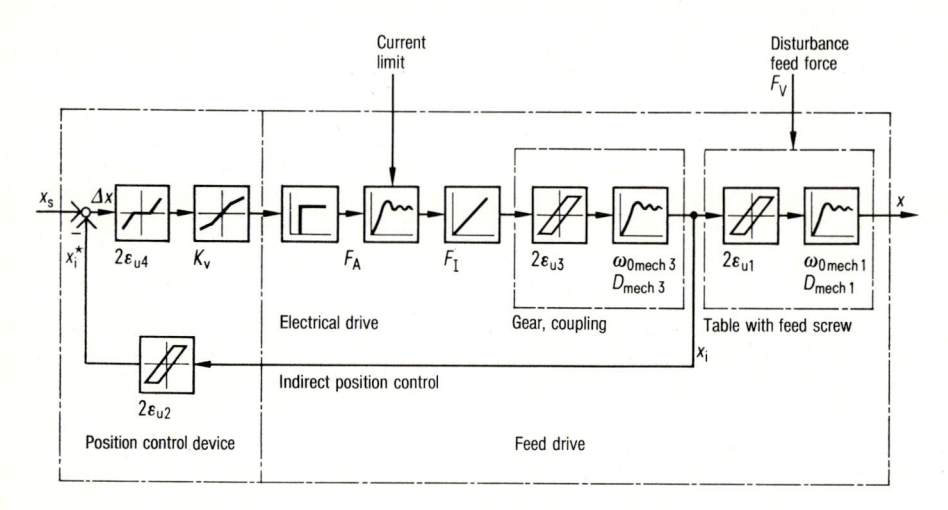

Fig. 1.23 Position control loop with non-linearities (feed screw drive)

section 4.2.1). The mass of the table with the feed screw and the bearing construction elements, has a nominal frequency $\omega_{0\,\text{mech}\,1}$ and a damping gradient $D_{\text{mech}\,1}$. The reversing error is assignated the value $2\varepsilon_{u\,1}$. For position control loops with indirect position control, this block lies outside the position control loop. It affects circumstancially the behavior of the control loop, only through reaction. It determines the accuracy on the work piece, through the value $2\varepsilon_{u\,1}$, because of permanent position control deviations.

The reversing error $2\varepsilon_{u\,2}$ characterizes the coupling of the position transducer to the feed screw. It can be kept very low, generally.

Gears also have a reversing error, and it is included in the reversing error $2\varepsilon_{u\,2}$. They exhibit the behavior of 2nd order mass oscillators with nominal angular frequency $\omega_{0\,\text{mech}\,3}$ and damping gradient $D_{\text{mech}\,3}$. Generally, $\omega_{0\,\text{mech}\,3}$ is significantly higher than $\omega_{0\,\text{mech}\,1}$, and has no effect on the position control.

The reversing error $2\varepsilon_{u\,3}$ lies within the position control loop. It can have a destabilizing effect on the position control behavior, and thus prevent high values to be set for the position loop gain K_v. It does not make possible any permanent position deviations.

The variable $2\varepsilon_{u\,4}$, active within the position control itself, is the insensitivity range, whose effect is especially apparent during digital signal processing.

The gain of the position control must be maintained linear throughout a wide range. Non-linear characteristic curves and uneven gains in the different feed axes cause contour deviations to occur during contour operations. With the help of a command value modifier, as the one described in chapter 5, reaching the current limit during accelerations and decelerations can be avoided, since reaching the current limits would lead to overshoots in the position control loop. On simple numerical controls, the position loop gain is reduced in the rapid traverse range, thus reducing the acceleration and braking of the feed drive.

The effects of non-linearities in the electrical drive can be strongly reduced with the aid of the subordinated speed control loop. Non-linearity outside the speed control loop is a property of mechanical transmission elements. The magnitude and the effect on the position control can be reduced only by designing the machine tool accordingly (see chapter 4).

For the sake of simplicity, further effects of the drive and machine have been omitted in the structure presented in fig. 1.23. More detailed block schemes for computer simulations are presented in the specialty literature [1.7], [1.8].

In practice, the measuring methods with which the properties of drive and machine parts can be evaluated, have been proved adequate to allow the evaluation of the position control loop behavior. Chapter 6 explains some of these procedures, and presents the measurement results obtained.

1.3 Position Loop Gain

1.3.1 Position Control Loop Model

If the behavior of a feed drive approaches that of a 2nd order delay element, the model for the corresponding position control loop can be presented as a block diagram, as in fig. 1.24. The dead time T_T has been omitted.

With a common damping gradient setting of $D_A = 0.5$, the command frequency response curve of this position control loop, according to equation (1.47) becomes:

$$F_{wL} = \frac{x_i(j\omega)}{x_s(j\omega)} = \frac{1}{1 + j\omega \frac{1}{K_v} + (j\omega)^2 \frac{1}{K_v \cdot \omega_{0A}} + (j\omega)^3 \frac{1}{K_v \cdot \omega_{0A}^2}}. \quad (1.47.1)$$

The following simulations can be executed on a computer with the scheme corresponding to this frequency response curve. The results can be used to evaluate the real behavior of position control loops.

An evaluation concerning the necessary position loop gain K_v is possible for the following operations:

positioning
drilling
turning
milling
thread cutting

It must be clarified how a position loop gain that has been chosen too high or too low, affects the operations mentioned above.

For this purpose, criteria will be established for the individual operation. These criteria will dictate the appropriate requirements for the position loop gain dependent on the feed, respectively contour velocity, the allowable contour

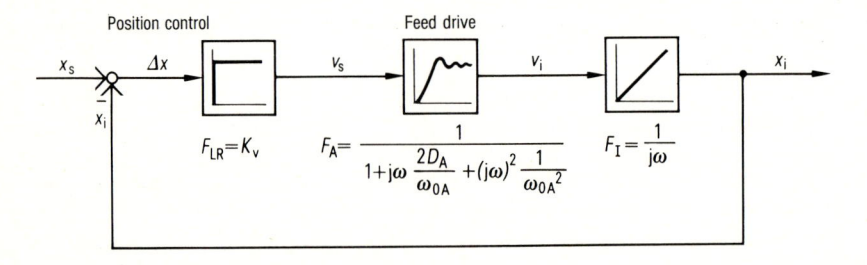

Fig. 1.24 Simplified block diagram of a position control loop

deviations for each operation, and the drive nominal angular frequency ω_{0A} (for calculating ω_{0A}, see 2.2.2.5). A linear position control loop is assumed here.

1.3.2 Settling Time and Overshoot Width at Positioning

The settling time T_{Aus} and the overshoot width $ü_a$ serve as criteria for setting the position loop gain K_v for positioning. The settling time T_{Aus} represents the time span necessary for the positioning itself. During the positioning process, the following error Δx between the position command and actual values inherently resulting from the construction of the position control loop, is reduced. This following error is always present when the position command value reaches its position end value. Hereby, depending on the setting of the position loop gain, an overshoot can occur; this is generally undesirable on a machine tool.

According to fig. 1.25, settling time T_{Aus} begins when the position command value has reached the desired end position value x_1. It ends, when the position actual value reaches the preset tolerance range $2e_x$ around the end position, and does not exceed it any more. The terms are:

Δx following error
$x_i(t)$ position actual value
$x_s(t)$ position command value

Curve 1 shows the positioning process without overshoot,
Curve 2 shows the positioning process with overshoot.

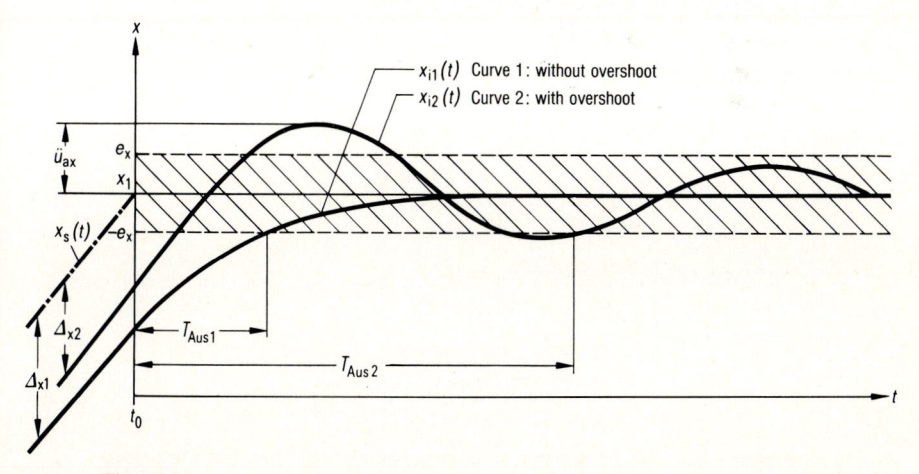

Fig. 1.25
Settling time T_{Aus} and overshoot $ü_{ax}$ for positioning with, and without overshoot

The effects of the position loop gain K_v on the settling time T_{Aus} and the overshoot width $\ddot{u}_{ax}$ at a given transient response tolerance $2e_x$, are presented as normalized curves in fig. 1.26.

The reference values are the feed rate v_V, and the nominal angular frequency ω_{0A}. The settling time and the overshoot width for a particular velocity v_V, can be determined from the curves presented in fig. 1.26, if the values of K_v, ω_{0A}, and $2e_x$ are known.

From fig. 1.26 we can derive:

Dependent on the transient response tolerance, the settling time is shortest for the values

$$0.3 \leq \frac{K_v}{\omega_{0A}} \leq 0.38$$

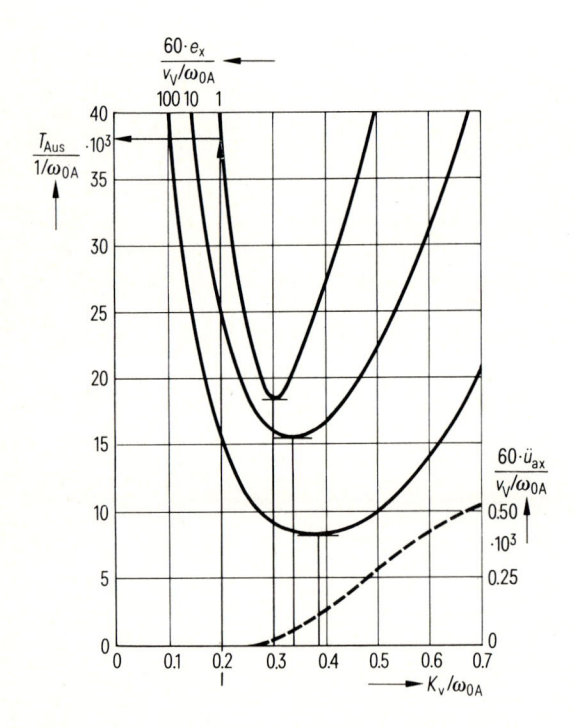

Fig. 1.26
Normalized settling time (—) for different transient response tolerances, and normalized overshoot width (-----). Plotted over K_v/ω_{0A}. (Numeric values are valid for the units given in the symbols summary. See the following example)

At high tolerances (e.g. 100), the curve at minimum is flat.
At low tolerances (e.g. 1), the minimum shows more pronounced.

For $\frac{K_v}{\omega_{0A}} \leq 0.3$, no overshoot occurs.

Conclusions:

▷ At low tolerances, the minimum is significant for the optimal K_v value for positioning. Short settling times result from high ω_{0A} values, and a K_v setting of K_v approximately $0.3\omega_{0A}$.
▷ Since overshoots are generally not permitted, a value of $K_v \geq 0.3\omega_{0A}$ cannot be set for a given ω_{0A}.
▷ With a given ω_{0A}, a setting $K_v \geq 0.2\omega_{0A}$ must be required, since otherwise T_{Aus} would become too large. (Longer non-productive times.)

For positioning, the requirements are thus that:

The nominal angular frequency of the drive ω_{0A} should be as high as possible, and the position loop gain $K_v = (0.2 \cdots 0.3)\omega_{0A}$.

Example of determining the settling time T_{Aus} according to fig. 1.26:

A position should be approached with a velocity of $v_V = 6$ m/min; the positioning process should hereby be considered terminated when the actual position value remains in the $2e_x$ range $= 0.002$ mm. The position loop gain is set at $K_v = 20$ s^{-1}, and the value of the nominal angular frequency is $\omega_{0A} = 100$ s^{-1}.

With these values, for the normalized position loop gain and the normalized transient response tolerance, we obtain:

$$K_v/\omega_{0A} = 0.2 \quad \text{und} \quad \frac{60 \cdot e_x}{v_V/\omega_{0A}} = 1$$

With these values, the normalized settling time derived from fig. 1.26, is:

$$\frac{T_{Aus}}{1/\omega_{0A}} \approx 38 \cdot 10^3$$

Then the settling time proper, has a value of approximately 380 ms.

If the positioning velocity v_V and the desired tolerance range $\pm e_x$ are given, the settling time. T_{Aus} or the required position loop gain K_v can be read out of fig. 1.27. The assumption here is, that the drive's nominal frequency $\omega_{0A} = K_v/0.3$.

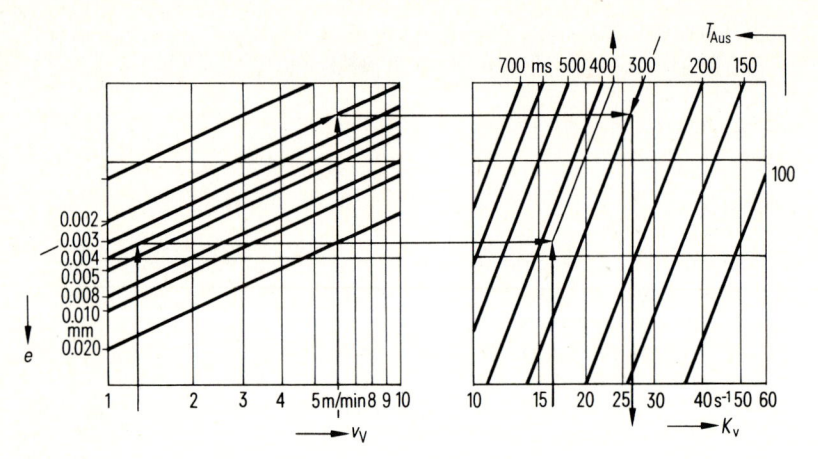

Fig. 1.27
Nomogram for determining the settling time T_{Aus} or the position loop gain (under the condition that $K_v/\omega_{0A}=0.3$).

Examples to fig. 1.27:

a) Given: $v_V = 6$ m/min, $e_x = 0.002$ mm, $T_{Aus} = 300$ ms
   Find: $K_v = ?$

   From the nomogram, the required position loop gain should have a value $K_v \approx 26$ s^{-1}. The nominal angular frequency required in this case, is $\omega_{0A} \geq 90$ s^{-1}.

b) Given: $v_v = 1.2$ m/min, $e_x = 0.004$ mm, $K_v = 16.6$ s^{-1}
   Find: $T_{Aus} = ?$

   From the nomogram, the settling time at positioning has a value of $T_{Aus} \approx 360$ ms.

   If a shorter settling time is required, the position loop gain must be set higher; this requires a higher drive nominal angular frequency. At $K_v/\omega_{0A} > 0.3$ overshooting occurs; at $K_v/\omega_{0A} < 0.3$, T_{Aus} increases.

1.3.3 Ramp Distance for Thread Cutting

For thread cutting on a lathe, the position command value x_s of the feed drive is commonly generated from the main spindle actual position, through a digital evaluation logic. Therefore, before the start of the cutting process, a ramp distance must be traveled. This ramp distance is necessary, in order to accelerate the feed drive up to the required feed rate. If thread cutting starts already during the acceleration phase, pitch errors occur because of the varying speed and following error.

The required ramp distance x_{an} results, when for a linearly increasing position command value x_s, the position actual value x_i remains within a tolerance range of $\pm e_x$ around a straight line, which shifts over time according to the following error. This shift over time is determined by the position loop gain.

The time is $t = \dfrac{1}{K_v}$.

This ramp process is shown in fig. 1.28.

The magnitude of the ramp distance depends on the value of the set position loop gain. When determining the ramp distance in relation with K_v/ω_{0A}, the feed rate v_V, and the tolerance range $2e_x$, the solution is similar to that obtained when determining the settling time at positioning (see fig. 1.29).

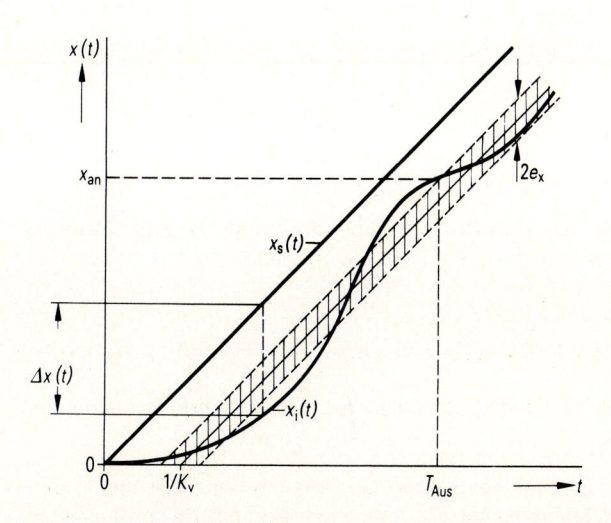

Fig. 1.28 Determining the ramp distance x_{an} at thread cutting

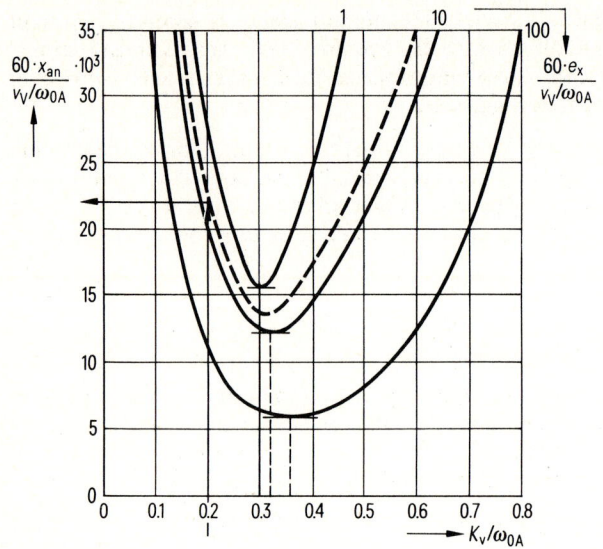

Fig. 1.29
Referenced ramp distance in dependence on K_v/ω_{0A} and on the referenced tolerance range $\dfrac{60\,e_x}{v_V/\omega_{0A}}$, at external thread cutting (numerical values apply as shown in the symbols summary. See examples following)

From fig. 1.29, we derive:

Depending on the magnitude of the settling tolerance, the minimal ramp distance is between

$$0.3 \leq \frac{K_v}{\omega_{0A}} \leq 0.36$$

At high tolerances, the curve obtained is flat at the minimum.

At lower tolerances, the minimum is more emphasized; thus the ramp distance is significantly stretched by a deviation from the optimum.

Conclusions:

▷ Short ramp distances are obtained with high ω_{0A}, and a K_v setting of approximately $0.3\,\omega_{0A}$.
▷ $K_v \geq 0.2\,\omega_{0A}$ should be adhered to, otherwise the ramp distances required would be too large.

For minimal ramp distances at thread cutting, the conditions that apply are: the drive nominal angular frequency ω_{0A} must be high, and the position loop gain must be in the range $K_v = (0.2 \cdots 0.3)\,\omega_{0A}$.

Example to fig. 1.29:

At a feed rate $v_V = 1.8$ m/min, a thread is to be cut with a maximal tolerance of $0.00 \cdots 0.002$ mm. Find the necessary ramp distance.

Given: $K_v = 20 \text{ s}^{-1}$, $\omega_{0A} = 100 \text{ s}^{-1}$, $v_V = 1.8$ m/min, $e_x = 0.002$ mm

It follows that:

$$K_v/\omega_{0A} = 0.2 \quad \text{and} \quad \frac{60 \cdot e_x}{v_V/\omega_{0A}} = 6.67$$

According to fig. 1.29, the necessary ramp distance is derived from

$$\frac{60 \cdot x_{an}}{v_V/\omega_{0A}} \approx 22 \cdot 10^3 \quad \text{to} \quad x_{an} \approx 6.6 \text{ mm}$$

If this value is too high, then K_v/ω_{0A} must be increased to the maximum 0.3.

The nomogram in fig. 1.30 is derived from fig. 1.29. In this case, the reference value is no longer the feed rate v_V, but the surface speed v_{Sch} on the cutting tool. With fig. 1.30 and values given for:

- v_{Sch} surface speed
- h_G thread lead
- d_G thread diameter
- e_x desired tolerance,

either the ramp distance x_{an} for a given position loop gain K_v,

or the required position loop gain K_v for a selected ramp distance x_{an}, can be determined.

Examples of use of the nomogram:

Thread M48 × 5   Starting values $K_v = 20 \text{ s}^{-1}$
Surface speed 40 m/min $\omega_{0A} = 100 \text{ s}^{-1}$
Max. deviation 0.01 mm

Find the necessary ramp distance x_{an}.

$$\frac{h_G}{d_G} = \frac{5}{48} \approx 0.1 \quad \text{and} \quad \frac{K_v}{\omega_{0A}} = 0.2$$

According to the ① lines, the ramp distance is $x_{an} \approx 3$ mm.

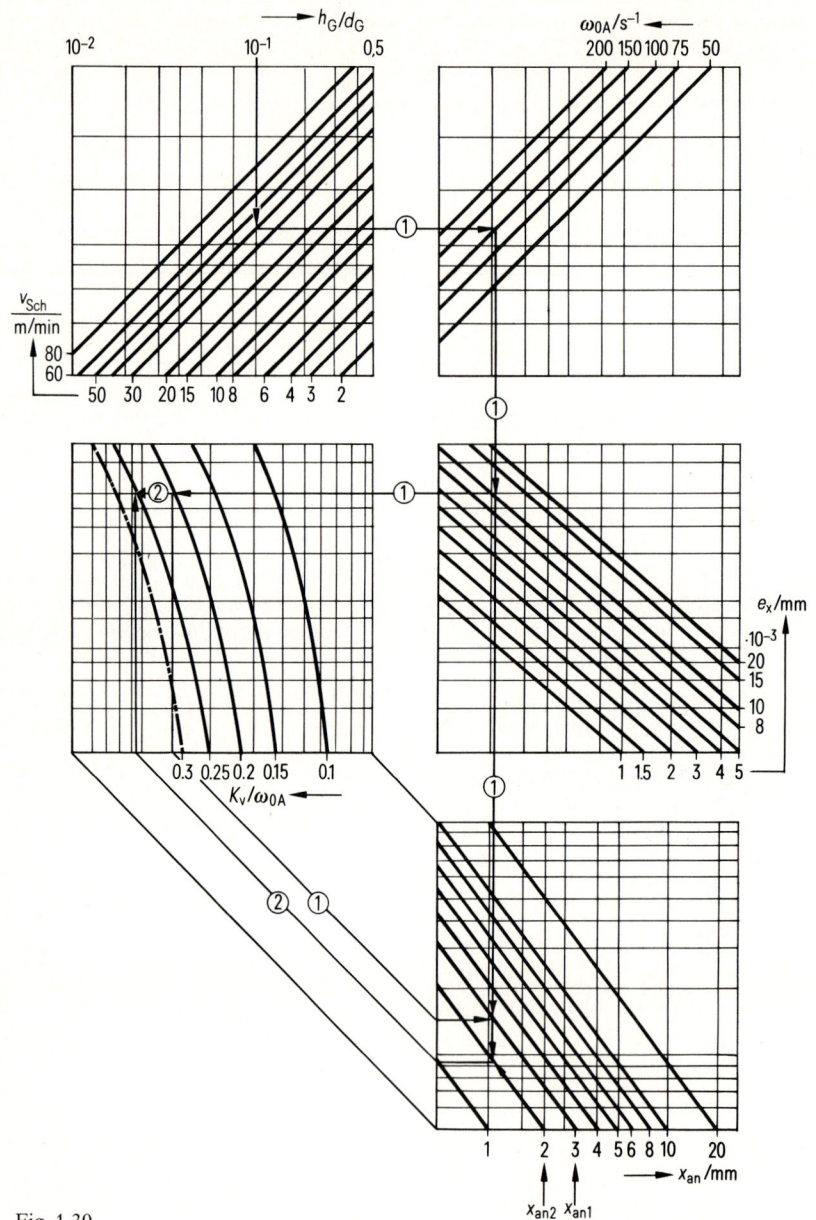

Fig. 1.30
Nomogram for determining the ramp distance x_{an} at thread cutting with given position loop gain, resp. determining the position loop gain K_v when the ramp distance x_{an} is given

If the ramp distance can be only $x_{an\,2} \approx 2$ mm, then, going backwards according to the ② line, we obtain for $K_v/\omega_{0A} = 0.25$; for this, the position loop gain K_v must be increased to $K_v = 25$ s^{-1}.

1.3.4 Contour Deviations on Corners

Contour deviations are a measure for the control quality of the position control loop. The relationship K_v/ω_{0A}, as is the case at positioning and thread cutting, is the deciding factor in contour deviations during contour directional changes (it is presumed that both axes are under the same dynamic conditions). The requirements on this ratio can be determined according to the contour deviations allowed [1.6].

A qualitative curve as in fig. 1.31 can be obtained for the course of the actual contour at traveling around a corner with no stop, with a corner angle β.

In this case, the contour distortion is characterized by the corner deviation e_E, respectively the overshoot width $\ddot{u}_a$:

The *corner deviation* e_E is the smallest distance from the actual contour to the corner point.

The *overshoot width* $\ddot{u}_a$ is the maximal deviation of the actual contour from the command contour.

The larger value of e_E and $\ddot{u}_a$ serves as characteristic value for the measurement of the position loop gain. The maximal corner deviation becomes:

$$e_{\max E} = \max \{e_E, \ddot{u}_a\}$$

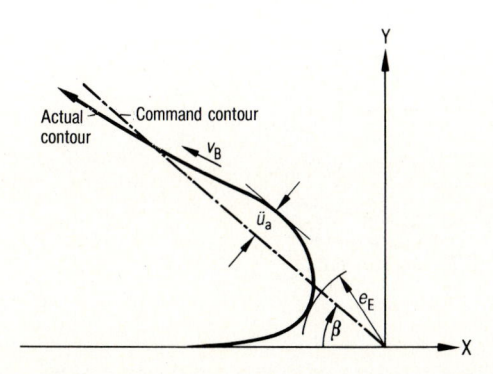

Fig. 1.31
Contour deviations at travel around a corner with no stop, with corner angle β

This characteristic value is presented in fig. 1.32, for corner angle $\beta = 30°$, $90°$ and $150°$, referenced over K_v/ω_{0A}.

From this, it follows:

▷ For $\beta = 90°$, the minimum of the maximal corner deviation is at $K_v/\omega_{0A} = 0.5$.
▷ $K_v/\omega_{0A} \leq 0.3$ should be adhered to, because otherwise the contour deviations increase considerably.

The referenced presentation in fig. 1.32 is possible, due to the fact that contour deviations at travel around corners are proportional to the contour velocity v_B, and inversely proportional to the nominal angular frequency of the drive ω_{0A}. The contour velocity v_B is the combination of the feed rates of the two axes X and Y. It is assumed that the command values x_s and y_s have not, in any way, been modified (see chapter 5).

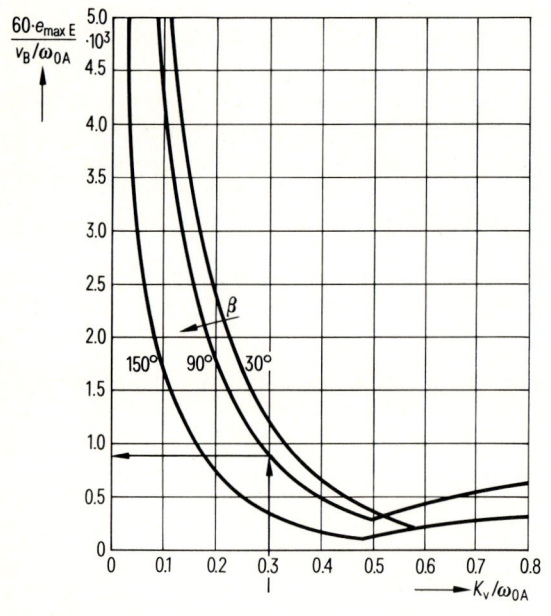

Fig. 1.32
Referenced maximal contour deviation at travel around corners in dependent relation with K_v/ω_{0A} and the corner angle β. (Numerical values apply according to the symbols summary units. See the following example)

Example to fig. 1.32:

Travel around a 90° corner with $v_B = 1$ m/min, with $K_v = 30$ s^{-1} and $\omega_{0A} = 100$ s^{-1}. From this, the resulting referenced position loop gain is $K_v/\omega_{0A} = 0.3$. From fig. 1.32, the maximal contour deviation thus is

$$\frac{60 \cdot e_{\max E}}{v_B/\omega_{0A}} \approx 0.9 \cdot 10^3$$

The actual deviation is then

$$e_{\max E} \approx \frac{0.9 \cdot 10^3 \cdot 1}{60 \cdot 100} \text{ mm} = 0.15 \text{ mm}.$$

This value represents a relatively large contour deviation.

If we presume an admissible deviation and consider $K_v = 0.3 \omega_{0A}$ as the set value, then the necessary position loop gain at a 90° corner is

$$K_v \approx 4{,}5 \cdot \frac{v_B}{e_{\max zul\, E}}$$

(where $e_{\max zul\, E}$ is the maximal allowed corner deviation).

With the values in the example above, at $v_B = 1$ m/min and $e_{\max zul\, E} = 0.05$ mm, $K_v = 90$ s^{-1} and $\omega_{0A} = 300$ s^{-1}.

Such values can be reached on electrical drives for K_v and ω_{0A} only with high quality devices and careful dimensioning. It further follows, that small contour deviations can be achieved only at low contour velocities, or with a command value modifier in the numerical control (see chapter 5).

1.3.5 Contour Deviations on Circles

For the test contour circle, a curve for the actual contour can be obtained qualitatively, as presented in fig. 1.33.

The measure used for the contour distortion is the maximal difference between the command and the actual radius

$$e_{\max Kr} = \max\left\{r - r_i\right\}$$

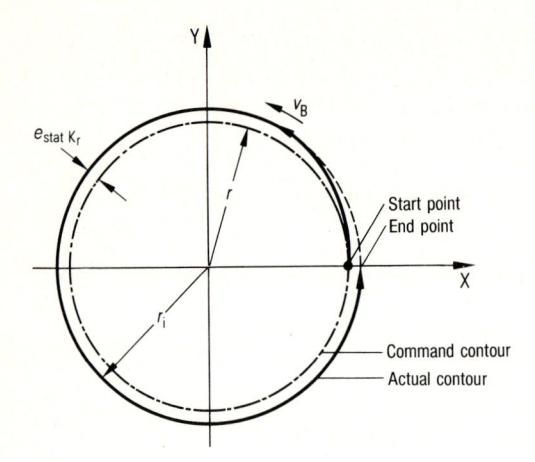

Fig. 1.33
Actual and command contour at traveling of circles. Starting point $x = r$, $y = 0$

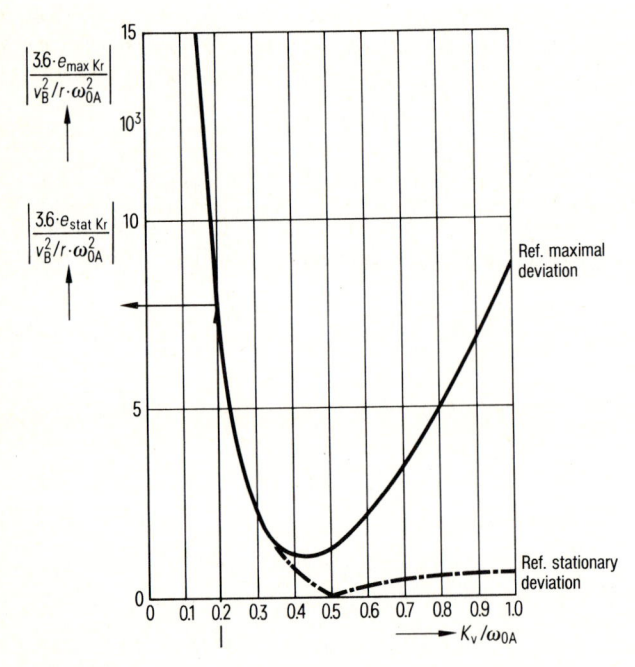

Fig. 1.34
Values of the referenced maximal and stationary deviations at traveling of circles, referenced over K_v/ω_{0A} (Numerical values apply for the units given in the symbols summary. See following example)

There are two different cases which dictate the procedure to be followed to determine this maximal deviation:

1st Case:

Prior to traveling the circle, the position control loop is in a settled (stationary) state.

In this case, the maximal contour deviation is the same as the stationary deviation $e_{\text{stat Kr}}$.

2nd Case:

The circle contour will be traveled from standstill.

In this case, in addition to the statinary deviation, errors due to settling action also occur. The deviations can be determined from the command frequency response curve of the position control loop, according to equation (1.47.1). For a contour feed rate in the range $v_B < 0.1 \, r \cdot \omega_{0A}$, common on machining with machine tools, we obtain curves as shown in fig. 1.34, for the values of the referenced maximal deviation $\dfrac{3.6 \, e_{\text{stat Kr}}}{v_B^2 / \omega_{0A}^2}$ over K_v / ω_{0A}.

From fig. 1.34, we derive:

At $K_v / \omega_{0A} \leqq 0.3$, the deviations $e_{\text{max Kr}}$ and $e_{\text{stat Kr}}$ are the same.

Minimal values for the contour deviation are obtained for:

$\dfrac{K_v}{\omega_{0A}} = 0.5$ for the 1st case, and

$\dfrac{K_v}{\omega_{0A}} = 0.45$ for the 2nd case.

The value of the maximal deviation is proportional to v_B^2 / r.

Example:

On a milling machine, a circle contour of radius $r = 100$ mm is to be cut with a contour velocity $v_B = 1$ m/min, starting from standstill. Find the maximal deviation $e_{\text{max Kr}}$.

Given values:

$K_v = 20 \text{ s}^{-1}, \quad \omega_{0A} = 100 \text{ s}^{-1}, \quad r = 100 \text{ mm}, \quad v_B = 1 \text{ m/min}.$

Since $K_v < 0.3\,\omega_{0A}$, and $v_B < 0.1\,r\omega_{0A}$ are met conditions, from fig. 1.34, we derive:

$$\frac{3.6 \cdot e_{\max Kr}}{v_B^2/r \cdot \omega_{0A}^2} = \frac{3.6 \cdot e_{\text{stat} Kr}}{v_B^2/r \cdot \omega_{0A}^2} \approx 7.6 \cdot 10^3$$

Thus $e_{\max Kr} = e_{\text{stat} Kr} \approx 0.0021$ mm.

This error is significantly below the finest fit tolerance IT5, which allows for a round fit of diameter $2r = 200$ mm, a deviation of 0.01 mm.

It follows from this:

▷ A $K_v = 5\omega_{0A}$ requirement is not appropriate.
▷ The stated requirement for positioning

$$0.2 \leq \frac{K_v}{\omega_{0A}} \leq 0.3$$

is also sufficient for traveling of circles in the contouring mode.

1.3.6 Disturbance Transient Response

The structure of the simplified position control loop of fig. 1.24 is presented as a block diagram in fig. 1.35, for the use in diagnosis of the disturbance behavior. It is presumed here, that the control behavior of the feed drive corresponds to that of a 2nd order delay element with a damping gradient $D_A = 0.5$. The 2nd order delay element of the drive's behavior is hereby substituted by the serial connection of two integral elements with feedback.

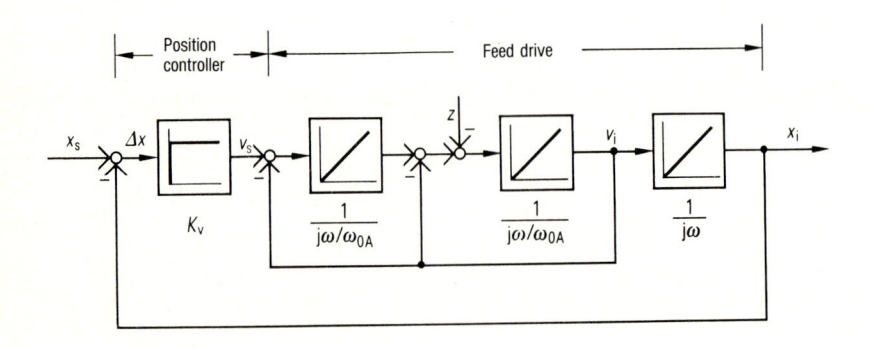

Fig. 1.35
Altered block diagram of the position control loop for determining the disturbance transient response

The disturbance frequency response curve is

$$F_z(j\omega) = \frac{x_z(j\omega)}{z(j\omega)} = \frac{1}{\omega_{0A}} \cdot \frac{-j\omega \dfrac{1}{\omega_{0A}}}{\dfrac{K_v}{\omega_{0A}} + j\omega \dfrac{1}{\omega_{0A}} + (j\omega)^2 \dfrac{1}{\omega_{0A}^2} + (j\omega)^3 \dfrac{1}{\omega_{0A}^3}}. \qquad (1.50)$$

In the range $0.2\omega_{0A} \leq K_v \leq 0.3\omega_{0A}$, the maximal position deviation, resulting from a jumping disturbance variable z (feed force F_V), can be calculated to approximately

$$x_{\max z} \approx \frac{F_V \cdot h_{Sp}^2}{i^2 \cdot 4\pi^2} \cdot \frac{1}{\omega_{0A}^2 \cdot J_{Ges}} \qquad (1.51)$$

Where

F_v	feed force
h_{Sp}	feed screw lead
i	gear ratio
J_{Ges}	total moment of inertia reflected onto the motor shaft

From the relationship (1.51), it follows:

A favorable disturbance transient response of the position control loop can be achieved at high nominal angular frequencies of the drive, and thus high position loop gains. High gear ratios also improve the disturbance transient response, but it should be kept in mind that within the feed range, the drive should not reach any limits (current, voltage limits) if contour distortions are to be avoided. The selected motor speed should not be too high, and possible disadvantages of high gear ratios should be taken into consideration (see section 4.4).

A high total moment of inertia also results in a better disturbance transient response, but ω_{0A} is affected by J_{Ges} (see section 2.2.2.6). For this reason, it should be checked in each individual case whether a mass increase will actually result in an improvement of the disturbance transient response.

1.3.7 Requirements for the Position Loop Gain

The results of a theoretical diagnosis of the setting of the position loop gain of single operating cases, with a simplified linear control model of the position control loop (see figs. 1.24 and 1.35), are summarized in table 1.2.

The values obtained for the position loop gain are theoretical values which, due to the non-linearities present in actual position control loops (e.g. reversing errors, dead time, and limits), cannot be reached in real situations. A smaller

Table 1.2
Diagnosis results concerning the requirements for the position loop gain

Criterion	Operating mode	Optimal value	Remarks
Settling time	Positioning	$K_v = 0.3\omega_{0A}$	No overshoot for $K_v \leq 0.3\omega_{0A}$
Ramp distance	Thread cutting	$K_v = 0.3\omega_{0A}$	$0.2\omega_{0A} \leq K_v \leq 0.3\omega_{0A}$
Contour deviation	Milling, turning	$K_v = 0.5\omega_{0A}$	
Disturbance transient response	Milling, turning	No relative optimum	Maximal deviation $x_{max\,z}$ dependent on the nominal angular frequency of the drive (see equation (1.51))

Conclusion: a high value for the nominal angular frequency ω_{0A} of the drive is desired

position loop gain results in a longer settling time at positioning, and a larger ramp distance at thread cutting; at contouring, the contour deviations will also increase.

Reversing errors, results of a too low machine stiffness and backlash, skewing, errors in the position measuring system, etc., can lead to larger errors than would be possible in a linear system, due to the dynamic behavior of the position control loop.

In summary, we can conclude:

▷ A favorable position control loop behavior can be expected at a high nominal angular frequency of the drive ω_{0A}, and with a position loop gain setting in the range of

$$0.2\omega_{0A} \leq K_v \leq 0.3\omega_{0A} \tag{1.52}$$

▷ Increasing the position loop gain beyond this range results in overshoots in the position control loop.

1.4 Bibliography

[1.1] Hütte I: Theoretische Grundlagen, 28. Auflage: 1. Abschnitt Mathematik, S. 76 und S. 162. Berlin, Verlag von Wilhelm Ernst u. Sohn, 1955

[1.2] Fröhr, F.; Ortenburger, F.: Einführung in die elektronische Regelungstechnik, 5. Auflage. Berlin und München, Siemens AG, 1981

[1.3] Pestel, E.; Kollmann, E.: Grundlagen der Regelungstechnik. Braunschweig-Wiesbaden, Vieweg Verlag, 1979

[1.4] Oppelt, W.: Kleines Handbuch technischer Regelvorgänge. Weinheim/Bergstraße, Verlag Chemie GmbH 1964

[1.5] Best, R.: Die Theorie und Anwendung des Phase–locked Loops, Anhang F. Stuttgart, AT-Fachverlag GmbH, 1976

[1.6] Augsten, G.; Boelke, K.; Schmidt, D.: Die Lageregelung an Werkzeugmaschinen. Abschn. 2, Strukturen von Lageregelkreisen. Stuttgart, Selbstverlag des Instituts für Steuerungstechnik der Werkzeugmaschinen und Fertigungseinrichtungen der Universität Stuttgart, 1972

[1.7] Wilharm, H.: Aufstellung und Einsatz von mathematischen Modellen für mechanische Systeme. Siemens Forschungs- und Entwicklungsberichte, Band 3, Nr. 5, Seite 281 D–287 D. Berlin-Heidelberg-New York. Springer Verlag, 1974

[1.8] Wilharm, H.: Erstellung des nichtlinearen Modells und Folgeregelung für eine Vorschubachse bei Werkzeugmaschinen. Frankfurt/Main, VDI/VDE-Aussprachetag „Nichtlineare Regelsysteme", 1979

Boelke, K.: Analyse und Beurteilung von Lagesteuerungen für numerisch gesteuerte Werkzeugmaschinen. Berlin-Heidelberg-New York, Springer Verlag 1977

Stute, G.; Breuer, F.: Regelung an Werkzeugmaschinen. München, Wien, Carl Hanser-Verlag, 1981.

DIN 19221, Februar 1981: Formelzeichen der Regelungs- und Steuerungstechnik

DIN 19226, Mai 1968: Regelungstechnik und Steuerungstechnik

DIN 19229, Oktober 1975: Übertragungsverhalten dynamischer Systeme

DIN 44300, März 1972: Informationsverarbeitung

DIN 5483, Februar 1974, Zeitabhängige Größen, Formelzeichen

DIN 5487, November 1967: Fourier-Transformation und Laplace-Transformation, Formelzeichen

DIN 5490, April 1974: Gebrauch der Wörter bezogen, spezifisch, relativ, normiert und reduziert

2 DC Motors for Feed Drives

2.1 Requirements for the Feed Drive

The requirements for the feed drive are determined by the desired dynamic behavior for the superimposed position control loop, and by the loads resulting from the operating process, as well as the frictional conditions in the feed unit. A distinction is made here between stationary and non-stationary operation.

2.1.1 Requirements for Stationary Operation

▷ The required torques for overcoming the frictional and working forces, must be provided.
Working forces of up to 20,000 N occur on common drives. These require a torque range of up to 40 Nm on the motor shaft. For lathes, a reserve of up to twofold the rated motor torque should be available, for minutewise overloads.
▷ A high speed stiffness $\Delta M/\Delta n$ is required, for achieving equal velocity with different machining forces.
▷ A speed range of at least 1:1,000 must be available. Control ranges of larger than 1:10,000 are common, which means that at a maximum speed, e.g. 1,200 min^{-1}, the feed motor must run smoothly and without noticeable waviness, at a speed of 0.1 min^{-1}.
▷ Positioning of smallest distance elements (also called position increments Δs) to approximately 1–2 μm, should be possible. For a DC servo motor, this represents an angular rotation of approximately 2–5 angular minutes.
▷ A linearity must exist between the velocity command value v_s and the velocity actual value v_i.
▷ There should be no insensitivity in the range of feed velocity zero.

2.1.2 Requirements for Non-stationary Operation

▷ Good command transient response which allows only minute contour deviations and sufficient surface quality, is necessary. For this, the nominal angular frequency ω_{0A} must be between 100 s^{-1} and 400 s^{-1}, and the damping gradient D_A should be about 0.5.
▷ Good disturbance transient response is necessary to ensure short settling times after load changes, and to keep temporary position deviations small.

▷ The coupling of the tacho generator used for speed feedback must be stiff on the motor shaft, in order to avoid oscillations in the speed control loop.

2.2 DC Shunt Motor

From the point of view of its characteristics and behavior, the DC shunt motor is, at present, the most suitable electrical servo motor. It has an only slightly sloped speed/torque characteristic curve, and thus high speed stiffness and favorable transient behavior. With the required current converter units, its speed can be altered over a wide range.

2.2.1 Construction Forms

There are permanent magnet-excited and electrically excited forms of DC servo motors. They are equally suitable for feed drives from the point of view of their control properties, and the nominal angular frequencies ω_{0A} which they can achieve. The following presentation omits the disc rotor servo motors and motors with bell shaped armatures, which are used as drives only under special circumstances. Chapter 7 summarizes the technical data for the permanent magnet-excited DC servo motors 1HU.., and those for the electrically excited DC servo motors 1GS... The permanent magnet-excited motors are designed for rated speeds of 1,000–3,000 min^{-1}. They can be mounted in part directly to the feed screws, and an intermediate gear is possible as an alternative. Section 4.4 shows in some examples where a reduction gear between motor and feed screw can be advantageous, and where direct coupling is preferable.

The electrically excited motors have rated speeds of up to approximately 6,000 min^{-1}. They are usually coupled through gears.

2.2.1.1 Permanenet Magnet-excited DC Servo Motors of the Series 1HU..

The distinguishing characteristics are:

▷ Direct coupling of the motor to the feed screw is possible.
▷ Rated speed is up to 3,000 min^{-1}, thus gears are also possible.
▷ Protection IP44, respectively IP54 (also available in the force ventilated version IP21 for special application) (the type of protection designated according to DIN 40050).
▷ High overload capability.
▷ Motor inbuilt optional devices, e.g. resolver gears, pulse coder, holding brake.
▷ Insensitive to demagnetization over a wide range.

From the standpoint of design, there is a distinction between the principle of flux concentration, and the shell magnet principle. Figure 2.1 shows both construction forms in crossection 2.1.

In the *flux concentration principle,* the magnetic circuit contains pole sheets of soft magnetic material. These cause the flux concentration, and permit a high degree of magnetic exploitation of the armature. They result in an even magnetic flux. Due to the form of the pole shoes, the magnetic field is shaped accordingly, so that the slotted armature can run in and out of the field smoothly. The pole shoes also represent a good magnetic conductor for feedback of the armature cross flow, which makes demagnetization of the main poles impossible, even at high overcurrents. Because of the square construction form, magnet plates that are easily manufactured, can be used.

In the *shell magnet principle,* the excitation poles are made out of bent magnet shell segments. Here, the smoothness with which the armature runs in and out at the main pole is achieved through the strength of the magnetization. The permanent magnet is magnetized more weakly at the ends. With this construction principle, reactions from the armature's cross flow to the excitation field are kept minimal, by providing a larger air gap. At very high overcurrents, there is a danger of a partial demagnetization; therefore, the current limit in the associated converter unit must prevent prohibited overcurrents. Shell

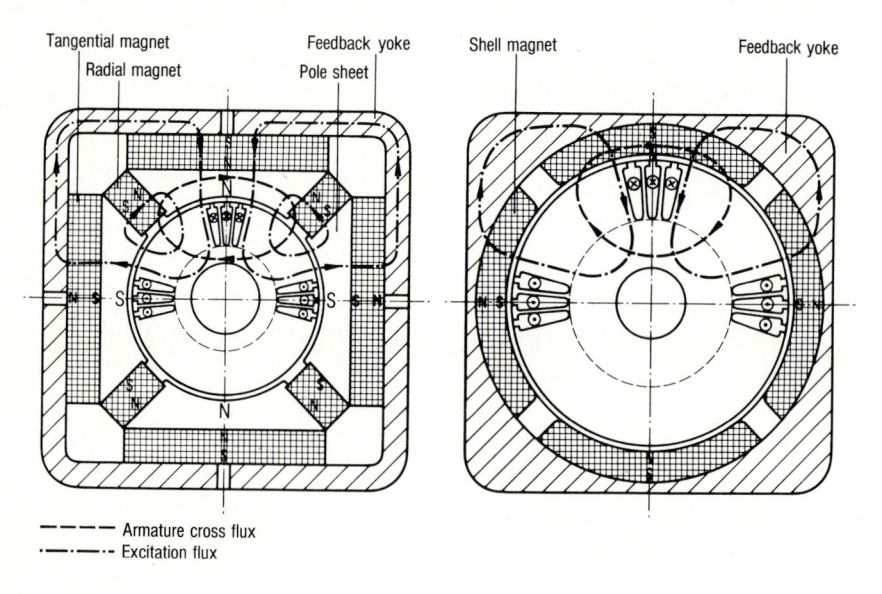

– – – – Armature cross flux
– · – · – Excitation flux

a) Flux concentration principle   b) Shell magnet principle

Fig. 2.1 Crossection through permanent magnet-excited DC servo motors

magnet motors are designed mostly with round crossection. A square construction offers the advantage of the feedback yoke being assembled out of stamping parts. The hight of the yoke is kept very small at the center of the flat sides; the excitation flux here is zero, thus producing a high magnetic resistance in the armature flow.

The magnet material for the permanent magnets used in both principles, is highly coercive ferric material in an aluminium-nickel-cobalt alloy, and lately also a combination of rare earth metals with cobalt. Each type of material has its own advantages and disadvantages on which we will not elaborate here [2.2].

Permanent magnet-excited DC servo motors are generally built without compensating poles. For this reason, when operating at high speeds, the commutating limit must be closely watched. The respective protection circuits of the current converter inhibit the flow of prohibited currents. Figure 2.2 shows as an example, the operating ranges for the DC servo motors of series 1HU3 1.., in the armature circuit version designed for a maximal speed value of $1{,}200 \text{ min}^{-1}$.

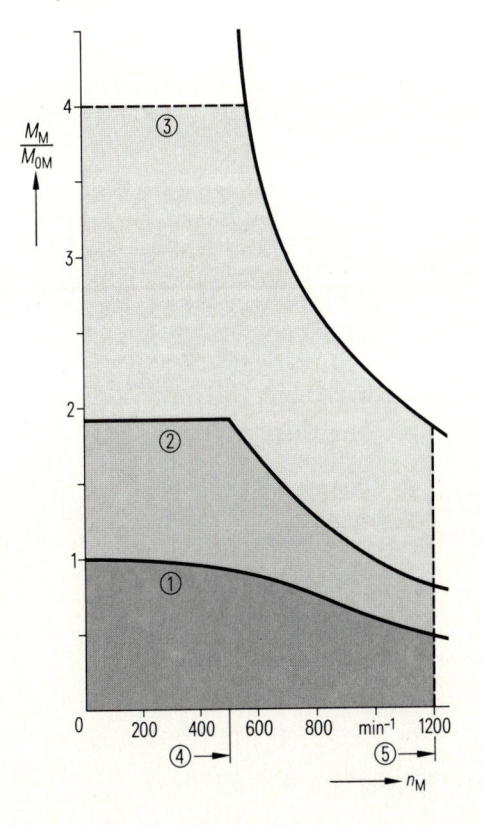

① Continuous duty S1
② Intermittent duty S3
③ Dynamic limit range with commutating limit curve
④ Feed range
⑤ Rapid traverse range

M_M Motor torque
M_{0M} Rated motor torque of DC servo motor
n_M Motor speed

Fig. 2.2
Operating ranges of the permanent magnet-excited DC servo motor 1HU3 1..

The motor torque is referenced to the rated motor torque M_{0M}. In the feed range, this torque is almost constant and can be supplied by the motor constantly, down to the lowest speeds. It is the determining factor for feed drives.

The individual ranges are characterized as follows (see VDE 0530):

① Continuous duty S1: With the exception of the standstill, all operating points M_M/n_M below this line are continuously possible. The limitation is given by the losses in the motor, and at higher speeds, by the commutating load. At longer standstills (longer than 5 min), approximately 50% of the rated torque is allowed (see technical data in chapter 7). In the feed rate range ④ up to approximately 500 min^{-1}, the torque permitted is approximately constant. For short term rapid traverse movements, the torque limited by the commutation is sufficient (range ⑤).

② Intermittent duty S3:
Higher torques are permitted for brief overloads. At higher speeds, the limitation is set mostly by the commutation.

③ Dynamic limit range:
This range is needed briefly for acceleration and deceleration processes. Out of economical considerations, when selecting the current converter, the current limit is chosen to be approximately four times the rated torque [2.3]. The commutating limit curve is simulated by a speed dependent function within the converter. This is given by the brush arcing still allowed for a brush lifetime of over 4,000 h.

The relatively high moment of inertia of permanent magnet-excited DC servo motors, as compared to that of the normally coupled external moment of inertia, allows the speed controller to be preset. The start-up requirement is thus kept low. It should be noted here, that the external moment of inertia, as reflected onto the motor shaft, should be only approximately 0.3 times the moment of inertia of the motor's rotor. If this is exceeded, limitations of the dynamic characteristic values must be traded, and an adjusted control setting becomes necessary (see section 2.2.2.6).

Due to their high heat storage capability, these motors are especially suited for feed drives which go through brief high, non-periodic overload occurrences. Such feed drives are used on numerically controlled machining centers, drill-presses, as well as milling machines and lathes. The mounting problems are not significant because of the closed rectangular construction form, so that today these motors are widely used.

2.2.1.2 Electrically Excited DC Servo Motors of the Series 1GS..

The distinguishing characteristics, are:

▷ Speed range is up to 6,000 min^{-1}.
▷ Mounted mostly in combination with gears.

▷ Mounted external ventilation, therefore protection type IP21.
(Closed construction form of type IP44 is possible for lower torques; protection type designated according to DIN 40050).
▷ Light weight, small dimensions.
▷ High power, wide range of constant torque.
▷ Linearity between current and torque, up to high overcurrents.
▷ Small mechanical and electrical time constants.

Figure 2.3 shows the basic difference between a commonly used DC motor, and the electrically excited DC servo motor. Both versions are shown as partial crossections, and have the same rated torque. The DC servo motor (right)

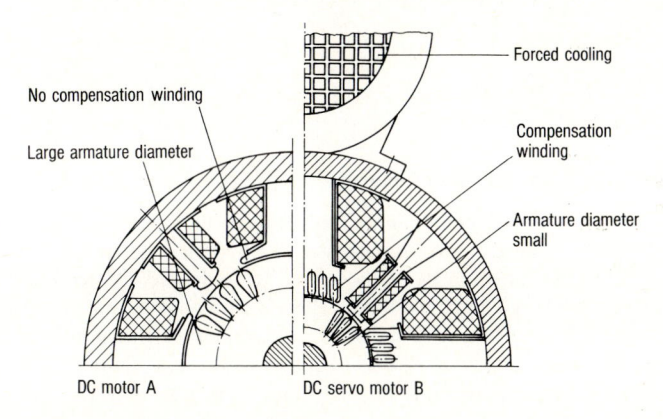

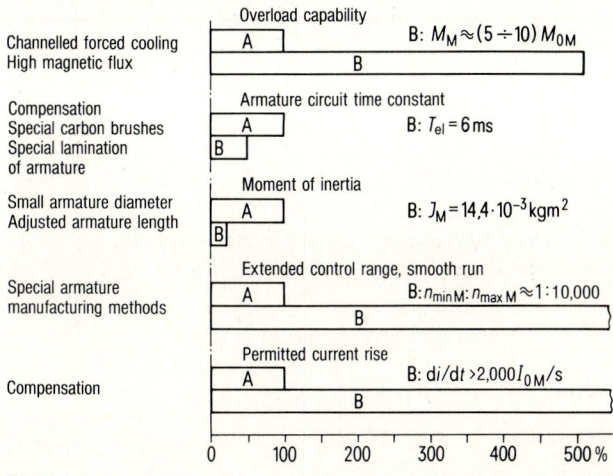

No compensation winding
Large armature diameter

Forced cooling
Compensation winding
Armature diameter small

DC motor A   DC servo motor B

Results of specific dimensioning measures

	Overload capability	
Channelled forced cooling High magnetic flux	A	B: $M_M \approx (5 \div 10) M_{0M}$
	B	
	Armature circuit time constant	
Compensation Special carbon brushes Special lamination of armature	A	B: $T_{el} = 6$ ms
	B	
	Moment of inertia	
Small armature diameter Adjusted armature length	A	B: $J_M = 14{,}4 \cdot 10^{-3}$ kgm²
	B	
	Extended control range, smooth run	
Special armature manufacturing methods	A	B: $n_{min M} : n_{max M} \approx 1 : 10{,}000$
	B	
	Permitted current rise	
Compensation	A	B: $di/dt > 2{,}000 I_{0M}/s$
	B	

0   100   200   300   400   500 %

Fig. 2.3
Comparison between a conventional DC motor and a DC servo motor

displays the compensation winding and the small armature diameter. The effects of these structures are presented at the bottom of the figure.

The operating ranges shown in fig. 2.2 for the permanent magnet-excited servo motor, are also valid qualitatively for the electrically excited DC servo motor. Due to the compensation poles used, the commutating limits are significantly higher, so that they generally can be disregarded when setting the current limit.

These servo motors can be advantageously used in situations where high torques are required at low weights, and where constantly high working torques with brief overloads occur. Machining tasks on turning automats are especially well handled with this type of motor. In addition, the dynamic requirements on large tracer mills are easier to achieve with these motors.

2.2.2 Drive Behavior

2.2.2.1 Differential Equation and Block Diagram

The substituted block diagram of the uncontrolled drive of fig. 2.4, is the basis for the analytical description of both stationary and dynamic behaviors. It is presumed here, that the impedances in the converter unit and the feeding line are added to the armature circuit resistance R_A and to the armature circuit inductivity L_A. The external moment of inertia J_{ext} combines all the moments of inertia of all mobile parts of the mechanical transmission elements reflected onto the motor shaft (see 4.2.5). The flux of excitation is constant over time.

The motor angular velocity is:

$$\omega_M = 2\pi \cdot n_M \tag{2.1}$$

as normalized dimensional equation, it is

$$\frac{\omega_M}{\text{rad} \cdot \text{s}^{-1}} = \frac{2\pi}{60} \frac{n_M}{\text{min}^{-1}} \tag{2.2}$$

The total moment of inertia reflected onto the motor shaft becomes:

$$J_{Ges} = J_M + J_{ext} \tag{2.3}$$

The basic equations for DC shunt motors at constant field excitation, are:

Voltage equation $\qquad u_A - e_M = R_A \cdot i_A + L_A \dfrac{di_A}{dt}$ \qquad (2.4)

Induced countervoltage $\qquad e_M = c_M \cdot \omega_M$ \qquad (2.5)

Torque equation $\qquad M_M = M_L + M_B = M_L + J_{Ges}\dfrac{d\omega_M}{dt}$ \qquad (2.6)

Motor torque $\quad M_M = c_M \cdot i_A$ (2.7)

Acceleration torque $\quad M_B = J_{Ges} \dfrac{d\omega_M}{dt}$ (2.8)

c_M is the motor constant; the selecting tables show usually two values for it:

$K_E \quad$ voltage constant, unit Vs/rad
(V/min^{-1} and V/1,000 min^{-1} are also common, see conversion factors in table 2.1)
$K_T \quad$ torque constant, unit Nm/A

In units Vs/rad and Nm/A, the values are identical:

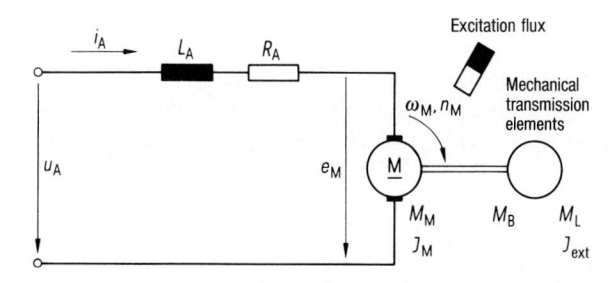

$$\dfrac{c_M}{\text{Vs/rad oder Nm/A}} = \dfrac{K_E}{\text{Vs/rad}} = \dfrac{K_T}{\text{Nm/A}} \qquad (2.9)$$

$\mu_A \quad$ time variable armature voltage
$e_M \quad$ time variable induced counter voltage in motor (another designation is "electro-motive force". When carrying opposite sign, this voltage is also known in the literature as source voltage, DIN 1323)
$i_A \quad$ time variable armature current
$R_A \quad$ total armature resistance of the feed drive, including brush contact resistance in the DC servo motor
$L_A \quad$ total armature inductivity of the feed drive
$\omega_M \quad$ motor angular velocity
$n_M \quad$ motor speed
$M_M \quad$ motor torque
$M_L \quad$ load torque reflected on motor shaft
$M_B \quad$ acceleration torque
$J_M \quad$ motor moment of inertia
$J_{ext} \quad$ external moment of inertia reflected on motor shaft

Fig. 2.4
Substituted block diagram of a drive with DC servo motor (permanent magnet-excited)

Table 2.1
Conversion table for the different units of the voltage constant K_E

Unit	Vs/rad	V/min^{-1}	V/1,000 min^{-1}
Vs/rad	1	0.1047	1.047 10^2
V/min^{-1}	9.549	1	1·10^3
V/1,000 min^{-1}	9.549·10^{-3}	1·10^{-3}	1

For the induced counter voltage, from equations (2.5) and (2.7), we obtain:

$$e_M = K_E \cdot \omega_M \tag{2.5.1}$$

and for the motor torque:

$$M_M = K_T \cdot i_A \tag{2.7.1}$$

The differential equation of the drive can be derived from equations (2.4)–(2.8). If we substitute in equation (2.4) the e_M of equation (2.5), and the i_A from equations (2.7) and (2.6), we obtain:

$$u_A - c_M \cdot \omega_M = R_A \left(\frac{M_L}{c_M} + \frac{J_{Ges}}{c_M} \cdot \frac{d\omega_M}{dt} \right) + \frac{L_A}{c_M} \cdot \frac{dM_L}{dt} + \frac{L_A \cdot J_{Ges}}{c_M} \cdot \frac{d^2\omega_M}{dt^2}$$

Dividing both sides by R_A and introducing the electrical time constant of the drive T_{elA}, with

$$T_{elA} = \frac{L_A}{R_A} \tag{2.10}$$

we obtain the equation:

$$\frac{u_A}{R_A} - \frac{c_M}{R_A} \cdot \omega_M = \frac{M_L}{c_M} + \frac{J_{Ges}}{c_M} \cdot \frac{d\omega_M}{dt} + \frac{T_{elA}}{c_M} \cdot \frac{dM_L}{dt} + \frac{T_{elA} \cdot J_{Ges}}{c_M} \cdot \frac{d^2\omega_M}{dt^2}$$

For convenience, the values of the equation are normalized as follows:

The motor angular velocity to the maximal motor angular velocity in idle

$$\omega_{maxM},$$

the armature voltage to the maximal armature voltage in idle

$$U_{maxA} = E_{maxM} = c_M \cdot \omega_{maxM} \tag{2.11}$$

and the load torque to the maximal torque of the drive with full armature voltage and the shaft blocked. This torque is designated as the short circuit torque of the drive

$$M_{StA} = I_{maxA} \cdot c_M = \frac{U_{maxA} \cdot c_M}{R_A} = \frac{\omega_{maxM} \cdot c_M^2}{R_A} \tag{2.12}$$

These normalized values are obtained by multiplying the differential equation by $R_A/U_{\max A}$:

$$T_{elA} \frac{J_{Ges}}{M_{StA}} \cdot \frac{d^2\omega_M}{dt^2} + \frac{J_{Ges}}{M_{StA}} \cdot \frac{d\omega_M}{dt} + \frac{\omega_M}{\omega_{\max M}} = \frac{u_A}{U_{\max A}} - \frac{M_L}{M_{StA}} - T_{elA} \frac{d\frac{M_L}{M_{StA}}}{dt}$$

Now, the mechanical time constant of the drive is defined as

$$T_{mechA} = \frac{J_{Ges} \cdot \omega_{\max M}}{M_{StA}} = \frac{J_{Ges} \cdot R_A}{c_M^2} \tag{2.13}$$

and we obtain the 2nd order differential equation for the DC motor

$$T_{elA} \cdot T_{mechA} \cdot \frac{d^2 \frac{\omega_M}{\omega_{\max M}}}{dt^2} + T_{mechA} \frac{d \frac{\omega_M}{\omega_{\max M}}}{dt} + \frac{\omega_M}{\omega_{\max M}}$$
$$= \frac{u_A}{U_{\max A}} - \frac{M_L}{M_{StA}} - T_{elA} \frac{d\frac{M_L}{M_{StA}}}{dt} \tag{2.14}$$

The solution of this equation for input voltage changes at constant M_L, results in the command response of the uncontrolled drive. The disturbance transient response is derived with constant voltage u_A, and a change in the load torque.

The behavior can be presented as a block diagram, which can be derived from the physical substituted block diagram of fig. 2.4 by using the basic equations (2.4)–(2.8). This block diagram is shown in fig. 2.5.

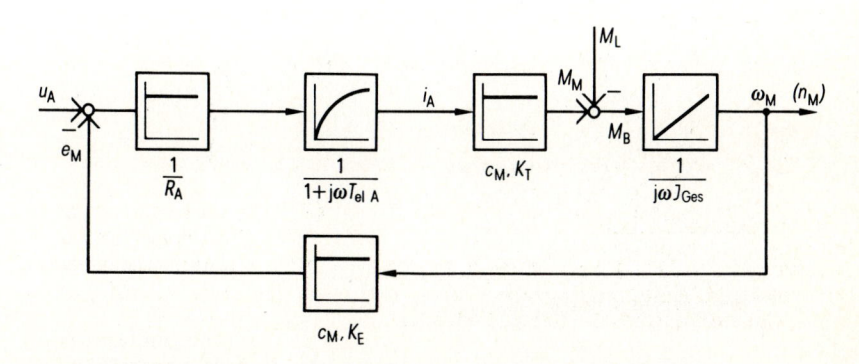

Fig. 2.5 Block diagram of the uncontrolled drive with DC servo motor

The blocks show the transient response functions. Behind the input block $1/R_A$, follows a 1st order delay element with the electrical time constant T_{elA}. The output value is the current i_A of equation (2.4). The next block represents the conversion of the motor torque M_M according to equation (2.7). The summation point shows equation (2.6) again, and the following integral element represents equation (2.8).

2.2.2.2 Stationary Behavior of the Uncontrolled Drive

In the stationary operation of the DC servo motor, the time derivations of equation (2.14) become zero. According to equation (2.6), at this point the motor torque equals the load torque. From equation (2.14), we then obtain:

$$\frac{M_L}{M_{StA}} = \frac{M_M}{M_{StA}} = \frac{U_A}{U_{maxA}} - \frac{\omega_M}{\omega_{maxM}} \tag{2.15}$$

and with equation (2.1)

$$\frac{M_L}{M_{StA}} = \frac{M_M}{M_{StA}} = \frac{U_A}{U_{maxA}} - \frac{n_M}{n_{maxM}} \tag{2.15.1}$$

Equation (2.15), respectively (2.15.1), describes analytically and in normalized form, the stationary characteristic curves of the drive with DC servo motor. The result is a characteristic curve field consisting of parallel straight lines, with the armature voltage U_A as a parameter (fig. 2.6). The operating ranges known from fig. 2.2, are represented by dotted lines. They show the physical limits of the DC servo motor, respectively of the entire feed drive, due to commutation, thermal losses, and current limit.

The speed stiffness of the uncontrolled drive is represented by the straight slope in the characteristic curve field of fig. 2.6. According to equations (2.15)

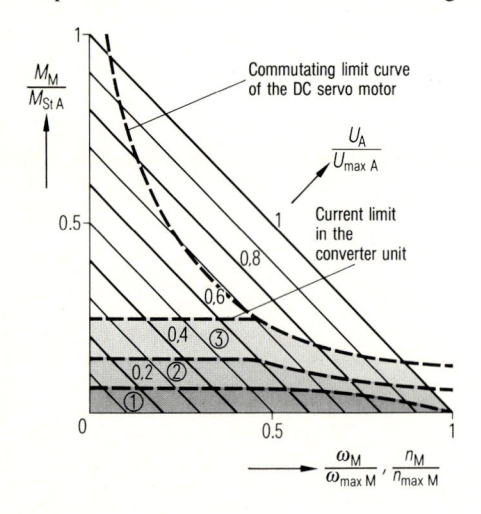

① Continuous duty S1
② Intermittent duty S3
③ Dynamic limit range

Fig. 2.6
Characteristic curve field of the uncontrolled drive with DC servo motor

and (2.12), for $U_A/U_{\max A}=\text{const.}$, it is

$$\frac{\Delta M_M}{\Delta \omega_M} = -\frac{M_{StA}}{\omega_{\max M}} = -\frac{c_M^2}{R_A} = -\frac{K_E \cdot K_T}{R_A} \qquad (2.16)$$

The adjusted dimensional equation for speed stiffness is obtained with equation (2.1)

$$\frac{\dfrac{\Delta M_M}{Nm}}{\dfrac{\Delta n_M}{\min^{-1}}} = -\frac{\dfrac{K_E}{V/1000\,\min^{-1}} \cdot \dfrac{K_T}{Nm/A}}{\dfrac{R_A}{\Omega}} \cdot 10^{-3} \qquad (2.16.1)$$

This speed stiffness should be as high as possible, so that when controlling load changes, the speed controller should not have to make large changes in its control angle.

Example:

The speed stiffness of an uncontrolled drive with a permanent magnet-excited DC servo motor 1HU3 104-0AD01, is: (the armature resistance $R_A = 0.6\,\Omega$, can be found in table 2.2, $K_E = 138\,V/1{,}000\,\min^{-1}$ and $K_T = 1.30\,Nm/A$).

$$\frac{\Delta M_M}{\Delta n_M} = -\frac{138 \cdot 1.30}{0.6} \cdot 10^{-3}\,Nm/\min^{-1} = -0.3\,Nm/\min^{-1}$$

The speed stiffness of an electrically excited DC servo motor 1GS3 107-5SV41 is: (the armature resistance $R_A = 0.7\,\Omega$ also found in table 2.2, $K_E = 30.8\,V/1{,}000\,\min^{-1}$, $K_T = 0.294\,Nm/A$).

$$\frac{\Delta M_M}{\Delta n_M} = -\frac{30.8 \cdot 0.294}{0.7} \cdot 10^{-3}\,Nm/\min^{-1} = -0.0129\,Nm/\min^{-1}$$

Even if the relationship of the maximal speeds of $1{,}200\,\min^{-1}$ to $6{,}000\,\min^{-1}$ is considered, the speed stiffness of drives with permanent magnet-excited DC servo motors is still about 4.6 times higher than that with electrically excited servo motors. Therefore, the permanent magnet-excited 1HU motor has a more favorable load change behavior.

The short-circuit torque defined in equation (2.12) is a mathematical value which can be calculated from the data list of the servo motors. From (2.12), we can obtain an adjusted dimensional equation for it by using equations (2.2) and (2.9)

$$\frac{M_{StA}}{Nm} = \frac{60}{2\pi} \cdot 10^{-3} \frac{\dfrac{U_{\max A}}{V} \cdot \dfrac{K_E}{V/1000\,\min^{-1}}}{\dfrac{R_A}{\Omega}} = \frac{\dfrac{n_{\max M}}{\min^{-1}} \cdot \dfrac{K_E}{V/1000\,\min^{-1}} \cdot \dfrac{K_T}{Nm/A}}{\dfrac{R_A}{\Omega}} \cdot 10^{-3} \qquad (2.12.1)$$

where $n_{\text{max M}}$ is the idle speed corresponding to the maximum current converter output voltage.

Example:

On a drive with the DC servo motor 1HU3 104-0AD01, at 200 V DC voltage the idle speed will be $n_{\text{max M}} \approx 1{,}350 \text{ min}^{-1}$, thus according to table 2.2 on page 90, $R_A = 0.6 \, \Omega$:

$$M_{\text{StA}} = \frac{1{,}350 \cdot 138 \cdot 1.30}{0.6} \cdot 10^{-3} \text{ Nm} = 404 \text{ Nm}$$

This represents approximately 16 times the rated torque.

At a maximal DC voltage of 200 V, the DC servo motor 1GS3 107-5SV41 has $n_{\text{max M}} \approx 7{,}000 \text{ min}^{-1}$ and, from table 2.2, $R_A = 0.7 \, \Omega$:

$$M_{\text{StA}} = \frac{7{,}000 \cdot 30.8 \cdot 0.294}{0.7} \cdot 10^{-3} \text{ Nm} = 90.6 \text{ Nm}$$

which represents approximately 13 times the rated torque.

2.2.2.3 Dynamic Behavior of the Uncontrolled Drive

The command transient response is given by the solution to equation (2.14) for a jumping intput voltage change. The solution for a jumping load torque change indicates the disturbance transient response. In the example of section 1.1.3.2, the solution was given for a voltage change when $T_{\text{el A}} \ll T_{\text{mech A}}$ and $M_L = 0$. We obtain for ω_M an exponential curve with the time constant $T_{\text{mech A}}$, the behavior of a 1st order delay element.

If $T_{\text{el A}}$ is not negligeable, the uncontrolled drive will behave like an oscillating element, also called a 2nd. order delay element. It is characterized by the damping gradient and the nominal angular frequency. The nominal angular frequency of an idle, uncontrolled DC servo drive ($M_L = 0$) can be determined from equation (2.14), through coefficient comparison with the basic equation (1.26.1), to be

$$\omega_{0A}^* = \sqrt{\frac{1}{T_{\text{mech A}} \cdot T_{\text{el A}}}} \qquad (2.17)$$

and for the damping gradient

$$D_A^* = \frac{1}{2} \sqrt{\frac{T_{\text{mech A}}}{T_{\text{el A}}}} \qquad (2.18)$$

(* represents the values of uncontrolled DC servo drives)

Figure 2.7. shows the transient response functions for a jumping change of the input voltage u_A in idle drive. Depending on the magnitudes of $T_{\text{el A}}$ and $T_{\text{mech A}}$, the uncontrolled drive will already display oscillating behavior. The damping gradient D_A^* determines the overshoot behavior.

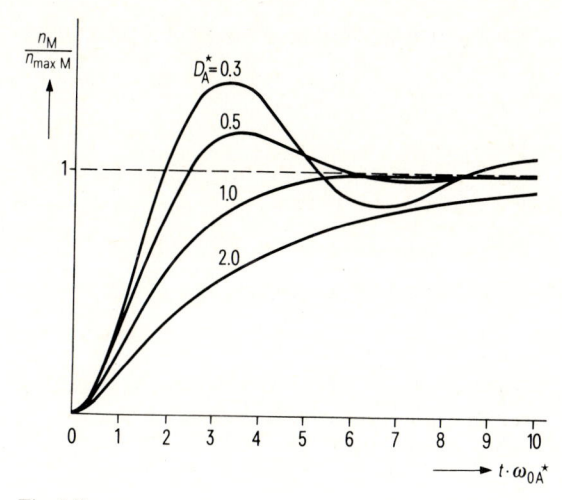

Fig. 2.7
Transient response functions of the idle, uncontrolled DC servo drive

The damping behavior of a drive can be improved through a constant frictional torque, e.g. with the friction of the slide guide. This is, for a feed drive, a desirable goal which should be kept in mind when designing the mechanical transmission elements.

Under the presumption that the drive is idle ($M_L = 0$), with equations (1.18) and (1.19) and with the ratio input variable/output variable determined in equation (2.14), we can determine the frequency response curve of the uncontrolled drive to be:

$$F_A^*(j\omega) = \frac{\dfrac{\omega_M}{\omega_{\max M}}(j\omega)}{\dfrac{u_A}{U_{\max A}}(j\omega)} = \frac{1}{1 + j\omega T_{\mathrm{mech\,A}} + (j\omega)^2 T_{\mathrm{mech\,A}} \cdot T_{\mathrm{el\,A}}} \tag{2.19}$$

With equations (2.17) and (2.18), we obtain

$$F_A^*(j\omega) = \frac{1}{1 + j\omega \dfrac{2D_A^*}{\omega_{0A}^*} + (j\omega)^2 \dfrac{1}{\omega_{0A}^{*2}}} \tag{2.19.1}$$

The frequency response curve of the idle, uncontrolled drive corresponds to the Bode diagram of fig. 1.14.

For calculations with list data of DC servo motors, the following adjusted dimensional equations are also stated:

Electrical time constant of the drive, from equation (2.10)

$$\frac{T_{elA}}{ms} = \frac{\frac{L_A}{mH}}{\frac{R_A}{\Omega}} \qquad (2.10.1)$$

Mechanical time constant of the drive, from equation (2.13)

$$\frac{T_{mechA}}{ms} = \frac{2\pi}{60} \cdot 10^3 \frac{\frac{J_{Ges}}{kg\,m^2} \cdot \frac{n_{maxM}}{min^{-1}}}{\frac{M_{StA}}{Nm}} = \frac{2\pi}{60} \cdot 10^6 \frac{\frac{J_{Ges}}{kg\,m^2} \cdot \frac{R_A}{\Omega}}{\frac{K_E}{V/1000\,min^{-1}} \cdot \frac{K_T}{Nm/A}} \qquad (2.13.1)$$

For example, for an uncontrolled, idle drive with the permanent magnet-excited DC servo motor 1HU3 104-0AD01, with the time constants in table 2.2, we derive, drive ①:

According to equation (2.17), the nominal angular frequency is

$$\omega_{0A}^* = \sqrt{\frac{1}{11.9 \cdot 10^{-3}\,s \cdot 10 \cdot 10^{-3}\,s}} = 97.7\,s^{-1}$$

and the damping gradient according to equation (2.18), is

$$D_A^* = \frac{1}{2}\sqrt{\frac{11.9\,ms}{10\,ms}} = 0.545$$

For a drive with the electrically excited DC servo motor 1GS3 104-5SV41 (drive ⑥ in table 2.2), we obtain:

The nominal angular frequency

$$\omega_{0A}^* = \sqrt{\frac{1}{24.3 \cdot 10^{-3} \cdot 10 \cdot 10^{-3}\,s}} = 64\,s^{-1}$$

and the damping gradient

$$D_A^* = \frac{1}{2}\sqrt{\frac{24,3\,ms}{10\,ms}} = 0,78$$

2.2.2.4 Dynamic Behavior of the Speed-controlled Drive

In order to ensure a good stationary and dynamic drive behavior, speed regulation is necessary. On an uncontrolled drive, this is accomplished with a proportional-integral acting control connected in front of it, and the motor speed

n_M is recorded by means of a tacho generator. The tacho generator voltage proportional to the actual speed n_i is fed back to the control input, to be compared to the command speed n_s proportional voltage.

The speed regulator controls the power section of a current converter which in turn supplies the armature voltage of the servo motor. Converter circuits have a delay time, which combined with the effects of the mechanical transmission elements, can be approximated as the dead time T_T of the drive system.

The frequency response curve of a proportional-integral acting control is, according to 2.4

$$F_{Rn}(j\omega) = K_{gn} \frac{1+j\omega T_{nn}}{j\omega T_{nn}}$$

K_{gn} and T_{nn} are the control parameters. They are designated:
K_{gn} gain
T_{nn} integral action time

For a speed controlled drive, under these assumptions, a block diagram as in fig. 2.8 can be produced.

The command frequency response curve of this control loop can be described, according to equation (1.35.1), as

$$F_A(j\omega) = \frac{n_i(j\omega)}{n_s(j\omega)} = \frac{v_i(j\omega)}{v_s(j\omega)} = \frac{1}{1+\dfrac{1}{F_{Rn} \cdot F_T \cdot F_A^*}} \qquad (2.20)$$

respectively

$$F_A(j\omega) = \frac{(1+j\omega T_{nn}) \cdot e^{-j\omega T_T}}{e^{-j\omega T_T} + j\omega \dfrac{T_{nn}}{K_{gn}}(1+K_{gn} \cdot e^{-j\omega T_T}) + (j\omega)^2 \dfrac{T_{nn}}{K_{gn}} \cdot \dfrac{2D_A^*}{\omega_{0A}^*} + (j\omega)^3 \dfrac{T_{nn}}{K_{gn}} \cdot \dfrac{1}{\omega_{0A}^{*2}}}$$

(2.20.1)

From this command frequency response curve equation, it can be seen that no damping gradient D_A and no nominal angular frequency ω_{0A} can be deter-

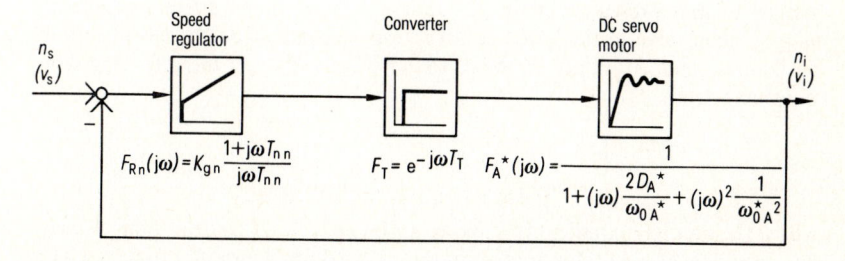

Fig. 2.8 Block diagram of the speed controlled DC servo drive

mined for the entire controlled drive. The total system must be adjusted so that it should behave in a manner approximate to that of a 2nd order delay element (oscillating element), with as high a nominal angular frequency as possible, and a damping gradient $D_A \approx 0.5$.

A high nominal angular frequency is required for the position control loop to reach a high position loop gain K_v, and thus achieve good dynamic properties (see section 1.3.7).

The frequency response curve of the so adjusted drive is, according to equation (1.41), without dead time effect

$$F_A(j\omega) = \frac{1}{1 + j\omega \dfrac{2D_A}{\omega_{0A}} + (j\omega)^2 \dfrac{1}{\omega_{0A}^2}}$$

and with dead time effect, according to equation (1.41.1)

$$F_A(j\omega) = \frac{e^{-j\omega T_T}}{1 + j\omega \dfrac{2D_A}{\omega_{0A}} + (j\omega)^2 \dfrac{1}{\omega_{0A}^2}}$$

Whether a dead time effect must be considered or not, is determined by the type of current converter. For line synchronized converters, there generally is a dead time to be considered (see section 2.2.2.7).

2.2.2.5 Calculating the Drive Nominal Angular Frequency

For the calculation of the drive nominal angular frequency, the mechanical and electrical time constants, $T_{\text{mech A}}$ and $T_{\text{el A}}$, of the drive, are necessary. $T_{\text{mech A}}$ is influenced by the external moment of inertia and by the sum of the armature circuit resistances (see equation (2.13)). $T_{\text{el A}}$ is affected also by the sum of armature resistances, and by the externally connected inductivities, e.g. circulating current and commutating reactors, determined by the converter circuit (see equation 2.10).

It can be shown through a computer simulation with a control loop structure similar to that presented in fig. 2.8, that for the nominal angular frequency $\omega_{0 \text{ max A}}$ achievable in the speed control loop, dependent on the electrical and mechanical time constants, in the range $1 \leq T_{\text{mech A}}/T_{\text{el A}} \leq 4$ the following relationship applies

$$\omega_{0 \text{ max A}} \approx \frac{1}{T_{\text{el A}}} \left(1 + \frac{1}{2 \cdot \dfrac{T_{\text{mech A}}}{T_{\text{el A}}}} \right) \tag{2.21}$$

where the damping gradient is assumed to be $D_A = 0.5$. The dead time element has been disregarded. This approximation is valid for the speed range in which the current limit is not yet being reached.

At velocity changes in the position control loops, in the range of machining velocities, attention must be paid that the subordinated control loops should not reach any limits. If this is allowed to happen, contour deviations will result (see section 1.2.3.4). If these limits are respected, the achievable nominal angular frequency $\omega_{\text{Grenz A}}$ will be determined by the set current limit and by the maximal necessary speed change. For this limit value of the nominal angular frequency of the drive, the following relationship applies:

$$\omega_{0\,\text{Grenz A}} \approx \frac{1}{T_{\text{el A}}} \cdot \frac{\dfrac{M_{\text{Grenz M}}}{M_{\text{St A}}} \cdot \dfrac{n_{\text{max M}}}{\Delta n_{\text{M}}}}{\dfrac{T_{\text{mech A}}}{T_{\text{el A}}} \cdot \dfrac{K_{\text{v}}}{\omega_{0\,\text{A}}}} \tag{2.22}$$

where

$M_{\text{Grenz M}}$ maximal achievable motor torque with current limit in the converter
$n_{\text{max M}}$ max. motor speed
$M_{\text{St A}}$ short-circuit torque of the motor (see equation (2.12))
Δn_{M} motor speed range required for machining, i.e. the max. speed change required by the motor without overshooting of limits and thus contour deviations
$T_{\text{el A}}$ electrical time constant of the motor (equation (2.10))
$T_{\text{mech A}}$ mechanical time constant of the motor (equation 2.13))
$K_{\text{v}}/\omega_{0\,\text{A}}$ required relationship between position loop gain and nominal angular frequency as determined in equation (1.52)

For a given drive, the values $M_{\text{Grenz M}}$, $M_{\text{St A}}$, $n_{\text{max M}}$, Δn_{M} and the time constants $T_{\text{mech A}}$ and $T_{\text{el A}}$, can be calculated.

We know from sec. 1.4, that a favorable behavior of the position control loop can be obtained with a position loop gain in the range $K_{\text{v}} = (0.2 \cdots 0.3)\,\omega_{0\,\text{A}}$. We can use this knowledge to simplify equation (2.22), and to present it graphically, in combination with equation (2.21). For DC servo drives, the ratios lie in the following range

$$\frac{M_{\text{Grenz M}}}{M_{\text{St A}}} \approx 0.1 \ldots 0.3$$

$$\frac{n_{\text{max M}}}{\Delta n_{\text{M}}} \approx 3 \ldots 5$$

If these values are substituted in equation (2.22), we obtain for the quotient

$$Q = \frac{\dfrac{M_{\text{Grenz M}}}{M_{\text{St A}}} \cdot \dfrac{n_{\text{max M}}}{\Delta n_{\text{M}}}}{\dfrac{K_{\text{v}}}{\omega_{0\,\text{A}}}} \approx 1 \ldots 7.5$$

and we can derive the limit value of the nominal angular frequency, from equation (2.22), to

$$\omega_{0\,\mathrm{Grenz\,A}} \approx \frac{1}{T_{\mathrm{el\,A}}} \cdot \frac{1\ldots 7{,}5}{\dfrac{T_{\mathrm{mech\,A}}}{T_{\mathrm{el\,A}}}} \tag{2.22.1}$$

Fig. 2.9 shows the product $\omega_{0\,\mathrm{max\,A}} \cdot T_{\mathrm{el\,A}}$ over $T_{\mathrm{mech\,A}}/T_{\mathrm{el\,A}}$, according to equation (2.21), and $\omega_{\mathrm{grenz\,A}} \cdot T_{\mathrm{el\,A}}$, according to equation (2.22). The curve $\omega_{0\,\mathrm{max\,A}} \cdot T_{\mathrm{el\,A}}$ shows the maximum possible nominal angular frequency; the curve sheaf $\omega_{0\,\mathrm{max\,A}} \cdot T_{\mathrm{el\,A}}$ indicates the limitation on $\omega_{0\,\mathrm{A}}$, determined by the current converter and the desired linear transfer range. $\omega_{0\,\mathrm{Grenz\,A}}$ can be influenced to a certain extent with the choice of the drive components.

For evaluating the achievable nominal angular frequency $\omega_{0\,\mathrm{A}}$ of a drive system, the smaller value of $\omega_{0\,\mathrm{max\,A}}$ or of $\omega_{0\,\mathrm{Grenz\,A}}$ is the deciding factor. Therefore, considering the current limit and the speed range, the following relationship applies:

$$\omega_{0\,\mathrm{A}} \approx \min \left\{ \omega_{0\,\mathrm{max\,A}},\ \omega_{0\,\mathrm{Grenz\,A}} \right\} \tag{2.23}$$

From fig. 2.9, it is apparent that at high $T_{\mathrm{mech\,A}}/T_{\mathrm{el\,A}}$ values, the size of the attainable nominal angular frequency $\omega_{0\,\mathrm{A}}$ is mostly determined by the selection of current converter with its limit current, and the achivable idle speed. At low values of $T_{\mathrm{mech\,A}}/T_{\mathrm{el\,A}}$, this ratio itself determines the limitation of the drive nominal angular frequency.

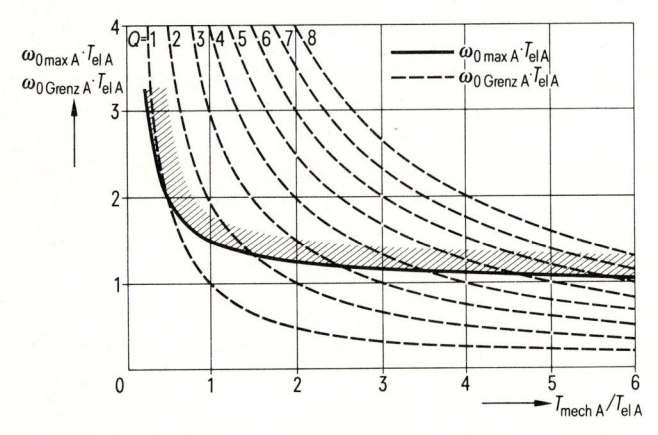

Fig. 2.9
Maximal nominal angular frequency $\omega_{0\,\mathrm{max\,A}} \cdot T_{\mathrm{el\,A}}$ and the limit angular frequency $\omega_{0\,\mathrm{Grenz\,A}} \cdot T_{\mathrm{el\,A}}$, as function of the ratio $T_{\mathrm{mech\,A}}/T_{\mathrm{el\,A}}$, with $D_{\mathrm{A}}=0.5$

In table 2,2. the attainable nominal angular frequencies of some DC servo drives are presented, based on the following presumptions:

▷ The torque is fixed by the current limit at four times the rated torque.
▷ The speed range for the machining is 40% of the idle speed, and so

$$\frac{n_{\max M}}{\Delta n_M} = 2{,}5$$

▷ The selected position loop gain is $K_v = 0.2\,\omega_{0\,A}$.
▷ The external moment of inertia J_{ext} reflected onto the motor shaft corresponds to medium values of feed slides.
▷ The external armature circuit resistor R_L contains the resistances of the overcurrent relay, wiring, and inductivities.
▷ A current dependent saturation is taken into consideration for the armature inductivity L_A; the value is therefore slightly lower than the sum of $L_M + L_D$.

The following drives are presented for comparison:

① Permanent magnet-excited DC servo motor 1HU3 104-AD01 with 6-pulse line synchronized thyristor converter, in circulating current-free anti-parallel connection
Rated torque   25 Nm
Rapid traverse speed   1,200 min^{-1}
Direct coupling onto feed screw

② Permanent magnet-excited DC servo motor 1HU3 076-0AF01 with transistor current converter
Rated torque   10 Nm
Rapid traverse speed   3,000 min^{-1}
Gear ratio   2.5

③ Permanent magnet-excited DC servo motor 1HU3 056-0AC01 with transistor converter
Rated torque   4.5 Nm
Rapid traverse speed   2,000 min^{-1}
Gear ratio   2

④ Permanent magnet-excited DC servo motor 1HU3 100-0AC01 with transistor converter
Rated torque   7 Nm
Rapid traverse speed   2,000 min^{-1}
Gear ratio   2

⑤ Permanent magnet-excited DC servo motor 1HU3 132-0AF01 with 6-pulse line synchronized thyristor current converter, circulating current-conducting cross connection
Rated torque   47 Nm
Rapid traverse speed   1,500 min^{-1}
Direct coupling onto feed screw

⑥ Electrically excited DC servo motor 1GS3 107-5SV41 with 3-pulse line synchronized thyristor converter, in circulating current-conducting anti-parallel connection
Rated torque   6.8 Nm
Rapid traverse speed   6,000 min^{-1}
Gear ratio   4

High nominal angular frequencies can be achieved with drives ①, ②, ④, and ⑥ in the configuration presumed. The position loop gain of the position control loop can be set at $K_v = 25$ s^{-1} – 30 s^{-1}. Drive ③ can be set up to a position loop gain of $K_v = 20$ s^{-1}. Here, the limitation is given by the low damping gradient of the system. Since $T_{mech\,A}/T_{el\,A}$ is smaller than 1, equation (2.21) does not approximate the real behavior accurately enough. Drive ⑤ applies to the large machine tool range, where the behavior is determined by the larger mass moments of inertia. The mechanical transmission elements determine the possible accelerations. Here, numerically controlled axes are moved with a position loop gain of $K_v = 13 - 16$ s^{-1}, for which the calculated nominal angular frequency is sufficient.

When calculating drives ②, ③ and ④, the fact was neglected that for transistor converters usually a current regulation dependent on the speed regulation is used (see 3.4.3). This makes the control loop structure differ from fig. 2.8. and thus, the conditions for deriving equations (2.21) and (2.22). The measurement results in chapter 6 will therefore show in part, better nominal angular frequencies than those calculated here. Furthermore, due to higher gain in the speed regulator, the nominal angular frequency of the drive $\omega_{0\,A}$ might be higher. For drive ②, $\omega_{0\,A}$ was determined by the limit angular frequency $\omega_{0\,Grenz\,A}$. If the speed ratio $n_{max}/\Delta n_M$ were reduced, from the maximal attainable nominal angular frequency $\omega_{0\,max\,A}$ we could then derive for the nominal angular frequency of the drive $\omega_{0\,A}$, the value $\omega_{0\,A} \approx 290$ s^{-1}. This value is consistent with the values measured in fig. 6.15, where for a set damping gradient of $D_A \approx 0.5$, the values measured for the nominal frequency were $f_{0\,A} \approx 50$ Hz.

Table 2.2
Calculated nominal angular frequencies for DC servo drives

Value formula	Drive	①	②	③	④	⑤	⑥
Rated torque M_{0M}	Nm	25	10	4.5	7	47	6.8
Limit torque M_{GrenzM}	Nm	100	40	18	28	188	27.2
Idle speed n_{maxM}	min^{-1}	1,350	3,700	2,300	2,370	1,750	7,000
Voltage const. K_E	V/1000 min^{-1}	138	53	75	80	111	30,8
Torque const. K_T	Nm/A	1.30	0.501	0.72	0.765	1.05	0.294

Table 2.2 (continued)

Value formula / Drive		①	②	③	④	⑤	⑥
Inner motor resistance R_{A+B}	Ω	0.450	0.273	2.42	1.044	0.114	0.53
External circuit resistance R_L	Ω	0.15	0.1	0.2	0.16	0.066	0.17
$R_A = R_{A+B} + R_L$	Ω	0.6	0.37	2.6	1.2	0.18	0.7
Drive short-circuit torque M_{StA} from equation (2.12.1)	Nm	404	266	48.4	121	1130	90.6
Ratio M_{GrenzM}/M_{StA}		0.247	0.15	0.372	0.232	0.166	0.3
Quotient $Q = \dfrac{\dfrac{M_{GrenzM}}{M_{StA}} \cdot \dfrac{n_{maxM}}{\Delta n_M}}{\dfrac{K_v}{\omega_{0A}}}$		3.1	1.88	4.65	2.9	2.08	3.76
Motor moment of inertia J_M	kgm²	0.028	0.0065	0.0022	0.0086	0.11	0.002
Reflected external moment of inertia J_{ext}	kgm²	0.0056	0.0015	0.0005	0.0014	0.06	0.001
$J_{Ges} = J_M + J_{ext}$	kgm²	0.034	0.008	0.0027	0.01	0.17	0.003
Mech. time const. T_{mechA} from equation (2.13.1)	ms	11.9	11.7	13.4	20.5	27.6	24.3
Motor inductivity L_M	mH	8.9	2	56.7	11.9	5.2	0.71
Reactor inductivity L_D	mH	–	–	–	–	2.5	10
Armature circuit inductivity L_A	mH	6	1.5	45	10	5.5	7
Electrical time const. T_{elA} from (2.10.1)	ms	10	4.05	17.3	8.33	30.6	10
Ratio T_{mechA}/T_{elA}		1.19	2.88	0.777	2.46	0.902	2.43
From fig. 2.9, the smaller value $\omega_{0\,maxA} \cdot T_{elA}$ or $\omega_{0\,GrenzA} \cdot T_{elA}$		1.42	0.65	1.64	1.18	1.55	1.21
Drive nominal angular frequency ω_{0A}	s^{-1}	142	160	95	142	50.7	121
Nominal frequency f_{0A}	Hz	22.6	25.5	15.1	22.6	8.07	19.3

2.2.2.6 External Inertia Effect

In the examples calculated in table 2.2, with the exception of drive ②, the achievable nominal angular frequency is determined from the relationship of the time constants $T_{\text{mech A}}/T_{\text{el A}}$. The magnitude of $T_{\text{mech A}}$ is affected by the external moment of inertia J_{ext} reflected onto the motor shaft.

If the ratio J_{ext}/J_M is introduced, equation (2.21) becomes:

$$\omega_{0\,\text{max A}} \approx \frac{1}{T_{\text{el A}}} \left[1 + \frac{1}{2\dfrac{T^*_{\text{mech A}}}{T_{\text{el A}}}\left(1+\dfrac{J_{\text{ext}}}{J_M}\right)} \right] \tag{2.21.1}$$

(where $T^*_{\text{mech A}}$ is the mechanical time constant of the drive without the external moment of inertia). From this, we can determine how the attainable nominal angular frequency will be reduced under the influence of the external moment of inertia. For this purpose we divide $\omega_{0\,\text{max A}}$ according to equation (2.21.1), by the nominal angular frequency $\omega^*_{0\,\text{max A}}$ of the drive without external moment of inertia, and we obtain

$$\frac{\omega_{0\,\text{max A}}}{\omega^*_{0\,\text{max A}}} = \frac{\dfrac{1}{T_{\text{el A}}}\left[1+\dfrac{1}{2\dfrac{T^*_{\text{mech A}}}{T_{\text{el A}}}\left(1+\dfrac{J_{\text{ext}}}{J_M}\right)}\right]}{\dfrac{1}{T_{\text{el A}}}\left(1+\dfrac{1}{2\dfrac{T^*_{\text{mech A}}}{T_{\text{el A}}}}\right)}$$

$$\frac{\omega_{0\,\text{max A}}}{\omega^*_{0\,\text{max A}}} = \frac{1}{1+\dfrac{J_{\text{ext}}}{J_M}} + \frac{\dfrac{J_{\text{ext}}}{J_M}}{1+\dfrac{J_{\text{ext}}}{J_M}} \cdot \frac{1}{1+\dfrac{T_{\text{el A}}}{2\,T^*_{\text{mech A}}}} \tag{2.24}$$

(* designates the values of the drive, without external moment of inertia)

In fig. 2.10, the relationship $\omega_{0\,\text{max A}}/\omega^*_{0\,\text{max A}}$ is shown over J_{ext}/J_M. The parameter is $T^*_{\text{mech A}}/T_{\text{el A}}$. It is apparent from the diagram, that external moments of inertia reduce a drive's possible nominal angular frequency more strongly, the lower the value of the ratio $T^*_{\text{mech A}}/T_{\text{el A}}$ is.

Example:

For drive ① of table 2.2, from equation (2.13.1)

$$T^*_{\text{mech A}} = \frac{2\pi}{60} \cdot 10^3 \frac{0.028 \cdot 1350}{404}\,\text{ms} = 9.8\,\text{ms} \quad \text{and thus} \quad \frac{T^*_{\text{mech A}}}{T_{\text{el A}}} = 0.98$$

With an external moment of inertia equal to the motor's moment of inertia, the attainable nominal angular frequency of the drive is thus only about $0.8 \cdot \omega^*_{0\,\text{mech A}}$ of that of the drive without external moment of inertia.

For drive ⑥, we obtain:

$$T^*_{\text{mech A}} = \frac{2\pi}{60} \cdot 10^3 \frac{0.002 \cdot 7000}{90.6} \text{ ms} = 16.1 \text{ ms} \quad \text{and thus} \quad \frac{T^*_{\text{mech A}}}{T_{\text{el A}}} = 1.61$$

Here, for an external moment of inertia equal to the motor's moment of inertia, $\omega_{0\,\text{max A}}$ is still about $0.87 \cdot \omega^*_{0\,\text{max A}}$.

We can gather from these examples that the external moment of inertia for drives with permanent magnet-excited DC servo motors of the 1HU... series should amount to about 0.3 times the moment of inertia of the motor itself. For electrically excited DC servo motors, the external moment of inertia can be approximately equal to the motor inertia. In that case, the achievable nominal angular frequency of drive $\omega_{0\,\text{max A}}$ will be only about 10% lower than the absolute maximal value $\omega^*_{0\,\text{max A}}$. (For acceleration drives, another calculation method applies, according to section 2.3.5.2.)

For drive ② of table 2.2, according to equation (2.22), the deciding factor for the attainable nominal angular frequency is $\omega_{0\,\text{Grenz A}}$. A higher nominal angular frequency can be reached in this case by sizing the converter for a higher current limit; this can compensate for the effects of a larger external moment of inertia.

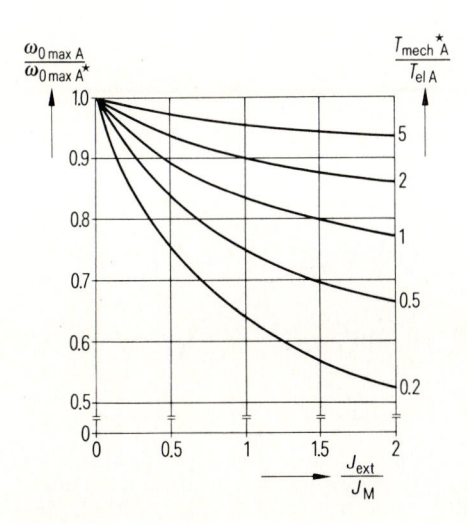

Fig. 2.10
Referenced nominal angular frequency $\omega_{0\,\text{max A}}$ dependently of the reduced referenced external moment of inertia J_{ext}

2.2.2.7 Dead Time Effect

The block diagram of the speed regulator of the feed drive presented in fig. 2.8 shows the effects of the converter and of the mechanical transmission elements, as a dead time element with the frequency response curve $F_T = e^{-j\omega T_T}$. The value of this dead time T_T depends on the system. Figure 1.19 shows the scheme of a position control loop with dead time element, and equation (1.41.1) indicates the frequency response curve of a drive with dead time element. The frequency response curve of a dead time element is presented in fig. 2.11.

The dead time element has a constant amplitude response curve, while the phase shift between input and output signals increases with increasing frequency. This means, that an additional phase shift occurs in the frequency response curve of the drive. The dead time always introduces an element of instability in the control loop. Therefore, depending on the magnitude of the dead time T_T, the position loop gain K_v must be reduced from its optimal setting accordingly.

Figure 2.12 shows dependently of the product $\omega_{0A} \cdot T_T$, the reduction of the referenced position loop gain K_v/ω_{0A}, starting from the values determined in equation (1.52), $0.2\,\omega_{0A} \leq K_v \leq 0.3\,\omega_{0A}$.

The mean dead times of electrical drives commonly found, are shown in table 2.3.

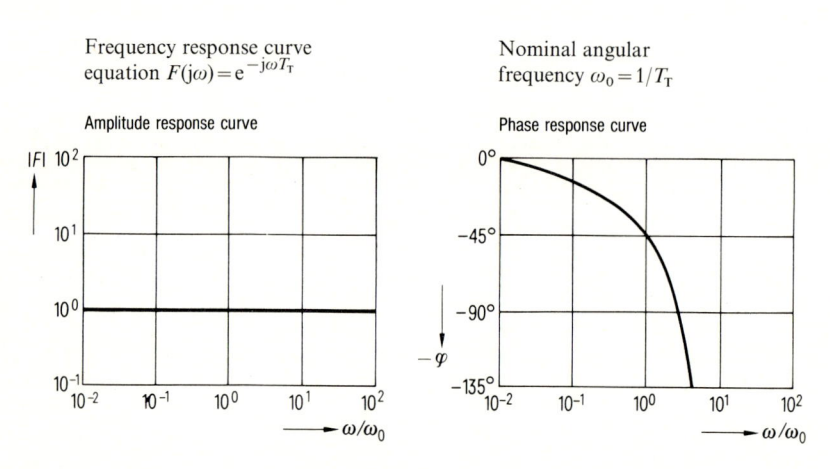

Fig. 2.11 Frequency response curve (Bode diagram) of a dead time element

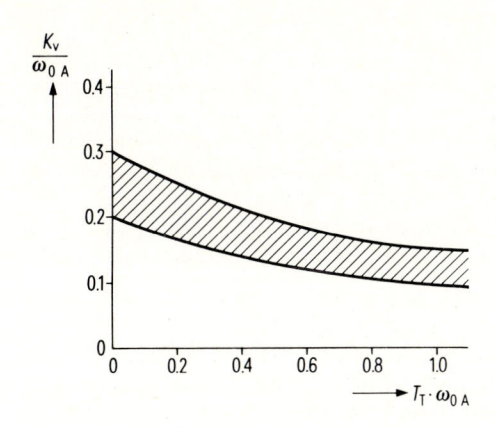

Fig. 2.12
Referenced position loop gain K_v/ω_{0A} dependently of the dead time T_T of the drive

[Graph: vertical axis K_v/ω_{0A} from 0 to 0.4; horizontal axis $T_T \cdot \omega_{0A}$ from 0 to 1.0; hatched band decreasing from about 0.2–0.3 at origin to about 0.15–0.17 at 1.0]

Table 2.1 Dead times of different drive systems (statistical means)

Current converter	T_T/ms
Transistor chopper	0.25
6-pulse thyristor converter with circulating current	1.67
3-pulse thyristor converter with circulating current	3.33
6-pulse thyristor converter circulating current-free	1.67 + approx. 6*)
2-pulse thyristor converter with circulating current	5

*) Additional reaction time through switch-over logic, at torque inversion

When these mean dead times are taken into consideration, e.g. for drive ① of table 2.2, the setting for the position loop gain becomes:

$$K_v \approx (0.2 \cdots 0.3)\, \omega_{0A} = 28 \cdots 43 \text{ s}^{-1}$$

reduced to

$$K_v \approx (0.1 \cdots 0.15)\, \omega_{0A} = 14 \cdots 21 \text{ s}^{-1}$$

(The additional reaction time should also be taken into consideration, since it does have to be traveled through at torque inversion.) Reactions of the mechanical transmission elements can further reduce this value.

2.3 Selection of DC Servo Motors

2.3.1 Methods of Calculation

The total feed drive consists of (see fig. 1.17, 1.18, and 1.23):

▷ DC servo motor,
▷ power section with speed regulator,
▷ mechanical transmission elements.

As emphasized in the preceeding section, besides the motor's characteristic data, the current converter and the mechanical transmission elements also affect the dynamic properties of the drive, and thus the behavior of the position control loop (see equation (2.22)).

The selection of an appropriate drive is a difficult one, since each part influences the effects of the other parts, and thus affects the entire servo system. If, for instance, a feed rate increase in a machine axis is desired, it can be accomplished by changing the motor speed, the gear ratio, or the feed screw lead. If the motor speed is increased, the total energy exchanged between the motor and the mechanical transmission elements at each acceleration and braking process, is increased by the square of the relationship of the new speed to the old one. The motor itself, however, has the same heat capacity, so that under circumstances, the next larger type of motor would have to be used. In that case, the higher moment of inertia of this larger motor increases in turn the necessary energy, and affects the heat that results.

Such interaction between changes in a variable and the behavior of the entire system, make extensive calculations necessary for the assembly of a suitable drive system, especially designed for the machine to be equipped.

An appropriate aid for these calculations is the computer. It allows the calculation of the dynamic and thermal behavior of the drive, and it can simulate the operational behavior. It thus makes possible the comparison of alternative designs and the appropriate selection, still in the conceptual stage. Corresponding computer programs are described in [2.5] and [2.6].

The following discussion presents the information which would allow, even without the use of a computer, the selection of feed motors. Questions about current converter units are treated in chapter 3, and the mechanical transmission elements are handled in chapter 4. For the motor itself, we must consider the static and dynamic requirements, and the heating behavior. A static dimensioning based on the projected working and frictional forces, is by no means sufficient for this purpose, for which the acceleration and deceleration processes are more important considerations.

2.3.2 Stationary Load

The requirements on the feed drive in the stationary mode are presented in section 2.1.1. The most important factors are the frictional forces in the slide guides, the bearing friction, the frictional losses in the feed screw nut, respectively between rack and pinion, and the losses in the gear, when present. In addition, we have to consider the feed force necessary for machining at the part.

Reactions of the mechanical system can affect the velocity control. The so-called stick-slip effect (see section 4.3.4.2) even at the best speed control, leads to a jerking motion of the feed unit. Gears and ball screw nuts preloaded too strongly, can also generate discontinuities in the motion, and these must be improved at the spot, since they are not affected by settings of the speed control in the converter unit. For wide control range, the temperature dependent drift of the speed control amplifier alters the command value. This error can be limited to a control range of $1:10^4$ to $< \pm 30\%$ (in reference to the lowest speed), by taking special measures (applying low-drift operational amplifiers and compensating resistors with low temperature coefficient). Numerical control permits automatic drift compensation and can therefore prevent position deviations resulting from drift errors.

The stationary part of the load contributes significantly to the heating of the motor. It can be determined by finding the following values:

▷ load torque M_L, reflected onto the motor shaft,
▷ motor speed,
▷ load duration, duty cycle.

Load torque M_L consists of ΣM_R, necessary to overcome the slide guide friction and the losses in the bearings and gear, and of M_V, which is required for the working force:

$$M_L = \Sigma M_R + M_V \tag{2.25}$$

ΣM_R sum of torques for friction and losses
M_V torque for the machining force

2.3.2.1 Friction and Losses

Figure 2.13 shows the machining forces and the torques on feed drive with gear and feed screw. The torque for friction and losses ΣM_R hereby consists of:

a) Friction in the slide guides.
   For the feed screw drive, the necessary torque reflected on the feed screw is

$$M_{RF} = \mu_F(v) \cdot \frac{h_{Sp}}{2\pi} \left[(m_W + m_T) \cdot g + F_{VT} \right] \qquad (2.26)$$

$\mu_F(v)$ speed dependent frictional factor of the feed drive slides (see 4.3.3.2)
$(m_w + m_r) \cdot g$ weight force of the guide (including the heaviest possible part) ($g = 9.81$ m/s^2, gravitational acceleration)
F_{VT} cutting force component, perpendicular to the slide (dependent on the type of cutting, see section 2.3.2.2)
h_{Sp} feed screw lead

The speed dependent frictional factors for the corresponding pairing of materials can be derived from the literature, or from the measured values.

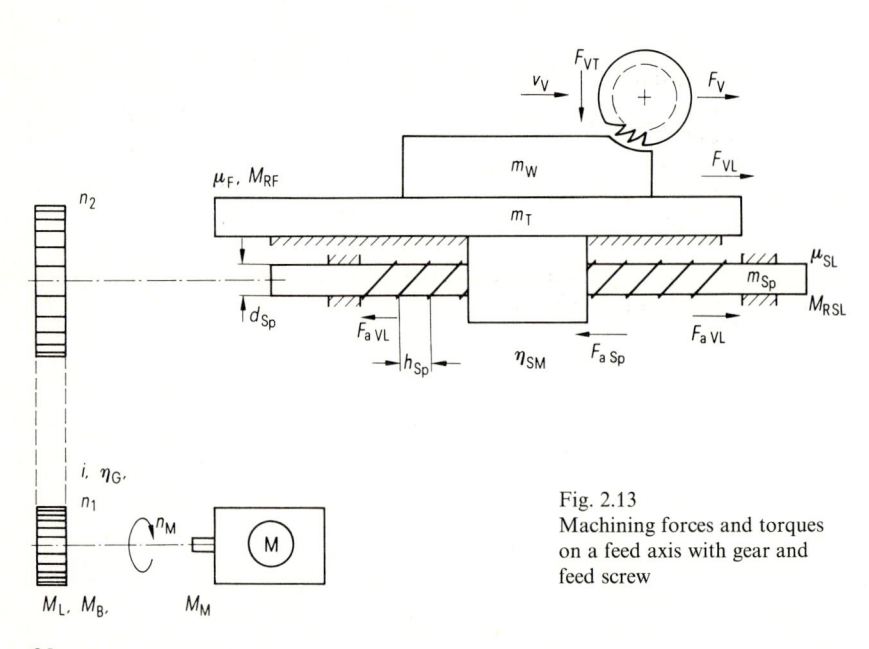

Fig. 2.13
Machining forces and torques on a feed axis with gear and feed screw

b) Frictional losses in the feed screw bearing.

For an axial roller bearing, the necessary torque reflected onto the feed screw will be, according to information of the INA company, approximately

$$M_{RSL} = \mu_{SL}(v) \cdot \frac{1}{2} d_{mL} \cdot F_{aVL} \tag{2.27}$$

For the commonly used combination of axial-radial bearings (see section 4.3.1.4), the portion contributed by the radial bearing to the total friction on the feed force load is so low, as to be disregarded.

$\mu_{SL}(v)$ speed dependent frictional factor of the feed screw bearing (see section 4.3.1.4, fig. 4.30)
F_{aVL} axial preload of the feed screw bearings in addition to feed screw preload (or the axial load due to the machining force F_{VL}, for bearings not preloaded)
d_{mL} mean bearing diameter

c) Frictional losses in the feed screw nut.

Feed screws with ball nuts are most commonly used nowadays. The frictional losses are measured through the efficiency coefficient η_{SM} (see 4.3.1.2.3). According to manufacturing data for ball screw drives, η_{SM} is approximately (data from SKF for transrol screws)

$$\eta_{SM} \approx \frac{1}{1 + 0.02 \cdot \frac{d_{Sp}}{h_{Sp}}} \tag{2.28}$$

d_{Sp} diameter of feed screw

For trapezoid thread feed screws, one can count on an efficiency coefficient of approximately 0.5–0.6.

d) Gear losses.

These losses are represented by the efficiency coefficient η_G, which, depending on the type of teeth and the gear ratio, can be between 0.8 and 0.95.

Thus for feed screw drives, the *torque for friction and losses* reflected onto the motor shaft, is:

$$\sum M_R = \frac{\dfrac{M_{RF}}{\eta_{SM}} + M_{RSL}}{\eta_G \cdot i} \tag{2.29}$$

i gear ration n_1/n_2 (for missing gears, η_G and i should be replaced by 1)

Figure 2.14 shows the machining forces and the torques for drives with rack and pinion. The torque for the friction of the slide guide can be determined as in equation (2.26)

$$M_{RF} = \mu_F(v) \cdot r_{Ri}[(m_W + m_T) \cdot g + F_{VT}] \tag{2.30}$$

r_{Ri} radius of pinion

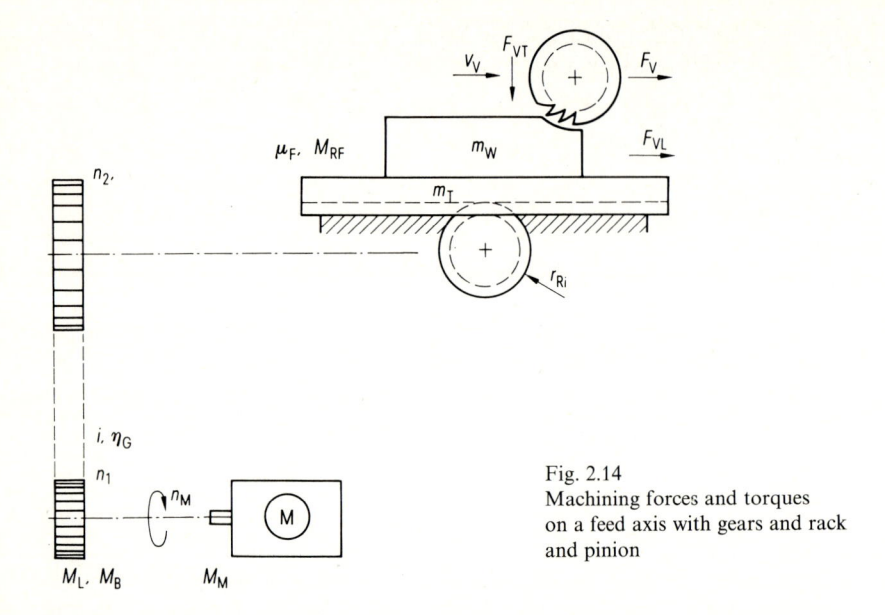

Fig. 2.14
Machining forces and torques
on a feed axis with gears and rack
and pinion

The frictional losses in the gear and between rack and pinion are given through the gear efficiency coefficient η_G, which can be counted on to be between 0.7–0.8, since a multiple stage gear must generally be used.

Thus the *torque for friction and losses* reflected onto the motor shaft *of a rack and pinion drive*, will be:

$$\sum M_R = \frac{M_{RF}}{\eta_G \cdot i} \tag{2.31}$$

2.3.2.2 Machining Force

For determining the torque for the cutting process, the cutting force must be known. It can be split into two components: one, acting in axial direction, is designated as machining force F_{VL}. The other acts perpendicularly to the guide and, in the following, is designated as F_{VT}.

The cutting force depends on the variables of the cut. It is affected by the cutting depth, surface speed, feed rate, tool geometry, the part-cutter combination of materials, pretreatment of the part, coolant-lubrication material used during machining, and by the wear state of the tool.

For the standard cutting case, many of these variables cannot be determined, and can have a wide range of variation. Therefore, it is recommended that in individual cases the corresponding literature be consulted [2.7].

For *feed screw drives,* the torque reflected onto the motor shaft necessary for the machining force, is:

$$M_V = \frac{F_{VL} \cdot h_{Sp}}{2\pi \cdot i \cdot \eta_G \cdot \eta_{SM}} \tag{2.32}$$

and for *rack and pinion drives,* respectively:

$$M_V = \frac{F_{VL} \cdot r_{Ri}}{i \cdot \eta_G} \tag{2.33}$$

For drives with ball screw drives without gear, for rough calculations, at 10 mm screw lead, the torque M_V can be calculated from

$$\frac{M_V}{Nm} \approx \frac{F_{VL}/N}{500} \tag{2.32.1}$$

Example:

A feed axis with a 500 kg slide mass and 1,000 kg maximal work piece weight is to be driven over friction guides, through a 2 m long ball screw with 40 mm diameter and 10 mm lead. A timing belt gear with a ratio of 2 must be used. The required machining force is 20,000 N. Find the stationary torque reflected onto the motor.

For the material pairing of cast iron/plastic, we assume that the frictional factor μ_F is approximately 0.08, and that the component of the cutting force perpendicular to the slide guide is $F_{VT} \approx 2,000$ N. The feed screw has bearings on both sides, and is preloaded so that with the axial-radial bearing combination used, only one part bearing is always loaded at high feed forces. We take the preload to be 6,000 N, and the frictional factor μ_{SL} to be 0.004; the mean bearing diameter is about 1.5–1.8 times the spindle diameter, and is taken to be 66 mm.

According to equation (2.26), the torque for friction of the slide guide is

$$M_{RF} = 0.08 \frac{10 \cdot 10^{-3} \, m}{2\pi} [(1,000 \, kg + 500 \, kg) \cdot 9.81 \, m/s^2 + 2,000 \, N] = 2.13 \, Nm$$

The torque for the frictional losses in a feed screw bearing with the frictional value presumed for $\mu_{SL} \approx 0.004$, according to (2.27), will be

$$M_{RSL_1} = 0.004 \cdot \frac{1}{2} \cdot 66 \cdot 10^{-3} \, m \cdot 6,000 \, N = 0.792 \, Nm$$

Thus for both bearings we obtain

$$M_{RSL} = 1.584 \, Nm$$

The efficiency coefficient of the ball nut is calculated according to equation (2.28)

$$\eta_{SM} \approx \frac{1}{1 + 0.02 \frac{40 \cdot 10^{-3} \text{ m}}{10 \cdot 10^{-3} \text{ m}}} = 0.926$$

The gear efficiency is taken to be $\eta_G = 0.9$.

Thus the torque for the sum of the frictional losses becomes, according to (2.29)

$$\sum M_R = \frac{2{,}13 \text{ Nm}/0.926 + 1.584 \text{ Nm}}{0.9 \cdot 2} = 2.16 \text{ Nm}$$

the torque for the machining force according to (2.32), is

$$M_V = \frac{20{,}000 \text{ N} \cdot 10 \cdot 10^{-3} \text{ m}}{2\pi \cdot 2 \cdot 0.9 \cdot 0.926} = 19.1 \text{ Nm}$$

and the stationary load torque, according to (2.25), amounts to

$$M_L = 2.16 \text{ Nm} + 19.1 \text{ Nm} = 21.26 \text{ Nm}.$$

2.3.3 Dynamic Load

Section 2.1.2 lists the requirements on a feed drive in the non-stationary mode of operation. The most important requirements are high acceleration and deceleration torques, and good command and disturbance transient response behaviors.

The values of ω_{0A} between 100 s^{-1} and 400 s^{-1} mentioned in section 2.1.2, require response times of about 35–8 ms. A high acceleration potential must be available without limitations in the feed rate range. The current limit set for the protection of the motors and for an economical dimensioning of the current converter unit, must be sufficiently high; a sufficient limitation value has been found to be 4–4.5 times the motor's rated torque M_{0M}. The motor must be capable of delivering this torque momentarily, and also be able to withstand it thermically.

The total torque M_M, that the motor must at any time be able to deliver, is the sum of the already described stationary load torque M_L and of the non-stationary part M_B, which is used to accelerate and decelerate the masses. According to equation (2.6), the torque balance is

$$M_M(t) = M_L(t) + M_B(t)$$

The drive does not react without delays to velocity changes requested through the velocity command value. The limitations and inertias immanent in the system affect the course of the velocity considerably, and thus also affect the acceleration and motor torque. By the same token, neither can load changes be controlled without delays. According to section 2.2.2.4, feed drives with position control loops are optimized so, that they approximate the behavior of a 2nd order delay element. The characteristic values are the drive nominal angular frequency ω_{0A}, and the damping gradient D_A which should be approximately 0.5.

For higher speeds, the current limit is reconstructed on the commutating limit. This results in a non-linearity in the drive system at larger control angles, which affects the calculation of the dynamic behavior. Furthermore, the effects of load changes on the course over time of the motor torque, velocity, and slide travel must also be taken into consideration.

As shown for the determination of the disturbance transient response behavior in fig. 1.35, a 2nd order delay element can also be represented as a serial connection of two integral elements with feedback. From equations (1.41) and (2.25), we can thus derive an extended substituted diagram of the controlled DC servo drive with load effect (fig. 2.15). In this block diagram we can determine the courses of $M_M(t)$ and $n_M(t)$ from the frequency response curve, with the help of a computer. Certain approximations and omissions must hereby be considered. For further information, consult the appropriate literature [2.8].

In order to have an approximate idea of the behavior within the position control loop from the machine data given, we can however, go through the following simple calculations.

From equations (2.8) and (2.2), we can derive the acceleration torque necessary for a particular speed difference. We obtain the adjusted dimensional equation

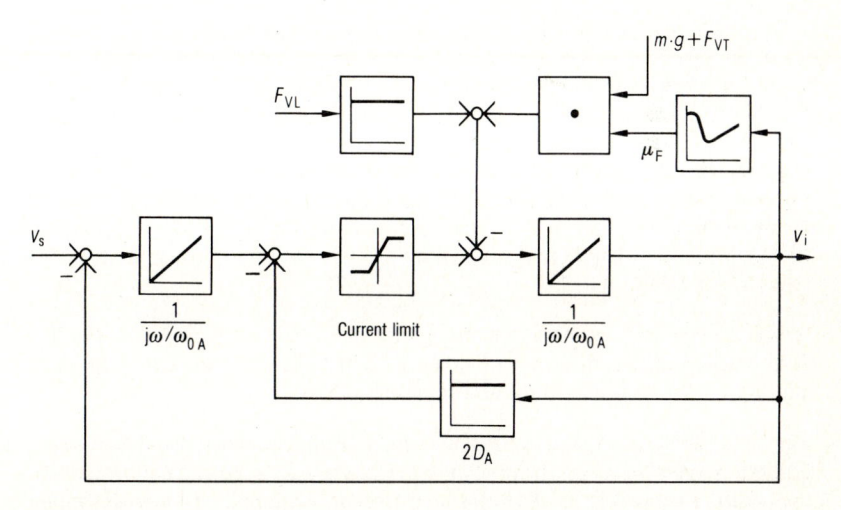

Fig. 2.15
Block diagram of the speed controlled DC servo drive with load effect

for the required acceleration torque

$$\frac{M_B}{\text{Nm}} = \frac{2\pi}{60} \cdot \frac{\frac{J_{\text{Ges}}}{\text{kg m}^2} \cdot \frac{\Delta n_M}{\text{min}^{-1}}}{\frac{t_H}{s}} \qquad (2.34)$$

and for the ramp-up time

$$\frac{t_H}{s} = \frac{\frac{J_{\text{Ges}}}{\text{kg m}^2} \cdot \frac{\Delta n_M}{\text{min}^{-1}}}{9{,}55 \cdot \frac{M_B}{\text{Nm}}} \qquad (2.35)$$

where:

J_{Ges} total inertia reflected onto the motor shaft $J_{\text{Ges}} = J_M + J_{\text{ext}}$
Δn_M speed difference for which t_H is to be calculated
M_B mean value of the available acceleration torque
t_H ramp time needed for speed difference Δn_M, at constant acceleration torque M_B

After calculating the stationary load torque according to equations (2.25), (2.29) and (2.32), respectively (2.31) and (2.33), a motor is selected. The ramp-up time t_H can be determined with the total moment of inertia and the selected rapid traverse feed rate. A first test run would establish whether the selected feed drive can be operated with a sufficiently high dynamic in the position control loop, and would require that this ramp-up time be consulted.

Since for the permanent magnet-excited DC servo motors of series 1 HU.. the course of the torque is controlled speed-dependently through the speed dependent current limit, the course of the acceleration torque during ramp-up processes is also variable. Calculations must therefore be done step by step for constant torques, and the time segments then added. The course of the maximal motor torque M_M can be derived from the motor data sheets (see the example of calculation), while keeping in mind the current limit of the selected converter unit.

A speed dependent current limit is not necessary for electrically excited DC servo motors. M_M is determined by the current limit set in the converter unit, and thus is constant over the entire ramp-up process.

For numerically controlled lathes and milling machines, generally a linear acceleration, respectively deceleration of 0.8–1 m/s², and a position loop gain in the position control loop of 1 (mm/min)/μm, are required. These values imply positioning times of approximately 400–300 ms. For these requirements, when designing a feed drive, as a rule of thumb a value of ≤100 ms for the ramp-up time to rapid traverse can be used for the selection of the current limit. For the exact appropriateness determination, it is necessary to calculate the drive angular frequency ω_{0A}. (See section 2.2.2.5.)

Example:

The drive calculated in the example of section 2.3.2.2, is computed for a motor 1 HU3 104-0AH01. The rapid traverse feed rate should be 10 m/min. The inertias must be calculated first (see section 4.3.4):

The linearly moved mass $m_W + m_T$ is 1,500 kg. According to equation (4.12), the moment of inertia reflected onto the feed screw is:

$$J_{T+W} = (m_W + m_T)\left(\frac{h_{Sp}}{2\pi}\right)^2 = 1{,}500 \text{ kg } \frac{1 \cdot 10^{-4} \text{ m}^2}{4\pi^2} = 3.8 \cdot 10^{-3} \text{ kgm}^2$$

From equation (4.10.1), the moment of inertia reflected on the ball screw is

$$J_{Sp} = 0.77 \cdot 10^{-12} \cdot \left(\frac{d_{Sp}}{\text{mm}}\right)^4 \cdot \frac{l_{Sp}}{\text{mm}} \cdot \text{kgm}^2 = 0.77 \cdot 10^{-12} \cdot 40^4 \cdot 2 \cdot 10^3 \text{ kgm}^2$$
$$= 3.94 \cdot 10^{-3} \text{ kgm}^2$$

For the gear wheel on the feed screw, including coupling and screw extension, we obtain

$$J_{Gt2} \approx 20 \cdot 10^{-3} \text{ kgm}^2$$

The moment of inertia of the gear wheel on the motor is estimated at

$$J_{Gt1} \approx 0.5 \cdot 10^{-3} \text{ kgm}^2$$

From this we obtain, according to equations (2.51) and (2.52), the external moment of inertia reflected onto the motor shaft

$$J_{ext} = J_{Gt1} + \frac{J_{Gt2} + J_{Sp} + J_{T+W}}{i^2} = \left(0.5 + \frac{27.74}{4}\right) \cdot 10^{-3} \text{ kgm}^2 = 7.44 \cdot 10^{-3} \text{ kgm}^2$$

The motor's moment of inertia is $J_M = 28 \cdot 10^{-3} \text{ kgm}^2$; the total moment of inertia is thus, according to equation (2.3)

$$J_{Ges} = J_{ext} + J_M = 35.44 \cdot 10^{-3} \text{ kgm}^2$$

The acceleration torque M_B is controlled dependently on the speed, through the converter unit. For this reason, we divide the acceleration or deceleration process into ranges for which we calculate the ramp-up times.

Figure 2.16 shows the set current limit in the current converter unit of the motor. We obtain, for instance, six sections, whereby the mean values for M_{Mn} are located so that each section creates equal areas above and under the current limit line.

From the previously calculated load torque M_L, only the portion concerning friction and losses $\Sigma M_R = 2.16$ Nm need be considered for the calculation of the acceleration torque M_B, since for ramping-up to rapid traverse, the machining force is not required. The remaining acceleration torque is

$$M_B = M_M(\mp)\Sigma M_R$$

(+ applies to deceleration, since the friction acts hereby as an additional braking force).

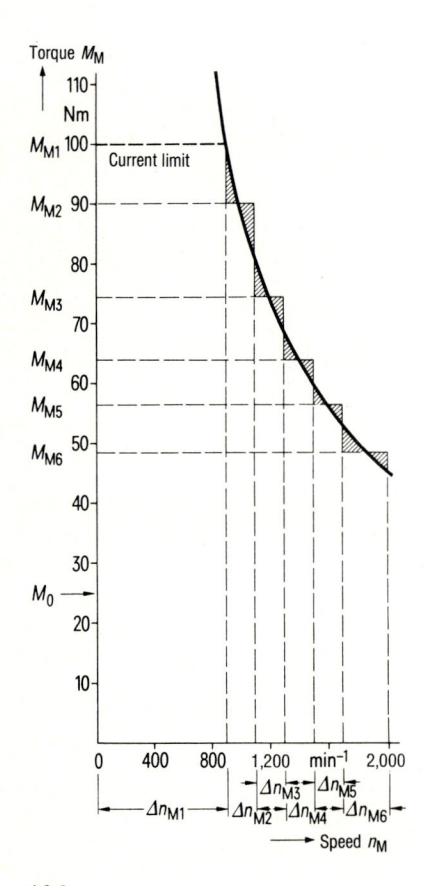

Fig. 2.16
Commutating limit for the 1HU3 104-0AH01 DC servo motor

106

The necessary partial ramp time segments are determined according to equation (2.35) and are summarized in table 2.4 below.

Table 2.4 Calculation of the ramp-up time of the feed drive

Section			1	2	3	4	5	6	Total ramp time
M_{Mn}		Nm	100	90	74.5	64	56.5	48.5	$\dfrac{t_H}{ms} = \sum\limits_{n=1}^{n=6} t_{Hn}$
$M_{Bn} = (M_{Mn} - \Sigma M_R)$		Nm	97.84	87.84	72.34	61.84	54.34	46.34	
Δn_{Mm}		$\min^{-1}$	900	200	200	200	200	300	
$\dfrac{t_{Hn}}{ms} = \dfrac{\dfrac{J_{Ges}}{kg\,m^2} \cdot \dfrac{\Delta n_{Mn}}{\min^{-1}}}{9.55 \cdot \dfrac{M_{Bn}}{Nm}} \cdot 10^3$			34	8.4	10	12	14	24	102.4

It appears from this that t_H will be larger than 100 ms. The drive however, can still be used in the position control loop, if the nominal angular frequency ω_{0A} suffices for the requirements. It can be calculated as shown in section 2.2.2.5, and the achievable position loop gain K_v must be compared with the requirements on the machine and with the machining program. A double check with table 2.2 results in a ω_{0A} of approximately $130\ s^{-1}$, and a K_v value of approximately $16\text{-}20\ s^{-1}$, even considering a dead time. These values suffice for this drive to be used without problems.

If $t_H \gg 100$ ms and ω_{0A} is too low, the external moment of inertia will have to be reduced, which can be accomplished, for instance, by selecting a different gear ratio. (In this last example, a 1HU3 078-0AF01 motor with a gear ratio of $i = 3$ could also be used.) Further information concerning the selection of gears can be found in section 4.4. Selecting a larger size motor will generally not accomplish the desired result, since its own larger moment of inertia must also be accelerated and decelerated, and the total power balance would be negatively affected.

By selecting a motor with a higher rated speed and the corresponding current converter, a shorter ramp-up time at rapid traverse will be ensured. The higher current limit and the more favorable speeds relationship $n_{maxM}/\Delta n_M$, increase the limit angular frequency $\omega_{0\,GrenzA}$ as calculated with equation (2.22). This improves the dynamic behavior of the feed drive, and the requirements for a ramp-up time at rapid traverse of less than 100 ms and for a high nominal angular frequency can be met in a better manner.

Such a selection is particularly favorable for high feed rates, e.g. at high speed milling or thread cutting on a lathe. A comparison should be made between the acceleration necessary for the required ramp distance and the acceleration that can be delivered by the drive (see the examples to figs. 1.29 and 1.30).

When accelerating to the rapid traverse and braking from it, reaching the current limit will not affect the work piece directly. During the acceleration process, the tool and the work piece are separated by a relatively large distance. Actual speed overshoots of the command value are usually not damaging, but for short distances, where there is an immediate switch from the ramp-up process into deceleration, it is disadvantageous to reach the current limit. For cases like this, it must be ensured that upon approaching the position during the positioning process, there is sufficient time for the position control to take charge of the drive again, for an overshoot-free positioning. This means that the current limit can be reached initially, but that upon approaching the selected position, it must be abandoned in time for positioning.

It can be double checked whether the drive has enough dynamic reserves for such acceleration processes, by means of a simple graphic procedure. The following example serves as an explanation.

Example:

For the drive calculated in the preceeding example, the total brake time during deceleration based on the current limit, can be computed according to table 2.5. As shown in fig. 2.17, the rapid traverse and the braking process calculated on page 107, are drawn sectionally from the calculated time segments t_{Hn}, resp. t_{Brn}, and the corresponding speed difference Δn_{Mn} is represented. (At acceleration, a delay time $T_u = 5$ ms, and at deceleration $T_u = 10$ ms have been assumed.) We obtain the dashed line ①; the overshoot lines drawn over and under it are estimations which depend on the setting of the speed controller.

In both diagrams, straight lines of constant acceleration can be shown. This acceleration a_w, is given by the numerical control through the command value

Table 2.5   Calculation of the braking time of a feed drive

Section		6	5	4	3	2	1	Total braking time
M_{Mn}	Nm	48.5	56.5	64	74.5	90	100	$\dfrac{t_{Br}}{ms} = \sum\limits_{n=1}^{n=6} t_{Brn}$
$M_{Bn} = (M_{Mn} + \Sigma M_R)$	Nm	50.66	58.66	66.16	76.66	92.16	102.16	
Δn_{Mn}	min^{-1}	300	200	200	200	200	900	
$\dfrac{t_{Brn}}{ms} = \dfrac{\dfrac{J_{Ges}}{kg\,m^2} \cdot \dfrac{\Delta n_{Mn}}{min^{-1}}}{9.55 \cdot \dfrac{M_{Bn}}{Nm}} \cdot 10^3$		22	13	11	10	8	33	97

modifier. (For NCs missing such a modifier, the acceleration is not constant, the ramp-up, respectively the braking process in the aim position, occur at the maximum acceleration $a_{i\,max} = K_v \cdot \Delta v_B$. More detailed explanations are presented in chapter 5, and from equation (3.11).)

According to equation (2.42), the acceleration of the machine tool's slide is

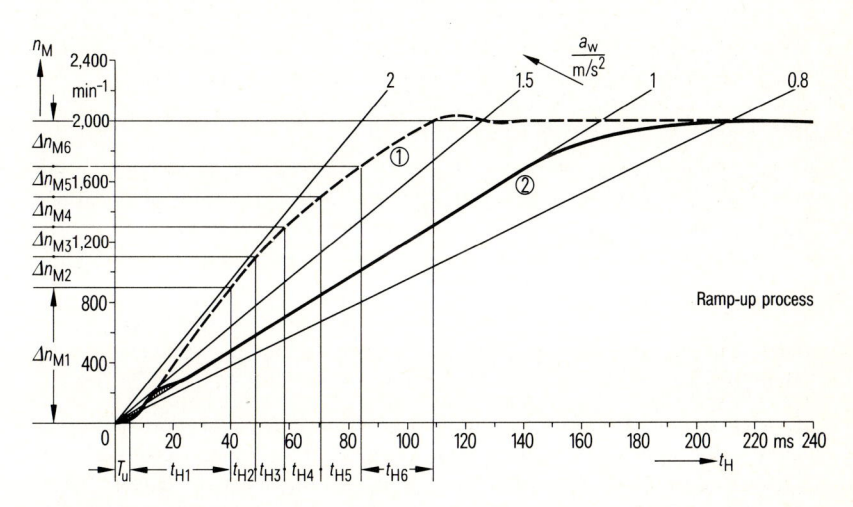

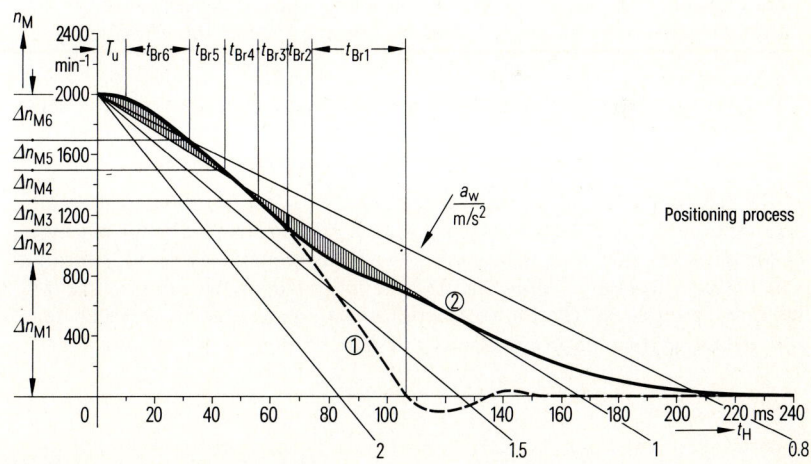

$$a_W = \frac{dv_V}{dt}$$

Fig. 2.17
Ramp-up and positioning processes of a DC servo drive
① --- only speed controlled
② — with position control and acceleration modifier $a_w = 1 \text{ m/s}^2$

From equations (2.46) and (2.48), the feed rate v_V can be written as

$$v_V = \frac{n_M \cdot h_{Sp}}{i}$$

n_M motor speed
i gear ratio
h_{Sp} feed screw lead

The slope of this straight line with constant acceleration, is thus

$$\frac{\Delta n_M}{\Delta t} \approx \frac{dn_M}{dt} = \frac{a_W \cdot i}{h_{Sp}}$$

Figure 2.17 shows the straight lines for $a_w = 0.8$; 1; 1.5 and 2 m/s².

If an acceleration of $a_w = 1$ m/s² is required, it should be recognized that during acceleration, a difference between the required and the possible accelerations can exist for only a short time. The resulting position difference will be controlled after 25 ms (shaded areas). The drive accelerates according to line ②.

During braking however, a larger difference is produced, and it is settled only after approximately 120 ms (shaded areas must be equal). The position controller can still take over the command and reach the end position without overshoot.

The limit for the possible acceleration is given during the positioning process at approximately $a_w \approx 1.5$ m/s², where the position difference resulting from having reached the current limit, cannot be compensated for any longer. (Since, while approaching the aim position, the diminishing following error in the position control determines the position command value, the exact course of the positioning curve cannot be determined by this method.)

The example shows that in the design state, an estimation of the possible acceleration values can already demonstrate whether the motor, gear ratio and moments of inertia are appropriately dimensioned or not.

If the feed drive is given higher requirements for the positioning time and acceleration, as would be the case for instance for punch and nibbling machines or press feeders, then an exact computation of the required acceleration torques will become necessary. After this, the thermal load of the servo motor will have to be checked. The above mentioned requirement for a ramp-up time $t_H < 100$ ms, can no longer be used as a design criterion.

2.3.4 Heating Behavior

2.3.4.1 Periodic Load Changes

Generally, load cycles cannot be stated for a machine tool. For feed drives in transfer units or, for instance, for press feeding equipment, we can however, construct a torque/time and a speed/time diagram for the machining of certain parts. An example of this is shown in fig. 2.18. With a rms calculation, it can hereby be double checked whether the motor meets the thermal behavior requirements.

The allowed motor torque M_{zulM} must be larger than the effective torque over the load period T_{Last}. Under the condition that the relationship M_M/I_M is linear, it can be expressed

$$M_{zul\,M} \geqq M_{eff\,M} = \sqrt{\frac{M_{M1}^2 \cdot t_1 + M_{M2}^2 \cdot t_2 + M_{M3}^2 \cdot t_3 + \ldots M_{Mn}^2 \cdot t_n}{T_{Last}}} \qquad (2.36)$$

For permanent magnet-excited DC servo motors, in the range of the speed dependent current limit, the torque corresponding to the static current limit is substituted, and the speed dependent reduction does not count, since current ripples result in additional losses (see for comparison e.g. M_{M1} in fig. 2.18).

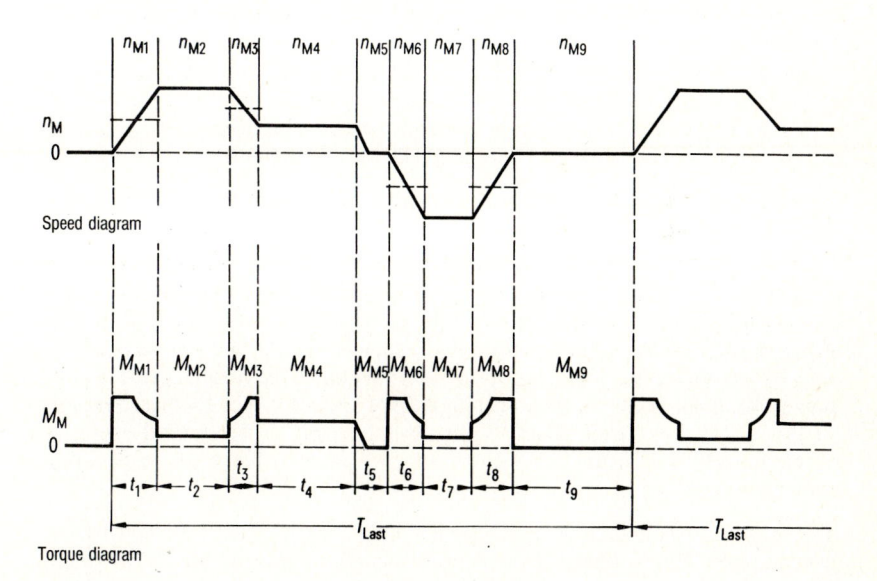

Fig. 2.18
Load cycle of a feed motor, during the machining of a work piece

It must be taken into consideration however, that for DC servo motors the allowed constant torque M_{zulM} is a speed dependent value (see line ① in fig. 2.2). For determining M_{zulM} we must therefore find the mean arithmetic values of the speed n_{mM} over the load period T_{Last}, and with it we can find the allowed constant torque for n_{Mn} from the selection tables of the motors. For acceleration and deceleration processes, the mean value between initial and final speed for n_{Mn} is taken as:

$$n_{mM} = \frac{n_{M1} \cdot t_1 + n_{M2} \cdot t_2 + n_{M3} \cdot t_3 + \ldots n_{Mn} \cdot t_n}{T_{Last}} \qquad (2.37)$$

If the DC servo motors are used mostly at higher speeds, the reduction in the allowed motor torque must be watched with particular attention; under circumstances, a larger motor may have to be selected. For current converter feeding, the form factor of the converter circuit must also be taken into consideration. Table 3.1 shows the form factors F_i of the most common converter circuits; equation (3.3) indicates the torque reduction. Since the commutating limits of DC servo motors, in dependence on form factors, are also lower, it is necessary for the determination of the allowed motor torque M_{zulM}, that the commutating limit curves for the given current converter feed should also be considered.

As a rule, calculating the M_{effM} is only sufficient as an approximation, since the heat transfer takes place on a machine tool through the flange and shaft, but other losses additionally present occur, due to e.g. settling processes and harmonics.

2.3.4.2 Overload Behavior

From the point of view of requirements set by the machine tool on a feed drive, there are two overload zones. The first range characterizes the short-term overload capacity for acceleration and deceleration processes. Here, the motors must be designed for the mechanical and magnetic forces, and for the acceleration values occurring. Accelerations of up to 60 times the normal gravitational acceleration can occur on the circumference of the rotor of a DC servo motor. In this range, the current limit of the thyristor unit will commonly react at approximately 4–4.5 times the rated current and will keep the acceleration forces within limits.

The second range characterizes overloads due to cutting, as for instance those occurring on lathes during drilling processes. Overloads of up to two times the rated torque are expected here. It is convenient in this range, to conduct a rms calculation as shown in section 2.3.4.1., and to design the corresponding converter unit with reactors and transformer accordingly.

Figures 2.19 and 2.20 show the overload curves of DC servo motors of series 1HU.. and 1GS... Depending on the preload of the motor, a more or less large overload is possible for a particular time span. It is hereby presumed that the motor is at operating temperature in the feed range, and additional iron losses are neglected.

In figs. 2.19 and 2.20:

Range A: Overload range for accelerations and decelerations up to 200 ms

Range B: Overload range for machining processes with 1.5 up to 2 times the rated current

$$\text{Preload factor} = \frac{M_{\text{eff M}}}{M_{0\,M}}$$

$M_{\text{eff M}}$ effective value of the motor torque
$M_{0\,M}$ rated torque of the motor

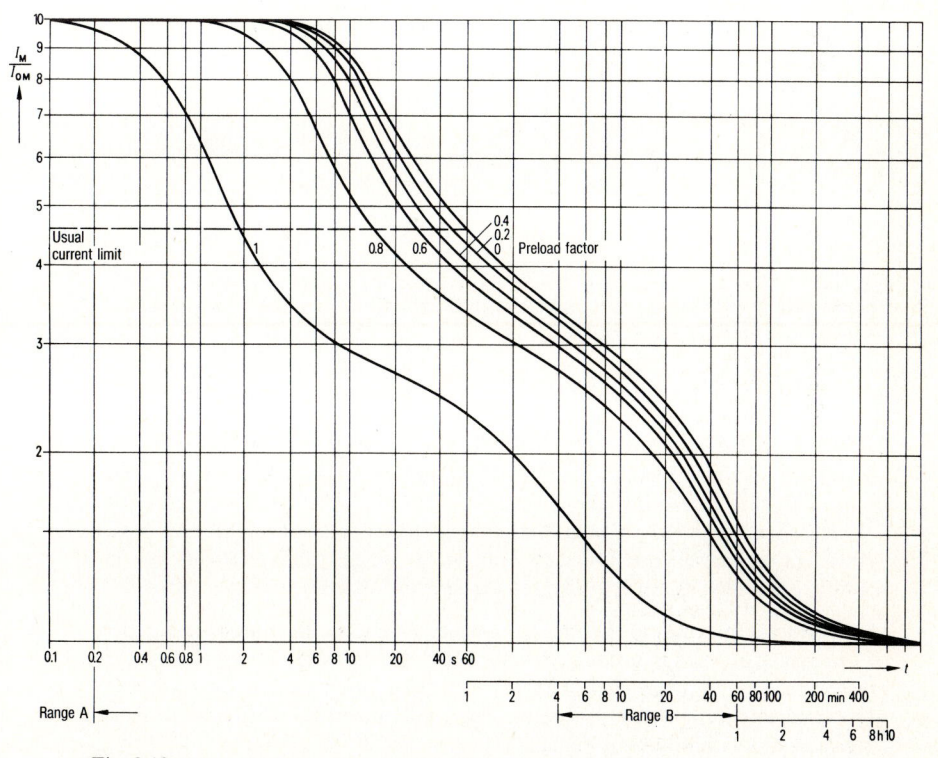

Fig. 2.19
Thermal overload capacity of permanent magnet-excited DC servo motor of the 1 HU3.. series, with DC feeding (form factor $F_i = 1$)

113

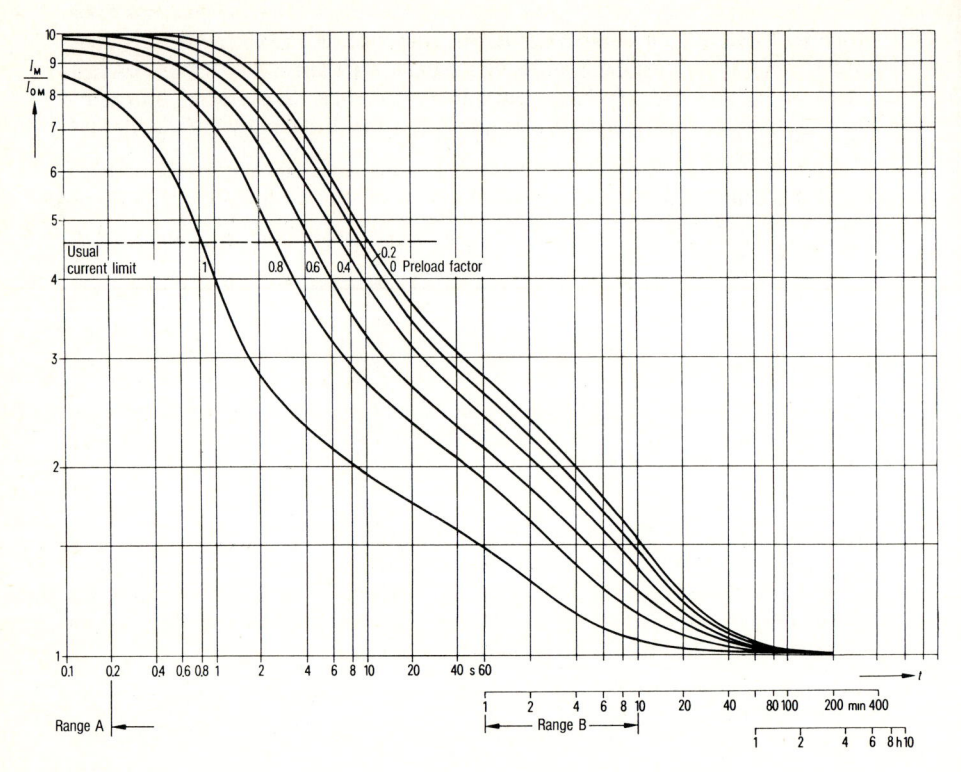

Fig. 2.20
Thermal overload capacity of the electrically excited DC servo motor of the 1GS3.. series, with DC feeding (form factor $F_i = 1$)

The thermal time constant of permanent magnet-excited DC servo motors of the 1HU3.. series amounts to about 120 min. For this reason, the overload range B is here higher by about a factor of 5 on the time axis than range B of electrically excited servo motors. This demonstrates the suitability of the permanent magnet-excited DC servo motors for longer term overloads in non-periodic processes, as already discussed in section 2.2.1.1.

In the short-term range A, besides the acceleration load, the magnetic reaction of the armature on the excitation field must also be considered. The motors are designed so that the peak currents occurring as part of the operation do not demagnetize the excitation poles. The limit shown for 10 times the rated current is still within the safe range.

For electrically excited DC servo motors, the thermal time constant for heating is approximately 60 min. Due to the mounted external blower, at continuous ventilation the cooling time constant is approximately 20 min. These motors are therefore better suited for periodic overloads. In the range A, the overload capacity is limited only by the commutating loads and mechanical forces.

According to VDE 0530 part 1, the operating modes of DC servo motors are assigned the letter S with a number annexed to it. The characteristic intermittent operation for feed drives, without effects of the starting process, is designated S3. It is characterized by the duty cycle ED in %, and by the cycle duration. A duty cycle commonly lasts 10 min. The selection tables of servo motors contain information for the overload capacity at 40% and 25% ED.

Figure 2.21 shows a scatter band of the allowed motor current I_M, in reference to the rated motor current I_{0M}, as it can be raised during intermittent duty cycle S3. The presumption is that the overloads occur in the feed rate range. The deviations will vary with the motor size. For the permanent magnet-excited DC servo motors of the series 1HU3 13. most especially, the overload values are somewhat lower. Exact values must be taken from the motor selection lists.

In the diagram of fig. 2.21, the effect of iron losses is disregarded, which is permitted when estimating the overload capacity of DC servo motors. For short term switch-ons, the effect of the 2nd thermal time constant, as described in section 2.3.4.3, can be seen. Besides, due to the high brush current density at the commutator, limitations of the overload capacity result here.

For a different load period and an alternating load, during a periodically repeated load time, a rms calculation in accordance with section 2.3.4.1 is to

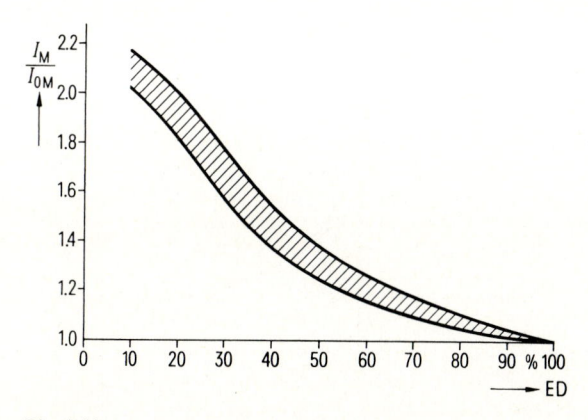

Fig. 2.21
Overload capacity in the intermittent operation S3 of DC servo motors of the 1HU3.. and 1GS3.. series, lasting load period 10 min, feed rate range

be conducted. More exact calculations are possible with a computer. During data input, the operation and pause times for the alternating load can be described better, and both thermal time constants of the motor can be taken into consideration.

2.3.4.3 Thermal Time Constants

As measurements [2.9] have proven, it is not sufficient to reconstruct the temperature behavior through an exponential curve with only one time constant. The evaluation of the conducted measurements generally reveals two time constants. One time constant, which is affected by the masses of the winding copper and iron, amounts to only a few minutes. It determines mainly the initial course of the heating process, and is thus the determining factor for short-term overloads. The second time constant is considerably larger, and determines the temperature course up to the point where a constant has been reached. It is affected by the heat transfer to the casing and by the heat losses due to the cooling air. The portion ϑ_1 of the overtemperature, resulting from the heating process with the smaller time constant, lies between 20% and 40% of the final overtemperature.

Figure 2.22 shows the course of the temperature with both time constants $T_{th\,1}$ and $T_{th\,2}$ for the heating, and $T_{th\,3}$, respectively $T_{th\,4}$ for the cooling process, and the summation of the two parts of the heating process.

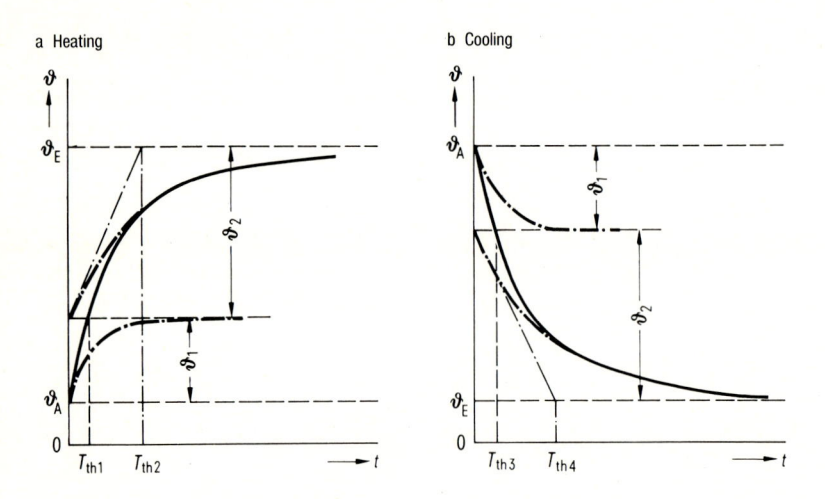

Fig. 2.22
Approximation of the heating process (a) and of the cooling process (b) through two exponential functions, for a DC servo motor

For self-cooling motors, T_{th2} and T_{th4} are not significantly different, while for forced cooling motors the differences between heating and cooling processes are considerable.

The evaluation of temperature curves of servo drives at high overloads as they occur in acceleration processes, is based on the adiabatic temperature behavior. This point of view describes a limit case which, when adopted, is always on the safe side of the thermal load capacity. The heat transfer capacity of the heated armature winding is hereby disregarded, and a linear temperature increase is presumed.

The limit value of the linear temperature rise is determined only from the heat storage capacity and the magnitude of the occurring losses. It is known as the so-called adiabatic temperature change slope $\Delta\vartheta_{adi}/\Delta t$. For the armature winding this is determined from the material constants, the overload current, and the effective crossection area of the winding. It holds that:

$$\frac{\Delta\vartheta_{adi}}{\Delta t} = \frac{I_{effM}^2}{\varkappa \cdot c_{WM} \cdot \rho \cdot q_M^2} \tag{2.38}$$

The effective crossection area of the winding q_M is determined from the number of armature branches, parallel wires, and the wire crossection.

$\varkappa$ specific conductivity, $Cu = 58 \dfrac{m}{mm^2 \cdot \Omega}$

c_{WM} specific heat $Cu = 388.5 \dfrac{J}{kg \cdot K}$

ϱ density $Cu = 8.9 \cdot 10^3 \dfrac{kg}{m^3}$

Substituted, the copper winding is

$$\frac{\Delta\vartheta_{adi}}{\Delta t} \approx 4.99 \cdot 10^{-3} \frac{I_{effM}^2}{q_M^2} \cdot \frac{K \cdot mm^4}{A^2 s} \tag{2.38.1}$$

Starting with the initial temperature of the winding ϑ_A, after the t_H acceleration or deceleration time, the following winding temperature results

$$\vartheta_E = \frac{\Delta\vartheta_{adi}}{\Delta t} \cdot t_H + \vartheta_A \tag{2.39}$$

Example:

A DC servo motor has a rated current of 20 A and a wire crossection of 1 mm². 4 armature branches are connected in parallel, so that the effective crossection $q_M = 4$ mm². Find the adiabatic temperature change ramp, at fourfold the rated current and 6-pulse thyristor feeding.

For a 6-pulse thyristor feeding, we calculate with a form factor of 1.05, so that at fourfold the current we have

$$I_{effM} = 1.05 \cdot 4 \cdot 20 \text{ A} = 84 \text{ A}$$

According to equation (2.38.1)

$$\frac{\Delta \vartheta_{adi}}{\Delta t} = 4.99 \cdot 10^{-3} \, \frac{84^2}{4^2} \, \frac{K}{s} = 2.2 \, K/s$$

This means that at four times the rated current, within 1 s a temperature increase of about 2.2 K is produced, and thus during occasional acceleration or deceleration processes, the motor will not heat significantly. At acceleration to higher speeds, the speed-dependent current limit of the motor may be taken into account, which will result in an even smaller temperature rise.

If equations (2.38) and (2.7) are compared, it becomes apparent that heating increases proportionally to the square of the current, while the torque is proportional to the current. According to equation (2.35), the ramp time is inversely proportional to the torque and thus to the current. This is the reason why the mentioned current limit of four to five times the rated current is considered economical.

The so-called acceleration drives on stamps, presses, or nibbling machines take over the feed movement of the work piece. They work with clock times of < 100 ms. Because of the heating and brush lifetime, the acceleration current can exceed only slightly the rated current of the motor. An exact checking of the heating and brush lifetime is here necessary.

2.3.5 Additional Optimization Criteria

A feed drive can be optimized according to different criteria, depending on the kind of gear ratio, feed screw lead, and motor that it has. Section 4.4 handles the advantages and disadvantages that should be considered concerning gears. Other optimization criteria are presented in the following sections.

2.3.5.1 Energy Content of Moved Masses

In a collision of the machine slide with the work piece, respectively with the tool, the energy of the moving masses must be absorbed by the bearings, the gear components, and the piece or the tool. The value of this energy should be as low as possible.

The energy content of the moving mass is

$$W = \frac{1}{2} J_{Ges} \, \omega_M^2 \tag{2.40}$$

W energy of moving masses
J_{Ges} total moment of inertia reflected onto the motor shaft
ω_M motor angular velocity, see equations (2.1) and (2.2)

A low energy content can be obtained by selecting the motor speed as low as possible, i.e. with as high a feed screw lead as possible.

2.3.5.2 Attainable Acceleration

This value has most special significance for drives on press feeders or nibbling machines, where the aim is towards the shortest positioning times. The faster the work piece can be accelerated, respectively braked, the longer the possible travel distance for a given time will be.

The angular acceleration of the motor shaft is

$$\alpha_M = \frac{d\omega_M}{dt} = \frac{M_B}{J_{Ges}} \qquad (2.41)$$

ω_M angular velocity from equation (2.1)
M_B acceleration torque from equation (2.6)
J_{Ges} total moment of inertia reflected on the motor shaft

The linear acceleration of the machine tool slide results from

$$a_W = \frac{dv_V}{dt} \qquad (2.42)$$

and is, for feed screw drives

$$a_W = \alpha_M \cdot \frac{h_{Sp}}{2\pi \cdot i} \qquad (2.43)$$

and for rack and pinion drives,

$$a_W = \alpha_M \cdot \frac{r_{Ri}}{i} \qquad (2.44)$$

v_V feed rate on the slide
h_{Sp} feed screw lead
r_{Ri} radius of the pinion
i gear ratio

From this, it cannot be derived that large feed screw leads, respectively large pinion diameters, and small gear ratios result in high linear accelerations. In equation (2.4), the external moment of inertia of J_{Ges} is contained in the denominator. It acts to reduce the acceleration, and for gears it appears in calculations reduced by the square of the gear ratio. The optimum is the gear ratio, where the reduced external moment of inertia equals the motor's moment of inertia, including the pinion on the motor shaft. The gear ratio for the highest acceleration is:

$$i_{opt} = \sqrt{\frac{J_2 + J_{Gt\,2}}{J_M + J_{Gt\,1}}} \qquad (2.45)$$

J_2 moment of inertia reflected on shaft 2 according to (2.49) resp. (2.50)
$J_{Gt\,2}$ moment of inertia of the gear wheel on shaft 2
J_M motor moment of inertia
$J_{Gt\,1}$ moment of inertia of the gear wheel on the motor shaft

For the designations of shafts, see the illustrations in table 2.7 on page 128.

For multiple stage gears, the moment of inertia of additional gear wheels reflected onto shaft 2, must be added to $J_{Gt\,2}$. Equation (2.45) can be derived from an energy analysis, as the one given e.g. in 2.10. The calculation of the mechanical time constant according to equation (2.13) also indicates this optimum. If we substitute hereby the maximal motor angular velocity by the maximal angular velocity of the shaft 2 $\omega_{max\,2}$, with the aid of equations (2.48), (2.2), (2.3), (2.51) (2.52) we obtain:

$$T_{mech\,A} = \frac{(J_M + J_{Gt\,1})\omega_{max\,2} \cdot i}{M_{St\,A}} + \frac{(J_2 + J_{Gt\,2}) \cdot \omega_{max\,2}}{M_{St\,A} \cdot i}$$

The first member of the sum is proportional to i, and the second to $\frac{1}{i}$.

From a graphic presentation as the one in fig. 2.23, we can derive the lowest mechanical time constant for a particular application, if the values for inertia, velocity, and the motor data are given.

The gear ratio we have determined can however, only by attained to the extent to which the speed range on the part of the drive, and the required speed range on the part of the machine allow it. On the other hand, i and i_{opt} do not have to coincide very closely, since most of the time the function $T_{mech\,A} = f(i)$ for $i \approx i_{opt}$ has a flat slope.

Equation (2.45) should not lead to increasing the external moment of inertia for adaptation. It is provided for calculating a favorable gear ratio when the moments of inertia are known. The sum of the moments of inertia should be kept as low as possible, in order to minimize the acceleration and deceleration energies, which are converted into heat losses in the drive system.

The influence of the external moment of inertia on the nominal angular frequency of the drive $\omega_{0\,A}$ has been explained in section 2.2.2.6. The conclusion

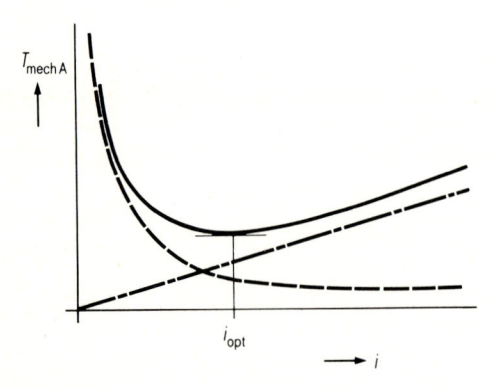

Fig. 2.23
Graphic determination of the gear ratio for the smallest mechanical time constant

drawn there about the value of the external moment of inertia also applies to regular, numerically controlled feed drives. The flat optimum of the gear ratio allows the selection of higher values for i. The advantage is, that in that case, the external moment of inertia has less influence on the speed regulator adjustment, and on the nominal angular frequency ω_{0A}.

2.3.5.3 Drive Comparisons

Depending on the motor selected, on the associated current converter unit, and on the possibly required gear, the costs of a feed drive will vary. By choosing different feed screw leads, gear ratios, and servo motors from the selection of possible versions, it is possible to design for the same task a feed drive at optimal cost, which will also meet the technical requirements.

Example:

For a feed slide, a feed force of about 12.5 kN is required, and the rapid traverse feed rate should be about 12 m/min. Table 2.6 shows the possible drive versions, with the values for the energy content of moving masses, attainable linear acceleration, cost comparisons (without gear) for a 6-pulse circulating current-free anti-parallel connection, or for a 3-pulse circulating current-free anti-parallel connection. (Friction, gear losses, and external moment of inertia have been included in the consideration. Timing belt gears were presumed for reductions.)

The following drives would be suitable for this task (the motor torques and final speed are given):

① Permanent magnet-excited servo motor 1HU3 104-0AD01 with direct coupling on a feed screw with a 10 mm lead.

② Permanent magnet-excited servo motor 1HU3 078-0AC01 with gear ratio $i=1.66$ to the feed screw with a 10 mm lead. The achievable feed force amounts to about 11.6 kN.

③ Permanent magnet-excited servo motor 1HU3 076-0AF01 with gear reduction $i=2.5$ to a feed screw with a 10 mm lead.

④ Permanent magnet-excited servo motor 1HU3 0780AC01 with direct coupling on a feed screw with a 6 mm lead. The achievable feed force amounts to approximately 11.6 kN.

⑤ Permanent magnet-excited servo motor 1HU3 108-0AD01 with direct coupling on a feed screw with 15 mm lead. The speed must be limited to 800 $\min^{-1}$; maximal attainable speed would be 18 m/min.

⑥ Electrically excited servo motor 1GS3 107-5SV41 with gear ratio $i=5$ to a feed screw with a 10 mm lead. The attainable feed force is 17 kN.

Table 2.6
Drives for a feed force of approximately 12.5 kN, and a rapid traverse of 12 m/min

Drive	Motor	$\frac{M_M}{Nm}$	$\frac{n_M}{min^{-1}}$	$\frac{W}{J}$	$\frac{a_W}{m/s^2}$	Cost. rel. % 6-puls.	3-puls.
① $h_{Sp}=10$ mm	1HU 3104	25	1200	364	3,5	100	112
② $i=1,66$ $h_{Sp}=10$ mm	1HU 3078	14	2000	250	4,7	88	98
③ $i=2,5$ $h_{Sp}=10$ mm	1HU 3076	10	3000	510	2,5	85	98
④ $h_{Sp}=6$ mm	1HU 3078	14	2000	290	4	88	98
⑤ $h_{Sp}=15$ mm	1HU 3108	38	800	210	6,1	121	138
⑥ $i=5$ $h_{Sp}=10$ mm	1GS 3107	6,8	6000	590	2,9	106	114

This comparison shows that drive ⑤ has the lowest energy content at the highest linear acceleration, but is also the most costly drive. Due to the relationship $T_{mech\,A}/T_{el\,A}$, it has the lowest nominal angular frequency $\omega_{0\,A}$. The calculation is demonstrated in table 4.6, on page 262.

The example shows that when the feed screw lead and the gear ratio are chosen freely, with different combinations of components, a well adjusted drive can be accomplished. The most suitable version is to be selected for each application. [2.12] shows such an optimization for the gear ratio i, and the feed screw lead, with the purpose to obtain the highest possible acceleration.

Considerations on the minimal ramp time possible for a feed drive are presented in [2.13]. The commutating limit is hereby a factor, as was shown in the example in section 2.3.3, and was also apparent from fig. 2.16.

2.3.6 Calculation Scheme

The flow chart in fig. 2.24 shows the sequence for the selection of a feed motor. The starting point is given by the technological and machine-related data:

Work piece mass m_W,
depending on the type of machine tool and on the axis concerned, m_W is not taken into account.

Feed table mass m_T,
includes the mass of all linearly moving parts.

Frictional factor μ_F between table and slide,
general values are:
for friction slides of cast iron/plastic approximately 0.08
for roller guides approximately 0.005

Cutting Force
Has two components: a machining force F_{VL} in the direction of the axis, and a component F_{VT} perpendicular to the slide guide, which must be taken into account when calculating the torque for the friction (if F_{VT} is not known, it is taken to be approximately 10–15% of the machining force).

Feed rate v_V.

Rapid traverse feed rate v_{Eil}

Load cycle data, if periodic loads are present.

Design data of the first draft:

Feed screw drive (fig. 2.13):		Rack and pinion drive (fig. 2.14):	
Feed screw lead	h_{Sp}	Pinion radius	r_{Ri}
Feed screw length	l_{Sp}	Moment of inertia of pinion	J_{Ri}
Feed screw diameter	d_{Sp}		
Feed screw moment of inertia (including coupling on gear or motor)	J_{Sp}	Gear:	
		Gear ratio (n_1/n_2)	i
Frictional factor of the feed screw bearing	μ_{SL}	Gear efficiency coefficient	η_G
		Moment of inertia:	
Bearing preload force	F_{aVL}	Entire drive	J_{Getr}
Feed screw nut efficiency coefficient	η_{SM}	For single stage gear	
		Gear wheel 1	J_{Gt1}
		Gear wheel 2	J_{Gt2}

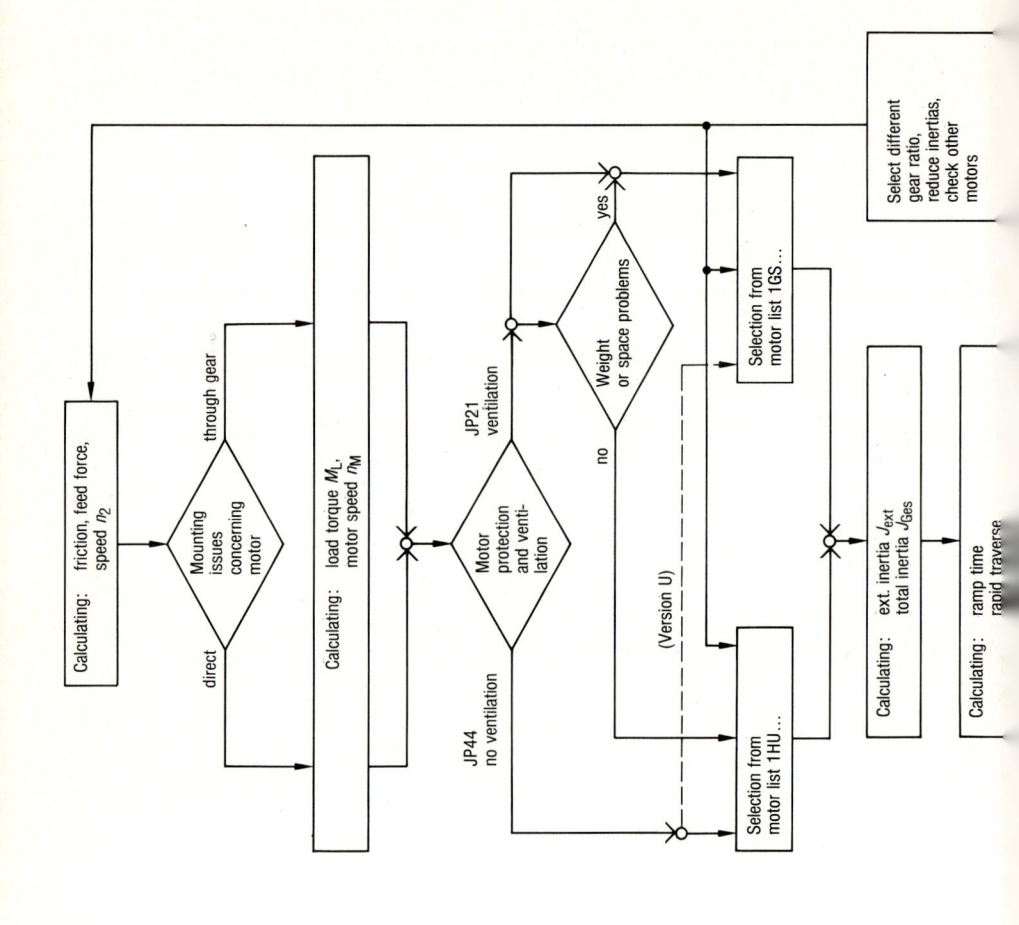

Fig. 2.24
Flow chart for the selection of DC servo motors for feed drives

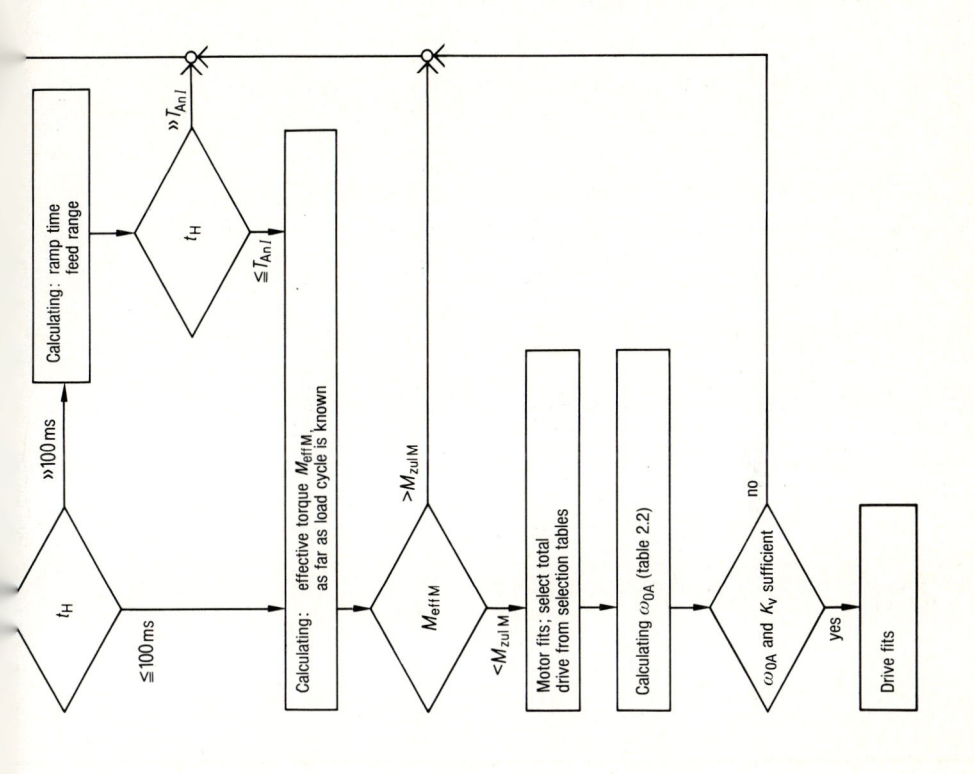

```
                                          ┌─ »T_AnI ──────────────────────────────────────────┐
                                          │                                                    │
  ┌──────────────┐   »100 ms   ╱ ╲       ╱ ╲                                                   │
  │ Calculating: │◄───────────╱ t \─────╱ ≤T \                                                 │
  │  ramp time   │            \ H /     \ AnI/                                                 │
  │  feed range  │             ╲ ╱       ╲ ╱                                                   │
  └──────────────┘              │         │                                                    │
                                │         │                                                    │
         ╱ ╲                    │   ┌─────▼──────┐   ╱ ╲       ┌──────────────┐  ┌──────────┐  ╱ ╲         ┌──────────┐
        ╱ t \  ≤100 ms          │   │Calculating:│  ╱ M \ <M_zulM│Motor fits;   │  │Calculating│╱ω_0A \  yes │          │
  ─────╱  H  ╲─────────────────►│   │effective   ├─╱ eff \──────►│select total  ├─►│ω_0A      ├╲and K_v╱────►│Drive fits│
       ╲     ╱                      │torque M_effM│╲  M  ╱       │drive from    │  │(table 2.2)│╲suff. ╱      │          │
        ╲   ╱                       │as far as    │ ╲ ╱         │selection tbls│  │          │ ╲  ╱         └──────────┘
         ╲ ╱                        │load cycle   │  │           └──────────────┘  └──────────┘  │no
                                    │is known    │  │>M_zulM                                    │
                                    └────────────┘  └──────────────────────────────────────────┘
```

125

The following steps are necessary for the selection of a suitable drive. (The formulas and symbols are summarized in table 2.7):

▷ The calculation of torques and speeds.

Load torque reflected on the motor shaft, according to equation (2.25)

$$M_L = \Sigma M_R + M_V$$

Sum of the *torques for friction and losses* reflected onto the motor shaft.

For feed screw drives according to equation (2.29)

$$\Sigma M_R = \frac{\frac{M_{RF}}{\eta_{SM}} + M_{RSL}}{\eta_G \cdot i}$$

For rack and pinion drives, according to equation (2.31)

$$\Sigma M_R = \frac{M_{RF}}{\eta_G \cdot i}$$

Torque for the machining force reflected onto the motor shaft.

For feed screw drives according to equation (2.32)

$$M_V = \frac{F_{VL} \cdot h_{Sp}}{2\pi \cdot i \cdot \eta_G \cdot \eta_{SM}}$$

For rack and pinion drives according to equation (2.33)

$$M_V = \frac{F_{VL} \cdot r_{Ri}}{i \cdot \eta_G}$$

The necessary *speeds* can be calculated from the desired feed rates, respectively desired rapid traverse rates.

For feed screw drives, the speed of motor shaft 2 (speed of the feed screw) is:

$$n_2 = \frac{v_V}{h_{Sp}}, \text{ and for rapid traverse } n_2 = \frac{v_{Eil}}{h_{Sp}} \qquad (2.46)$$

For rack and pinion drives, the speed of the 2nd shaft (speed of the pinion) is:

$$n_2 = \frac{v_V}{2\pi \cdot r_{Ri}}, \text{ resp. for rapid traverse } n_2 = \frac{v_{Eil}}{2\pi \cdot r_{Ri}} \qquad (2.47)$$

Depending on the gear present, the speed of the motor is

$$n_M = n_1 = i \cdot n_2 \qquad (2.48)$$

▷ Selection decision concerning the permanent magnet-excited servo motor of the series 1HU.., or the electrically excited servo motor of the series 1GS...

Selecting the proper size motor according to chapter 7, respectively from the technical data.

▷ The calculation of the moments of inertia.

The *moment of inertia of linearly moved masses* is:

According to equation (4.12), for feed screw drives, the inertia reflected onto shaft 2 is

$$J_{T+W} = (m_W + m_T)\left(\frac{h_{Sp}}{2\pi}\right)^2$$

and the sum of the moments of inertia reflected onto shaft 2 (feed screw) is

$$J_2 = J_{T+W} + J_{Sp} \qquad (2.49)$$

For rack and pinion drives, according to equation (4.13) reflected onto shaft 2:

$$J_{T+W} = (m_W + m_T) \cdot r_{Ri}^2$$

and the sum of the moments of inertia reflected onto shaft 2 (pinion) is

$$J_2 = J_{T+W} + J_{Ri} \qquad (2.50)$$

The *external moment of inertia* reflected on the motor becomes:

$$J_{ext} = \frac{1}{i^2} J_2 + J_{Getr} \qquad (2.51)$$

Without gear, $J_{Getr} = 0$, and $i = 1$, and the moment of inertia of the coupling must then be added to J_2. For single stage gears, the *gear moment of inertia* is

$$J_{Getr} = J_{Gt1} + \frac{J_{Gt2}}{i^2} \qquad (2.52)$$

and thus, according to equation (2.3), the *total moment of inertia* becomes

$$J_{Ges} = J_M + J_{ext}$$

▷ The calculation of the ramp-up time to rapid traverse, according to equation (2.35).

$$\frac{t_H}{s} = \frac{\dfrac{J_{Ges}}{\text{kg m}^2} \cdot \dfrac{\Delta n_M}{\text{min}^{-1}}}{9{,}55 \cdot \dfrac{M_B}{\text{Nm}}}$$

For permanent magnet-excited servo motors, the acceleration torque M_B is speed-dependent, and for this reason it must be calculated step-by-step (see example of section 2.3.3., and fig. 2.16). The time given, $t_H \leq 100$ ms, is a pragmatically determined value.

Table 2.7 Summary of the calculation formulas for the feed drive

Feed screw drive	Motor, without gear

Concept and units:

Work piece mass	m_W/kg
Feed slide mass	m_T/kg
Guide frictional factor	μ_F
Feed force	F_V/N
machining force (axis direction)	F_{VL}/N
slide direction component	F_{VT}/N
Feed rate	v_V/m/min
Feed screw lead	h_{Sp}/m
Feed screw length	l_{Sp}/m
Mean feed screw bearing diameter	d_{Sp}/m
Feed screw moment of inertia	J_{Sp}/kg m²*
Bearing frictional factor	μ_{SL}
Bearing preload force	F_{aVL}/N
Efficiency of feed screw nut	η_{SM}
Gravitational acceleration	$g = 9{,}81 \, m/s^2$

Concepts and units (motor):

Motor torque	M_M/Nm
Motor speed	n_M/min
Motor moment of inertia	J_M/kgm²
Gear moment of inertia	$J_{Getr} = 0$
Gear ratio	$i = 1$
Gear efficiency	$\eta_G = 1$

Load torque
Acceleration torque
External moment of inertia
(reflected on motor shaft)

Torques:

Friction bearing slide guide:

$$M_{RF} = \mu_F \cdot \frac{h_{Sp}}{2\pi} [(m_W + m_T) \cdot g + F_{VT}] \quad (2.26)$$

Friction feed screw bearing:

$$M_{RSL} = \mu_{SL} \frac{1}{2} d_{mL} \cdot F_{aVL} \quad (2.27)$$

Torque for friction and losses:

$$\sum M_R = \frac{\dfrac{M_{RF}}{\eta_{SM}} + M_{RSL}}{\eta_G \cdot i} \quad (2.29)$$

Torque for machining force:

$$M_V = \frac{F_{VL} \cdot h_{Sp}}{2\pi \cdot i \cdot \eta_G \cdot \eta_{SM}} \quad (2.32)$$

→ Stationary load torque
Acceleration torque

Ramp time:

M_M is speed dependent. Thus M_B of sections Δn must be determined. The sum of the ramp times of each section gives the total ramp time

*) incl. bearing and coupling   **) incl. shaft and coupling

Motor, with gear	Rack and pinion drive
Shaft 2: n_2, J_2, J_{Gt2} i, η_G, J_{Getr} Shaft 1: n_1, J_1, J_{Gt1}, M_L, M_B, J_{ext} Motor: n_M, J_M, M_M	(diagram with v_V, F_{VT}, F_V, m_W, F_{VL}, m_T, μ_F, M_{RF}, n_2, r_{Ri}, J_{Ri})
Concepts and units: Gear ratio — i Motor speed — n_1/min^{-1} Gear speed — n_2/min^{-1} Moment of inertia (on shaft 1) — J_{Getr}/kgm^2 Moment of inertia gear wheel 1 — J_{Gt1}/kgm^2** Moment of inertia gear wheel 2 — J_{Gt2}/kgm^2 Gear efficiency — η_G M_L/Nm M_B/Nm $J_{ext}/\text{kg m}^2$	**Concepts and units:** Work piece mass — m_W/kg Feed slide mass — m_T/kg Guide frictional factor — μ_F Feed force — F_V/N machining force (axis direction) — F_{VL}/N slide direction component — F_{VT}/N Feed rate — $v_V/\text{m/min}$ Pinion radius — r_{Ri}/m Pinion moment of inertia — $J_{Ri}/\text{kg m}^2$ Gravitational acceleration — $g = 9{,}81 \text{m/s}^2$
$M_L = \sum M_R + M_V$ (2.25) $M_B = M_M \mp M_L$ from (2.6) ($-$ acceleration, $+$ deceleration) $\dfrac{t_H}{s} = \dfrac{\dfrac{J_{Ges}}{\text{kg m}^2} \cdot \dfrac{\Delta n}{\text{min}^{-1}}}{9{,}55 \cdot \dfrac{M_B}{\text{Nm}}}$ (2.35)	Friction bearing slide guide: $M_{RF} = \mu_F r_{Ri} [(m_W + m_T) \cdot g + F_{VT}]$ (2.30) Torque for friction and losses: $\sum M_R = \dfrac{M_{RF}}{\eta_G \cdot i}$ (2.31) Torque for machining force: $M_V = \dfrac{F_{VL} \cdot r_{Ri}}{i \cdot \eta_G}$ (2.33)

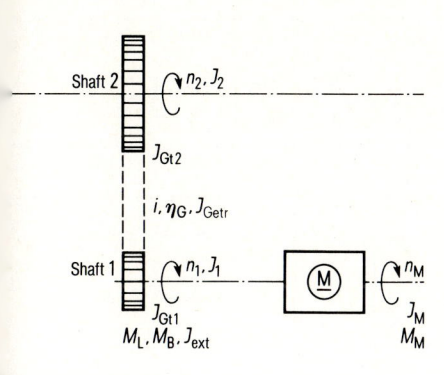

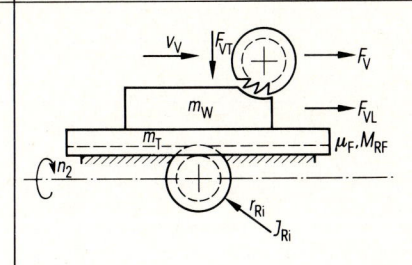

Table 2.7 continued

Feed screw drive	Motor
Speeds:	
Feed screw speed: $n_2 = \dfrac{v_V}{h_{Sp}}$ resp. $\dfrac{v_{Eil}}{h_{Sp}}$ (2.46)	→ Motor speed:
Moments of inertia:	
Linearly moved masses: $J_{T+W} = (m_W + m_T)\left(\dfrac{h_{Sp}}{2\pi}\right)^2$ (4.12) Sum, reflected on feed screw: $J_2 = J_{T+W} + J_{Sp}$ (2.49)	→ External moment of inertia: For single stage gear: Total moment of inertia (reflected onto motor shaft):

		Rack and pinion drive	
$n_M = n_1 = i \cdot n_2$	(2.48) ←	Pinion speed: $n_2 = \dfrac{v_V}{2\pi \cdot r_{Ri}}$ resp. $\dfrac{v_{Ei1}}{2\pi \cdot r_{Ri}}$	(2.47)
$J_{ext} = \dfrac{1}{i^2} \cdot J_2 + J_{Getr}$	(2.51) ←	Linearly moved masses: $J_{T+W} = (m_W + m_T)\, r_{Ri}^2$	(4.13)
$J_{Getr} = J_{Gt1} + \dfrac{J_{Gt2}}{i^2}$	(2.52)	Sum, reflected onto pinion: $J_2 = J_{T+W} + J_{Ri}$	(2.50)
$J_{Ges} = J_{ext} + J_M$	(2.3)		

▷ The calculation of the ramp time to feed rate.

In the feed range, the current limit should never be reached, thus the acceleration torque depends on the current rise, and is variable. The current rise time depends on the type of current converter, and on control setting. Table 3.1 on page 173 shows the approximate current rise times $T_{an\,I}$ to the current limit, for the individual current converters. The ramp time to the feed rate, can also be calculated from equation (2.35), and should not be significantly higher than this current rise time. Hereby, the mean value with approximately 50% of the torque for the current limit, is used for the determination of the acceleration torque.

If the ramp time to the feed rate is found to be significantly higher, then contour distortions can be expected in position controlled feed drives. By setting a reduced position loop gain in the position control loop, or by using a command value smoothing modifier for acceleration limitation, these contour distortions can be avoided (see chapter 5).

▷ The calculation of the effective torque (rms torque), is made according to equation (2.36), in as far as the load cycle is known.

▷ The selection of a current converter unit.

After the decision concerning the converter unit with the associated reactors has been made (see section 3.5), the achievable nominal angular frequency $\omega_{0\,A}$ can be double checked according to table 2.2 on page 90, and the so determined position loop gain K_v can be worked out.

2.3.7 Summary of Calculation Formulas

Table 2.7, and figs. 2.13 and 2.14, contain the data necessary for the determination and the sizing of a feed motor. More detailed explanations can be found in sections 2.3.2 and 2.3.3.

2.4 Bibliography

[2.1] Volkrodt, W.: Ferritmagneterregung bei größeren elektrischen Maschinen. Siemens Zeitschrift, 49 (1975), Heft 6, Seite 368 bis 374

[2.2] Fahlenbach, H.: Moderne Dauermagnete und ihre Grundlagen. radio mentor electronic, 1972, Heft 7, Seite 329 bis 331 und Heft 8, Seite 374 bis 376

[2.3] Boelke, K.: Strombegrenzung bei numerisch gesteuerten Werkzeugmaschinen. HGF Kurzberichte, Industrieanzeiger 95 (1973), Heft 16, Seite 305 bis 306

[2.4] Fröhr, F.; Ortenburger, F. (siehe Literaturhinweis [1.2])

[2.5] Phleps, R.D.: Computer assisted Servo Selection for NC-Machine Tools. Control engineering, Sept. 1974, Seite 143–144

[2.6] Ackermann, U.; Böbel, K.-H.: Rechnerunterstützte Auswahl von elektrischen Vorschubantrieben. wt-Zeitschrift für industrielle Fertigung, 66 (1976), Heft 3, Seite 129 bis 136

[2.7] Viktor, H.: Schnittkraftrechnungen für das Abspanen von Metallen. wt-Zeitschrift für industrielle Fertigung, 59 (1969), Heft 7, Seite 317 bis 327

[2.8] CAD-REKONE, Seminarunterlagen CAD-Seminar 24.–25. Juni 1976. Selbstverlag des Instituts für Steuerungstechnik der Werkzeugmaschinen und Fertigungseinrichtungen der Universität Stuttgart

[2.9] Stof, P.: Nachbildung des Erwärmungsverhaltens von Vorschub-Gleichstrommotoren. HGF Kurzberichte, Industrie-Anzeiger 97 (1975), Heft 25, Seite 490 bis 491

[2.10] Szalay, F.: Die Optimierung der Getriebeübersetzung bei Beschleunigungsantrieben. Werkstatt und Betrieb 109 (1976), Heft 2, Seite 69 bis 72

[2.11] Stute, G.; Stof, P.: Untersuchung über die Verwendbarkeit von Gleichstromantrieben als Vorschubantriebe für numerisch gesteuerte Werkzeugmaschinen. Frankfurt, VDW Forschungsbericht 1003, 1971

[2.12] Arafa, H.: Optimale Auslegung von Vorschubantrieben für Nachform- und numerisch gesteuerte Werkzeugmaschinen. wt-Zeitschrift für industrielle Fertigung 66 (1976), Heft 3, Seite 137 bis 141

[2.13] Berger, Th.: Die Dynamik als Gebrauchseigenschaft von Stellmotoren an Werkzeugmaschinen. Elektrie 32 (1978), Heft 9, Seite 493 bis 495

Kümmel, F.: Elektrische Antriebstechnik. Berlin-Heidelberg-New York, Springer Verlag 1971

Hesselbach, J.: Einsatz eines Mikrorechners zur Bestimmung des Zeitverhaltens von Vorschubantrieben. HGF Kurzberichte, Industrieanzeiger 101 (1979), Heft 2, Seite 24 bis 25

DIN 1323, Februar 1966, Elektrische Spannung, Potential, Zweipolquelle, Elektromotorische Kraft, Begriffe

DIN 1304, Allgemeine Formelzeichen

VDE 0530 Bestimmungen für umlaufende elektrische Maschinen

3 Current Converters for DC Servo Motors

3.1 Notions and General Overview

3.1.1 Requirements

Current converters must be understood in terms of: the power section for the energy supply of the motor – including the control circuits and the speed and current regulators. The input variable represents the speed set value, and the output variable is the motor voltage U_A.

Drives for position controlled machine tool feed drives are normally undelayed, reversible 4-quadrant drives. Such a drive can deliver energy in the drive mode, as well as accept, i.e. feed back into the line, energy during braking. Figure 3.1 shows the four operating ranges in the torque-speed plane:

Energy supplying = driving mode: I. and III. quadrant
Energy accepting = braking mode: II. and IV. quadrant

The dotted line at 4 M_M/M_{0M} designates the current limit of the current controller; the full line represents the commutating limit of the motor (see also fig. 2.2).

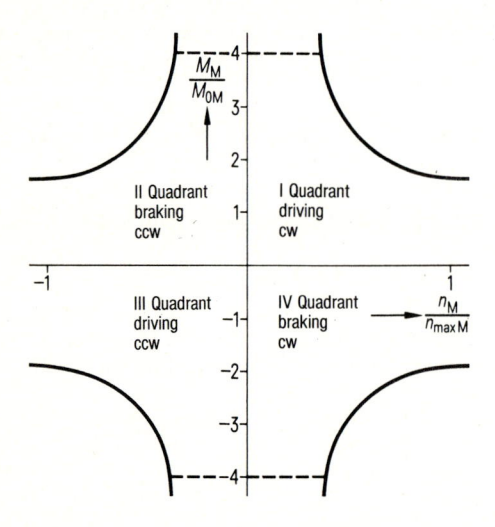

Fig. 3.1 Operating ranges of a 4-quadrant feed drive

The transition from one quadrant to another should be as free of delay as possible, in order to keep the drive response time for changes in the command value, short. During a dead time in the current converter, the drive is not controlled within the position control loop, so that corrections of position or velocity are not possible.

The current converter has a significant influence on the transfer behavior of the feed drive as a whole. Some of the requirements listed in section 2.1 can be affected by its design:

▷ the speed stiffness and the linearity between the velocity command value and the actual velocity, by a proportional-integral speed regulator,
▷ the wide speed range and the good disturbance transient behavior, by circuits acting adaptively within the control part,
▷ the high drive nominal angular frequency ω_{0A} and the high overcurrent allowed, by appropriate dimensioning of the power section.

Today, in practical application, the current control circuits used have transistors and thyristors. The DC voltage supplied to the servo motor is generated by the standard 3-phase AC lines, but is adjusted for amplitude and polarity by the appropriate circuit arrangement and control. The speed regulation, current regulation, or current limit preceed the control and power sections.

Circulating current-conducting and circulating current-free circuits, are common types of thyristor circuits. *Circulating current-conducting* circuits have the better dynamic behavior. The two part current converter combination is controlled so, that a circulating current is always flowing. For set value changes, especially at low speeds, the motor current can be increased rapidly, so that the speed change requirement can be quickly controlled. The delay time in the current converter is short.

Circulating current-free current converters need a command stage, which – depending on the polarity of the torque requirements – enables either one or the other part of the current controller. The switch-over pause which results must be kept short with some appropriate circuits; it causes a reduction in the drive nominal frequency of the speed control loop. For common position loop gains of approximately $1 \frac{\text{mm/min}}{\mu\text{m}}$, reaction times of $t \approx 6\text{-}8$ ms are possible, without affecting the behavior of the superceeding position control loop too negatively. The current-free circuits have modest requirements on transformer and reactors, so that they can be compactly and economically built into feed drives.

Circuits with *circulating current suppression* should also be mentioned. For these, a circulating current is flowing at a control for output voltage zero. This current decreases toward zero, as the output voltage is increased. The part of the current converter not in use for the moment, is inhibited by pulse suppression. Such current converters combine the good dynamic properties of circulating current-conducting drives, with the low requirements on reactors of the circulating current-free ones.

Decisive factors for selection of the appropriate circuit are:

ω_{0A} nominal angular frequency achievable
T_T delay time (approximated as dead time)
F_i form factor
P_{VSR} losses
η_{SR} efficiency coefficient
costs

The proper current converter circuit must be selected according to the requirements of the machine tool and the machining task. Table 3.1 on page 173 shows the nominal values of usual circuits, and the position loop gains achievable.

Extensive explanations concerning the achievable nominal angular frequency ω_{0A} of a drive are given in sections 2.2.2.5 and 2.2.2.6.

3.1.2 Dead Time

There is a time lapse before the drive reacts to a step in the set speed value. This period depends mainly on the control timing possible for the current converter circuit (see fig. 1.9).

For line synchronized current converters, the thyristors switch periodically with the line frequency. For free-running current converters, up to a certain frequency, the electronic switches can be switched on and off randomly. This particular switch frequency is designated as pulse frequency. Line frequency and pulse frequency are combined under the term f_N.

The second characteristic variable of a current converter circuit is the pulse number p_{SR}. It indicates the current pulses delivered by the circuit per line or pulse frequency period.

For current converter circuits, one calculates with a mean dead time, also called statistical dead time. It is the mean value between the maximal and minimal values possible for the dead time. The mean dead time can be calculated from the line or pulse frequency, and the pulse number:

$$T_T = \frac{1}{2 f_N \cdot p_{SR}} \quad (3.1)$$

Fig. 2.12 shows the effect of a dead time in a control loop, on the possible position loop gain.

3.1.3 Form Factor

For direct currents with superimposed alternating currents, the ratio of the total rms value I_{eff} to the arithmetic mean value I_d is known as the form factor F_i

$$F_i = \frac{I_{\text{eff}}}{I_d} \tag{3.2}$$

The torque of the DC servo motors is proportional to the arithmetic mean value I_d; the losses are proportional to the squared rms value I_{eff} of the armature current. At constant torque, increasing current ripple means increasing losses in the motor. The squared form factor is thus a measure of the additional thermal load.

For current converter circuits, the form factor of the output DC depends on the pulse number, and the momentary control angle, i.e. the ratio between the mean value of the DC voltage and the maximal possible voltage. The value of F_i is also determined by the size of the inductivity in the armature circuit. Table 3.1 on page 173 shows average values of installed units.

The selection tables for DC servo motors in chapter 7 show the torques for a form factor of 1.05. Thus, the torque permitted for current control supply is:

$$M_{\text{zul M}} \approx 1.05^2 \cdot M_{0\,M} \frac{1}{F_i^2} \tag{3.3}$$

The operating mode of a feed drive on a machine tool is not continuous duty at rated torque. Part changes and programmed load alternations result in pauses which are disregarded for the normal sizing, according to torque and thrust, of the feed servo motor. Based on this practical knowledge, usually no reduction of the admissable torque is necessary for 6-pulse and 3-pulse circuits. For 2-pulse current converters, the form factor must be improved with smoothing reactors, or the torque must be accordingly reduced.

The commutation of the DC servo motor is also affected by the current ripple. A higher commutation voltage in the armature coil short circuited by the brush, results when the peak values are higher in relation to the torque producing arithmetic mean values. For higher form factors, the commutation limit curves given in the selection lists must be reduced accordingly.

For extreme loads of DC servo motors, these effects on the allowable torque and commutation, must be in any case taken into consideration with regard to the motor life time. Computational considerations cannot include all influences. Extensive tests under real life conditions should therefore be conducted before any large scale application. There are some tips regarding easy ways to conduct tests, in chapter 6, under "test procedures".

Current ripples also cause frequency-dependent cyclic magnetizing losses in the magnetic circuit of the armature. They can be minimized through appropriate motor design, i.e. isolated lamination, and use of high quality dynamo sheet metal. These losses are more apparent at the high pulse frequency of transistor servo controllers, and result in additional heat losses in the iron parts. However, with a better form factor, the current heat losses are lower, so that the additional iron losses will usually not lead to a torque reduction.

For fast reacting feed drives, motor current ripple also causes speed oscillations superimposed on the mean speed value, which lead to oscillations of the coupled mechanical elements. Mechanical resonance frequencies in an unfavorable position relative to the frequencies of the superimposed AC, can lead to disturbing resonance phenomena. This is particularly true of the base frequency. For line synchronized thyristor converters, the frequency of the fundamental oscillation is proportional to the product of the line frequency f_N and the pulse number p_{SR} of the current converter.

$$f_1 = f_N \cdot p_{SR} \tag{3.4}$$

For transistor servo controllers, working in switching mode, the frequency of the fundamental oscillation of the current is given by the pulse frequency. It is typically so high, as to be outside the range of mechanical resonance frequencies.

Resonance frequency of gears and sheet metal covers should especially be kept away from the critical frequency. A high inertia of the DC servo motor, including mounted pinion or coupling, is here advantageous, since it limits the amplitude of the angular oscillation. Disadvantages of high inertias are commented upon elsewhere (2.2.2.5). The drive nominal angular frequency must not be reduced too much by measures designed to increase the inertia.

The line synchronized current converter causes a distortion of the input AC. The resulting harmonic currents do not contribute to power transportation, and reduce the power factor of the drive. Depending on the pulse number of the current controller, only harmonic oscillations of certain ordinal numbers result

$$v = p_{SR} \cdot k \pm 1 \quad (k = 1, 2, 3 \ldots) \tag{3.5}$$

The rms value of a harmonic current is approximately $1/v$ of the rms of the fundamental oscillation. The harmonic spectra of the different converter circuits with their frequencies and rms values, can be found in additional literature [3.1].

3.1.4 Efficiency

The choice of current converter circuit influences the losses in the motor. Generally, a minimal amount of heat loss on the machine parts is desired. The coefficient of thermal expansion of steel is $12 \times 10^{-6}/K$, i.e. at 1 K temperature change, a 1 m steel part expands by 12 µm. The heat loss flowing from the motor causes changes in the preload of the ball screws, and dimensional changes in the guide ways. Both can considerably affect the accuracy of the part.

For precision feed axes, the servo motor should therefore be either provided with a safe heat loss removal system, e.g. forced ventilation, or should not be sized too tightly. For this reason, the motors should not be mounted into closed machine cavities.

The problem of efficiency becomes important for larger drives. Permanent magnet-excited servo motors are especially suitable, since there are no losses due to electrical exitation, and thus the efficiency is increased considerably [3.2]. Current converters with high pulse number and therefore form factors nearing 1, are more efficient, and are from this stand, preferable.

The efficiency of a drive can be derived from the output power P_{ab}, and the sum of the losses ΣP_V

$$\eta_A = \frac{P_{ab}}{P_{ab} + \sum P_V} \tag{3.6}$$

The losses of a converter circuit-fed drive comprise:

> Transformer losses,
> lead, fuse, commutating reactor losses,
> current converter losses (thyristors, CSE circuits),
> circulating current reactor losses (or smoothing reactor where applicable),
> overcurrent relay losses,
> motor losses.

The power section losses are for the most part current dependent. The servo motor losses depend on the torque and the speed. Therefore, the efficiency characteristics are load dependent, and can be shown in a diagram, where the current control angle is shown as the ratio $n_M/n_{max\,M}$. Fig. 3.2 shows as an example, an efficiency diagram of a large permanent magnet excited DC servo motor with a 6-pulse current converter circuit with circulating current. In this example, a considerable part of the losses is caused by the transformer.

Efficiencies of other current converter circuits are similar to the one in this example. In the feed range, values of approximately $\eta_A = 0.6$–0.7 can be expected. The efficiency of the servo motor alone is approximately 0.8.

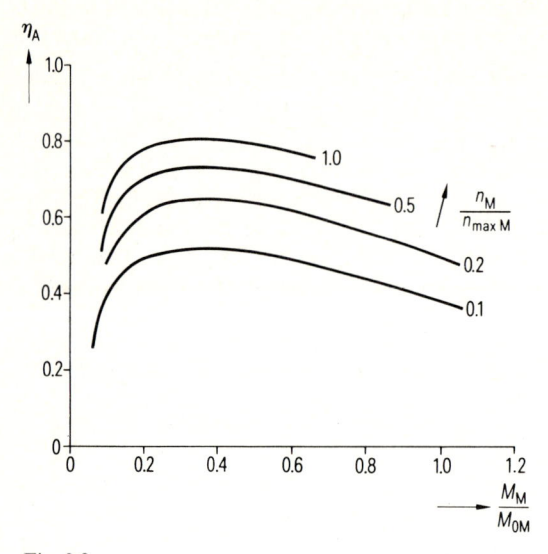

Fig. 3.2
Efficiency of a current converter-fed feed drive 1HU3 138–0AF01, with 6-pulse current converter circuit with circulating current

The output power of a feed servo motor is calculated as the product of torque and angular velocity. The adjusted dimensional equation with the speed is

$$\frac{P_{ab}}{\text{kW}} = 0.1047 \cdot 10^{-3} \frac{M_M}{\text{Nm}} \cdot \frac{n_M}{\text{min}^{-1}} \tag{3.7}$$

M_M and n_M are two related values of the speed/torque diagram.

Stating the power of a feed drive is generally irrelevant, because in the feed range, the machine tool requires high torques at low speeds. The relevant term is the available torque.

3.1.5 Costs

The aim is always towards an economical feed drive. The current controller is selected according to the part to be machined (see fig. 3.23). Installation and wiring costs, as well as space requirements in the cabinet must also be considered. Often, equipment uniformity is also a deciding factor for the user. Training of personnel in design, shops, and service creates high costs. Inventory of different systems appears in the negative in a cost calculation. Drive systems with current converters using the same boards thus offer cost efficient solutions (see also section 2.3.5.3).

3.2 Transistor DC Choppers

3.2.1 Applications

Transistor DC choppers are the ideal servo drives actuators. The transistors are commonly used in the switching mode, at frequencies between 1 kHz and 10 kHz (pulse width modulation). This implies almost no dead band, and a form factor of approximately 1. For small feed drives, analog transistor servo controls can also be used. In these cases, the limiting factor are the losses in the power section, which under continuous control may exceed the outputted real power.

The transistor DC chopper is used extensively. The steady development efforts on the power transistors have increased their reliability, and especially the output power. An economical limit is given by the necessary acceleration current of approximately 4–4.5 fold the rated current of the motor. The applications are generally limited because of the missing energy feedback into the line. Nowadays, feed drives are equipped with transistor choppers up to around 30 Nm.

Transistor DC choppers are used advantageously for fast and accurate turning and drilling machines. The dynamic data allow a trouble free adaption to the mechanical transport components. There are no difficulties due to resonance in machine parts. The position loop control can reach a position loop gain of up to $K_v \approx 80$ s^{-1}.

Multiple axes drives can be economically assembled by incorporating several systems into one unit. The common functions and the power supply for the power and control circuits need be provided only once per unit.

3.2.2 Principle of Operation

Figure 3.3 shows the basic diagram of the bridge circuit used. A rectifier feeds from the three-phase AC line into a DC bus. The buffer capacitor C can supply stored energy for acceleration, and can accept energy, as long as the motor must absorb mechanical energy during braking. This buffer capacitor is thus working as a generator and supplies electrical energy. It must be sized in such a way that the DC bus voltage changes only by little. The motor can be controlled selectively for clockwise or counterclockwise rotation, and can be accelerated or braked, by controlling two diagonally opposed transistors (T1–T3 or T2–T4).

The magnitude of the armature voltage U_A, and thus the speed n_M, is determined by either the more or less complete opening of the transistor pairs for analog controllers, or by pulse width modulation of the transistors acting as switches for a switched transistor chopper.

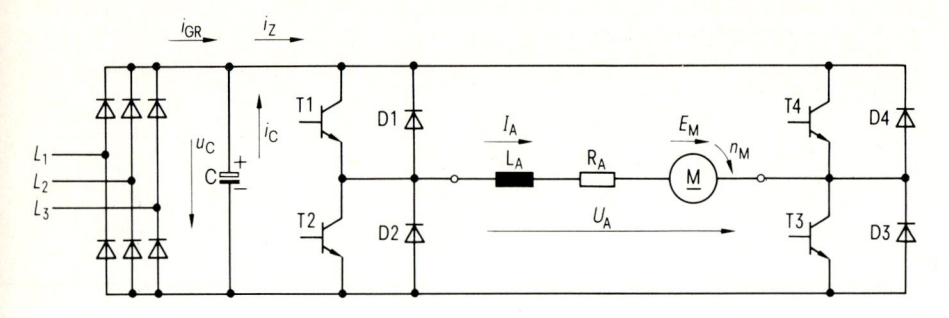

Fig. 3.3
Basic diagram of a transistor chopper in bridge circuit (T1 and T3 cross activated)

Two energy storages are necessary to operate a transistor controller in all four quadrants:

▷ A large buffer capacitor C, which maintains the voltage U_C constant on the input terminals, and is capable of accepting energy to store and to deliver.
▷ A load inductance, which smoothes the motor current and acts as an energy buffer; this is especially important during the braking mode. At high working frequencies of the controller, the armature inductivity L_M of the motor is generally sufficient.

The individual modes of the drive can be shown with the help of the basic diagram of fig. 3.3 and the pulse grid in fig. 3.4.

3.2.2.1 Driving, Clockwise, I. Quadrant

The voltage and current vectors shown in fig. 3.3 designate this operation mode. Figure 3.4 shows the pulse grid.

At time t_0, both transistor switches T1 and T3 are on. (The pulse grid shows logic "0" for the switched on transistor.) The armature voltage of the motor is positive $u_A = U_C$, and current i_A flows through the motor via T1, L_A, R_A, and T3.

At time t_1, switch T1 is opened. The motor current commutes from T1 to the diode D2, and is not flowing through the DC bus any more, but circulates in the lower half of the bridge from T3 through D2, L_A, R_A, motor, and back to T3 (free wheeling). At this point, the motor voltage u_A and the DC bus current i_z jump to zero.

At time t_2, the situation is the same as it was at time t_1. At time t_3, switch T3 opens. The motor current commutes now to diode D4, and circulates in the upper half of the bridge (circuit T1, L_A, R_A, motor, D4, T1), whereby motor voltage and DC bus current jump back to zero again. At time t_4, switch T3 is closed, and a new switch cycle begins.

Driving, clockwise

Driving, counter-clockwise

Braking, clockwise

Braking, counter-clockwise

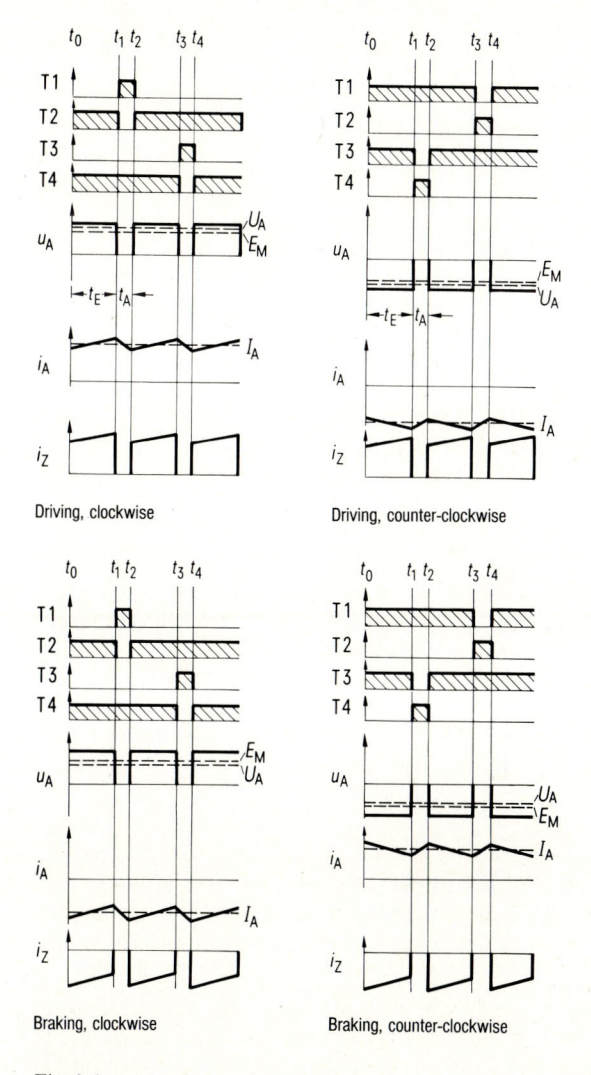

Fig. 3.4
Pulse grid, current, and voltage diagrams for the four operating ranges

The mean value of motor voltage U_A depends on the ratio between the switch-on time t_E, and the switch-off time t_A. During switch-on time t_E, the energy is derived from the DC bus, while at time t_A the current is driven by the energy stored in the inductance. The motor thus maintains a positive product from voltage U_A and current I_A during t_E and t_A, and thus converts electrical energy into transmitted mechanical energy. The voltage induced in the motor E_M is reduced, compared with the motor voltage U_A, by the amount of voltage drop inside the motor ($I_A \cdot R_A$).

3.2.2.2 Driving, Counterclockwise, III. Quadrant

The situation is similar for the "driving the motor counterclockwise" mode of operation. Here, the T2 and T4 transistors switch so that at the load there are negative voltage pulses, and the current flows through the motor in the opposite direction. In the free wheeling phases, the load current also circulates alternately through the lower and upper halfs of the bridge. During switch-on time t_E, the driving motor current is conducted by transistors T2 and T4. The switch-off time t_A is the free wheeling phase.

3.2.2.3 Braking, Clockwise, IV. Quadrant

The mean values of the motor current I_A and the armature voltage U_A must be of opposite signs for a back flow of actual power out of the motor circuit. If the current heat losses on R_A are disregarded, the control pulse pattern for current control braking looks the same as that for current control driving, except for the current flow polarity.

On the lower part of fig. 3.4, it is presumed that the mean value of the armature voltage U_A is reduced as compared to the voltage induced in the motor E_M. The direction of the current within the motor circuit is reversed. This transition from driving to braking is the result of a decrease in t_E, and an increase in t_A. At time t_1, switch T2 is closed. Voltage E_M drives a current through R_A, L_A, T2 and D3. Energy is stored in the inductance L_A:

$$W_{LA} = \frac{1}{2} L_A \cdot I_A^2 \tag{3.8}$$

At time t_2, T2 is switched off, and current i_A commutes over to diode D1, flows into DC bus, charges the capacitor, and returns through D3 to the motor. The voltage induced in the inductance and the induced motor voltage are in series; their sum is larger than the voltage delivered from the rectifier. Energy is fed back into the DC bus and stored in the capacitor:

$$W_C = \frac{1}{2} C \cdot U_C^2 \tag{3.9}$$

The voltage in the DC bus increases.

If now, at time t_3, the transistor T4 is switched on, the armature current will flow in the upper circuit through R_A, L_A, D1 and T4, and energy will again be stored in L_A. This energy will in turn be fed into the DC bus at time t_4, via diodes D1 and D3. This process is repeated periodically. (T1 and T3 switch-on at times t_2 and t_4 is of no consequence, since the current must flow against the direction of conduction, and is taken over by diodes D1 and D3.)

The time period t_A is the interval for storing energy in the inductance; period t_E is the interval of energy feedback into the DC bus. The mean values of motor current and motor voltage have opposed polarity. The motor is braked with mean constant torque, because actual power is fed back into the DC bus.

3.2.2.4 Braking, Counterclockwise, II. Quadrant

The sequence of events in braking of counterclockwise rotating motors parallels that involved in braking clockwise rotating motors.

Energy is stored in inductance L_A during time t_A, by switching transistor T3 on. As soon as, at time t_2, T3 is switched off, the current commutes to the free wheeling diodes D2 and D4, and flows back into the DC bus.

Here too, the motor current and voltage are of opposite signs, therefore an energy backflow and thus the generator mode of the drive is established.

3.2.3 DC Bus

Since the buffer capacitor is limited in the amount of energy it can store, and the DC bus voltage may not increase too much because of the transistors, it must be ensured – through a voltage supervising device – that this voltage is kept within limits. For large braking energies, a switched resistor stage should be connected to the DC bus to convert the energy into heat.

Since the DC bus is usually shared by several feed drives, when in driving mode the axes draw their energy from the DC bus. In the basic presentation of fig. 3.4, the voltage drops over the transistors, the diodes, and the motor inner resistance, including leads and over current detector, have also been disregarded. In reality however, a considerable amount of the mechanical energy is here converted into heat.

Actual experience with transistor converters has demonstrated that normal deceleration events do not result in an excessive voltage increase. The over voltage limiter switch mentioned above is necessary only for vertical axes without mechanical or hydraulical counter balance, or large inertias with no considerable friction.

Another possibility, is to use an inverting rectifier – which would feed the braking energy back into the line – to feed the DC bus. This circuit is normally not used, out of economic reasons.

3.3 Line Synchronized Thyristor Controllers

3.3.1 Generalities

Current converters with thyristors, because of their advanced technological level and practicality, are the most commonly used control elements. The most common versions used nowadays, are the 2-, 3-, and 6-pulse converters for circulating current-conducting circuits, and the 6-pulse version for circulating current-free circuits. A higher pulse circuit has smaller statistical dead times, and reacts faster dynamically.

As a rule, the 3-pulse circulating current-conducting converter circuit suffices for the dynamic requirements of the position control loop. For especially high demands, it is preferable to use the 6-pulse circulating current-conducting circuit, but for simple machines, the 2-pulse circulating current-conducting circuit can easily meet the requirements. The 6-pulse circulating current-free circuit has certain limitations in its use at high dynamic requirements, due to the additional reaction time necessary for torque reversion. It is, however, capable of delivering the necessary accuracy and dynamic over a wide range of applications. A more detailed explanation is to be found in the descriptions of the individual circuits.

The necessary 4-quadrant operation requires two groups of converters, since because of the semi-conducting properties of the thyristors, the current can only flow in one direction. These two groups are connected in anti-parallel or cross connection, thus allowing the DC servo motor to operate in all four quadrants. Figure 3.5 shows the principle of the arrangement.

An *anti-parallel connection* is a parallel connection on the DC side, of two converters whose thyristors conduct in opposite directions. Both converters lie on the same input line. It is therefore unlikely that a transformer should become necessary, only a secondary coil is required. An auto transformer can be used.

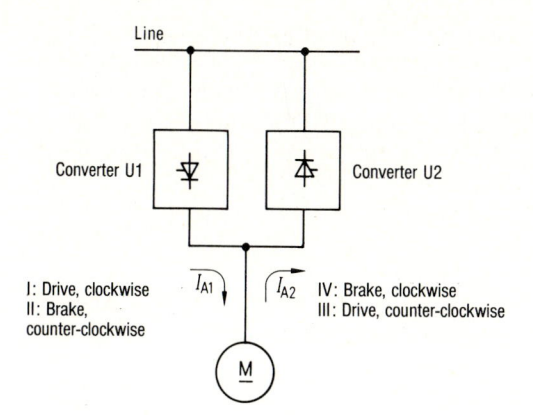

Fig. 3.5
Scheme of the reversing current converter

Line — Converter U1 — Converter U2

I: Drive, clockwise
II: Brake, counter-clockwise

I_{A1} I_{A2}

IV: Brake, clockwise
III: Drive, counter-clockwise

M

In a *cross connection,* each of the two current converters must be connected to its own transformer secondary coil; these transformers are more expensive to manufacture. Both converters are, here also, connected in opposite directions. Characteristic of this circuit is the fact that the systems are separated, connected only on the DC side.

3.3.2 Rectifier and Inverter Operation

For thyristors, the start of the current conducting time when the voltage on the anode, as compared to the cathode, is positive, can be controlled with a firing pulse. By comparison with the natural current commutating time of an uncontrolled rectifier, the positive voltage time area can be thus sectioned, and the mean value of the DC voltage can in this manner be controlled. The shift of the firing time, measured in electrical degrees, is designated as *control angel* α.

If inductivities are present in the DC circuit, the current can continue to flow even when the momentary value of the voltage becomes negative. The thyristor continues to conduct. When the control angle increases, the mean value of the DC voltage can also become negative; this reversion indicates a reversion of the energy direction, since the DC current direction, due to the semiconductive properties of the thyristor, cannot reverse.

In the control range, the converter works in *rectifier operation,* up to the point where the polarity of the DC mean value is reversed; at higher control angles, it works in *inverter operation.* It should be noted that inverter operation is only possible when the DC circuit of the converter contains an energy source.

Figure 3.6 shows the control range and control characteristic curves of a 6-pulse current converter. (The curve of the phase angle controlled voltage is presented, for the example of a 3-pulse converter, in fig. 3.11).

The control characteristic curve through the point $\alpha = 90°$, applies in the case of infinitely high inductivity in the DC circuit. The control curve terminating at $\alpha = 120°$, applies in case of a ohmic load of the 6-pulse current converter. Inbetween, lie the control curves for actual properties of the armature circuit resistances and inductivities. The mean value of the output voltage runs, in the ideal case ($L \to \infty$), according to a cosine function

$$U_A = U_{di} \cdot \cos \alpha \qquad (3.10)$$

U_{di} is the ideal idle voltage of the converter.

The limit of rectifier mode 1, and inverter mode 2 shifts, depending on the values of inductivity and resistance, within the range $\alpha = 90°$ el to $120°$ el.

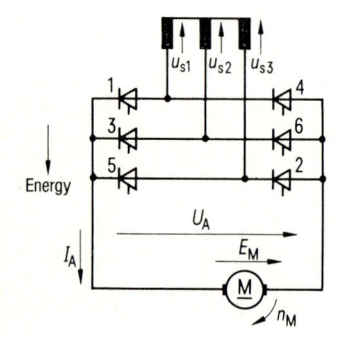

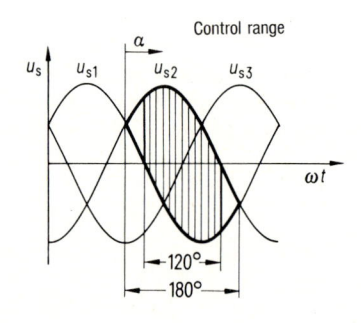

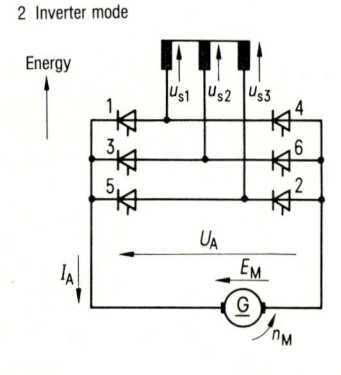

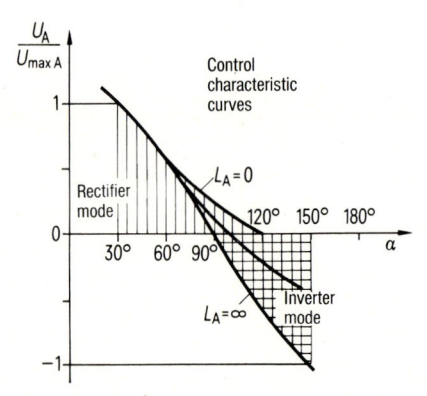

Fig. 3.6
Control range and control characteristic curves of a 6-pulse converter

In practice, because of the finite duration of the current commutation from one to the next thyristor, the maximum alternating control angle must be limited to approximately $\alpha = 150°$. This means that the maximal attainable inverter voltage is about 14% smaller than the maximum DC voltage. In order to prevent the counter-voltage induced in the motor from becoming larger than the inverting mode voltage, for 4-quadrant drives, the DC control angle should be limited to $\alpha = 30°$. If this measure is not taken, the current converter will not be able to control any longer the armature current. Thus, between the rectifier limit α_G and the inverter limit α_W, there is a usable control range of 120° el.

If the converter works in the rectifier mode, and if the converter voltage is higher than the induced machine voltage, the electrical energy is then derived from the line, and it is converted into kinetic energy, through the DC machine running as a motor. If the converter works in the inverting mode however, then, as in the example in fig. 3.6, a current flow with a unidirectional current converter, is only possible when the induced voltage, unlike in the rectifier mode, is reversed and larger than the converter voltage. (In fig. 3.6, the turning direction is shown reversed, for this reason.) In this case, the DC machine acts as a generator, and the kinetic energy of the mechanical system is fed back into the line as electrical energy, through the converter. For a 4-quadrant drive with reversing current converter, the second converter takes over the function of the inverter and vice-versa, so that the reversed energy flow is possible in both turning directions.

3.3.3 Circulating Current-free Connection

In a circulating current-free circuit, only one converter conducts the current, while the control pulses of the other converters are blocked. A command stage of the electronical control activates the appropriate current converter, depending on the required torque direction. With the electronical components available today, reaction times of about 6 ms can be reached, so that the reaction times which they thus cause in the closed control loop, have only small effect.

A disadvantage is, that the converter must deliver every value inbetween current 0 and the rated current, so that the pulsating current range is also included in the control angle range. Figure 3.7 shows the current forms for pulsating and non-pulsating operations of a 6-pulse current converter.

The ranges of the pulsating and non-pulsating current are determined by the resistance and inductivity in the armature circuit, as well as by the pulse number and control angle of the converter. Low pulse number and armature circuit inductivity result in large operating ranges with pulsating armature current.

In the pulsating range, a different structure of the control system appears. The electrical armature circuit time constant T_{elA} is no loger effective, therefore an optimized controller does not work optimally in this range. The response time of the current can only be kept constant, if the control structure is also

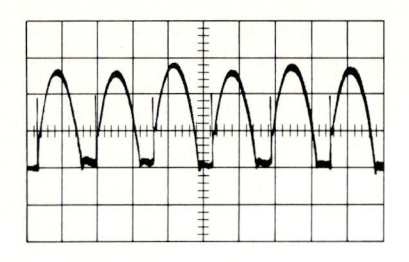

Pulsating operation 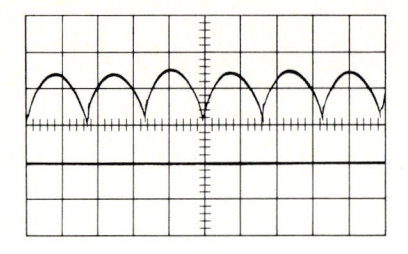   Non-pulsating operation

Fig. 3.7   Current forms of a thyristor converter

adapted. This can be accomplished with adapted modified controllers, as it is done especially in the case of spindle drives for machine tools [3.3], [3.4].

This operation type is not so significant in the case of feed drives, for which subordinated current controllers are used infrequently, because the constant basic friction of the machine usually provides the basic load current. The control is optimized for this basic load. If necessary, the behavior of the speed regulator, which becomes more critical as the load is increased, can be damped through a current dependent gain alteration.

Many numerically controlled feed axes, especially those on simpler production machines, are nowadays equipped with circulating current-free drives. This is justified from the point of view of the costs, by the low requirements on the current converter, the transformer, and the reactors.

3.3.4 Circulating Current-conducting Connection

This type of connection is characterized by the fact that between the two connected current converters flows a balancing current, also called circulating current. As shown in fig. 3.5, both connected converters are controlled continuously. It is ensured through the control device, that the control angle of, for instance, the rectifier mode of converter U1 corresponds to the inverter mode of converter U2. The converter U2 is in the inverter mode. Due to the counter-acting control angle of the two current converters, the mean values of both output voltages are equal, and they are opposed, because of the antiparallel connection. They cancel each other out, for this reason.

Because, however, the momentary values of the phase controlled voltages are not equal, a balancing current will flow between converters U1 and U2.

This circulating current is limited through circulating current reactors. The magnitude of the circulating current can be set with a circulating current command value. The converters are loaded additionally, due to the fact that the circulating current is added to the motor current.

The circulating current offers important advantages. One of them is, that the current in the converter can be kept outside the pulsating range at higher pulse numbers, through the basic load. For this purpose, the circulating current is set at approximately 10–15% of the rated current of the motor, and deteriorations in the control dynamic will be avoided. Another advantage is, that because the converters are continuously controlled, at any speed change a current commutation occurs immediately. The transition from driving to braking is almost delay-free. No additional switch-over time affects the dynamic of the speed control loop.

Circulating current-conducting connections have higher demands on the control circuit, transformer, and circulating current reactors. They have more favorable control behavior, and are used on high quality numerically controlled feed drives extensively.

3.3.5 6-Pulse Circulating Current-conducting Cross Connection

This type of connection is shown in fig. 3.8, as the power section. A transformer with two secondary coils is necessary for each feed axis. This provides the galvanic isolation, and prevents balancing currents between the two current converters.

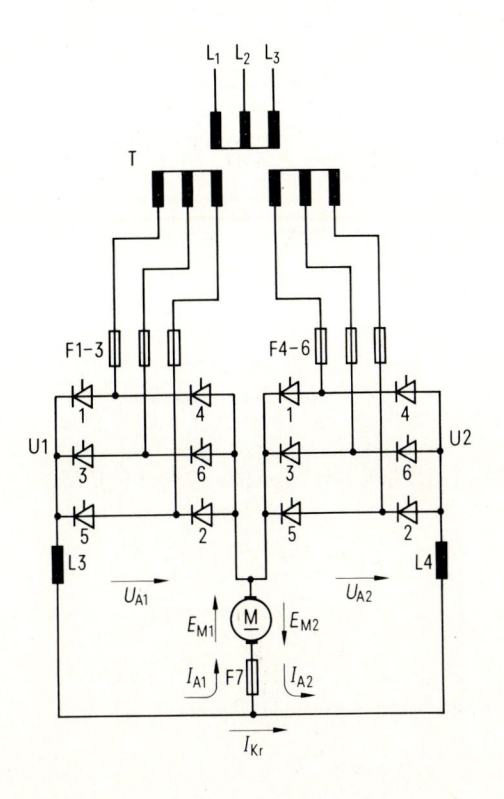

T	Transformer
F1–7	Rectifier fuses
U1, U2	Thyristor bridge connections
L3, L4	Circulating current reactors
M	DC servo motor
I_{A1}	Armature current for driving clockwise and braking counterclockwise
I_{A2}	Armature current for driving counterclockwise and braking clockwise
I_{Kr}	Circulating current
U_{A1}	Armature voltage for driving clockwise
U_{A2}	Armature voltage for driving counterclockwise
E_{M1}	Motor EMF in clockwise direction
E_{M1}	Motor EMF in counterclockwise direction

Fig. 3.8
6-pulse circulating current-conducting cross connection (B6C)X(B6C)

The current directions drawn in fig. 3.8 designate the possible operating modes. A better overview is given in fig. 3.9, which is a presentation of the 4 quadrants:

Driving Clockwise, I. Quadrant:

Converter U1 is controlled for rectifier operation, converter U2 for inverter mode. Current I_{A1} flows in the left half of the circuit, and drives the motor clockwise. Current I_{A2} equals zero.

In addition, the circulating current I_{Kr} flows from converter U1, through the circulating current reactors, to converter U2.

Driving Counterclockwise, III. Quadrant:

Converter U2 is controlled for rectifier mode, and converter U1 for inverter mode. Now current I_{A2} flows and drives the motor counter-clockwise, while current I_{A1} is zero. Circulating current I_{Kr} flows as before.

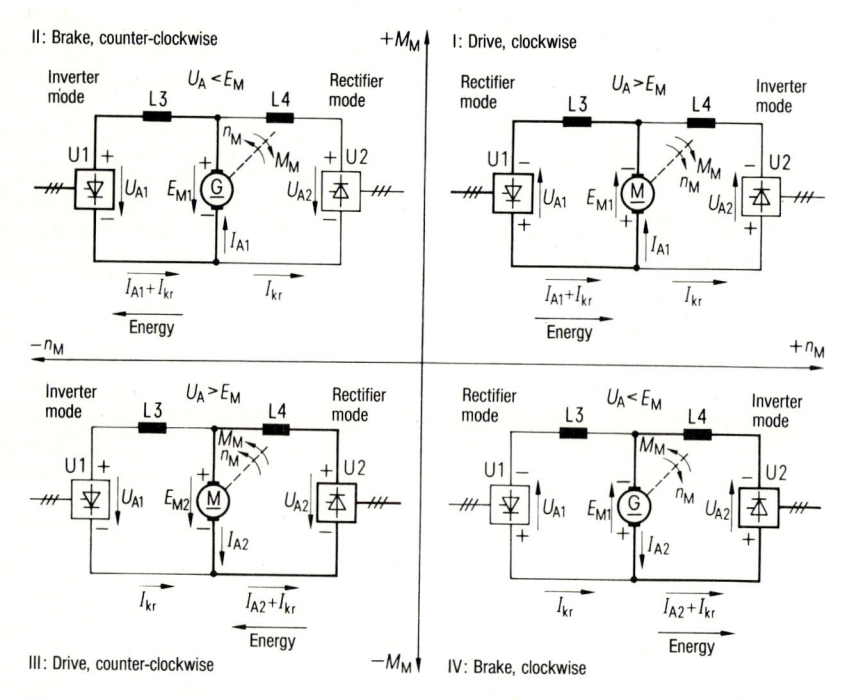

Fig. 3.9
4-quadrant operation of the 6-pulse circulating current-conducting cross connection

Braking Clockwise, IV. Quadrant:

By reducing the rectifier control angle of converter U1, current I_{A1} is brought to zero, since the motor EMF E_{M1} becomes now larger than the current converter voltage U_{A1}, and no current can flow against the semiconducting direction of converter U1. Simultaneously, voltage U_{A2} of the constantly controlled inverter U2 is reduced, and thus the motor EMF E_{M1} drives a current in the direction I_{A2}, against this inverter voltage. The resulting negative torque brakes the motor. Energy is fed back into the line through the current converter U2. The current is maintained within the limits permitted, through current control or a current limitation.

Braking Counterclockwise, II. Quadrant:

According to the braking mode in clockwise direction, the current converter U1 works here in the inverter mode. The motor EMF E_{M2} drives a current in the direction I_{A1}, which brakes the motor, and feeds energy back into the line through converter U1.

Since the energy of the mechanically moved masses is fed back into the line, no power limitations exist in these current converters, as would occur in transistor choppers. The form factor and dead time are favorable, and the circulating current makes possible feed drives of high dynamic quality. This type of circuit is preferred for dynamically demanding machine tools, and for torques $M_M >$ 30 Nm. For smaller torques, the demand for current converter transformers and circulating-current reactors, is too high. The position loop gain that can be reached in the position control loop, can be up to ca. 50 s^{-1}.

The alternating currents superimposed onto the motor current and the circulating current, have a basic oscillation that is six times the line frequency, thus generally higher than the mechanical resonance frequencies. This ensures a smooth run on the machine tool.

The circulating current flows between the two current converters. By comparison to the 3-pulse circulating-current conducting circuit, this can become a disadvantage when the frictional characteristics of the feed unit are unfavorable (see section 3.3.6). Stick-slip effects are less effectively controlled, and despite the high nominal angular frequency, the positioning behavior can, under circumstances, be inferior to that obtained with a 3-pulse circulating current-conducting circuit. Stick-slip effects occur when the feed unit requires a breakaway torque that is larger than the frictional torque of the motion, and thus it acquires a jerky motion at low feed rates (see 4.2.4).

3.3.6 3-Pulse Circulating Current-conducting Anti-parallel Connection

On the line side, this connection requires a loadable neutral point. It is therefore common to use a current converter transformer in Dzn0 connection. Several feed axes can be connected in parallel to the same transformer, and then they have to be decoupled through commutating reactors. Figure 3.10 shows the power section with both converters U1 (thyristors 1,3,5) and U2 (thyristors 4,6,2).

The demand on current converter semiconductors is only half of that of 6-pulse circulating current-conducting circuits. The current and voltage relationships for the motor are the same, as are the control angle ranges of the two converters U1 and U2. This type of connection can also feed mechanical braking energy back into the line, and can drive the motor in all four quadrants, as shown in fig. 3.9.

Fig. 3.11 shows the course of the DC mean values of different control angles α. It is presumed that the current is not pulsating, i.e. that there is a high inductivity in the DC circuit. At a control angle $\alpha = 30°$, which is the smallest possible control angle for a 4-quadrant drive, the corresponding DC mean

T	Transformer with loadable neutral
F1–3	Rectifier fuses
L1	Commutating reactors
U1, U2	Thyristor middle point connection
L3, L4	Circulating current reactors
M	DC servo motor

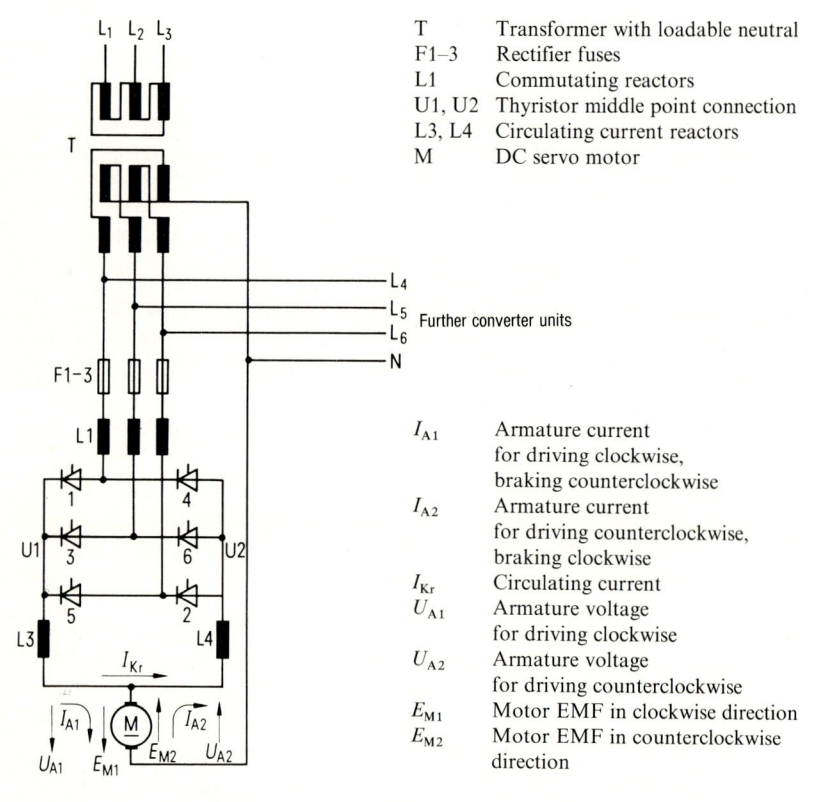

L4
L5 Further converter units
L6
N

I_{A1}	Armature current for driving clockwise, braking counterclockwise
I_{A2}	Armature current for driving counterclockwise, braking clockwise
I_{Kr}	Circulating current
U_{A1}	Armature voltage for driving clockwise
U_{A2}	Armature voltage for driving counterclockwise
E_{M1}	Motor EMF in clockwise direction
E_{M2}	Motor EMF in counterclockwise direction

Fig. 3.10
3-pulse circulating current-conducting anti-parallel connection (M3C)A(M3C)

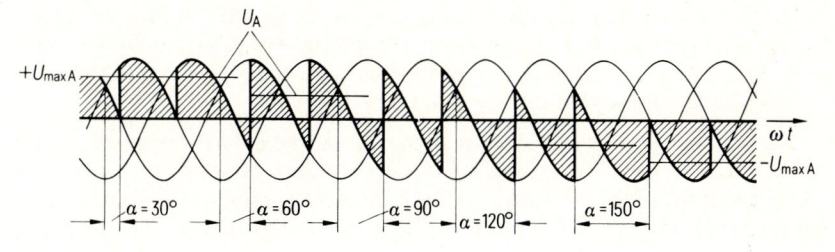

Course of the unsmoothed DC voltage

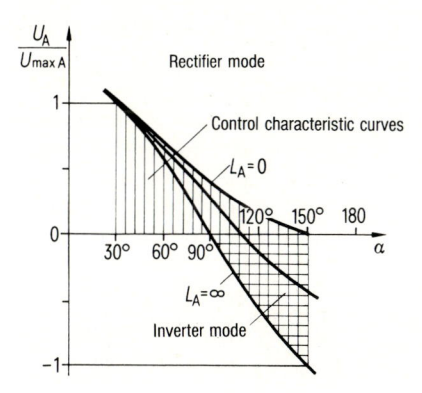

Fig. 3.11
Output voltage and characteristic curves of the 3-pulse current converter

value is at its maximum. As the control angle is increased, the DC mean value decreases, reaches zero when $\alpha = 90°$, and then increases again with reverse polarity when the control angle $\alpha > 90°$. The output voltage of the current converter can thus be set continuously, from a maximal positive to a maximal negative value.

At pure ohmic loads, the characteristic curves shift the rectifier mode up to $\alpha = 150°$. The characteristic curves that lie inbetween, represent the real condition of mixed-purely resistive/inductive load.

The way in which the circulating current is derived, is different from that of a 6-pulse cross connection. In the latter case, this current flows exclusively between the two converters, while in the 3-pulse circuit a significant part of it flows through the motor. Fig. 3.12 shows the currents dependent on the control angle, for a standing motor. Here, the term circulating current also includes the AC current that flows through the motor at standstill, towards N (neutral).

When the circulating-current setting is small (fig. 3.12a), which corresponds to a control angle of e.g. $\alpha = 130°$, a pulsating AC current results in the motor circuit. The two currents do not intersect, thus there is no circulating current between U1 and U2.

If a larger circulating current is set, e.g. a control angle of approximately $\alpha = 100°$ (fig. 3.12b), the currents overlap; the AC current in the motor takes on the shape of a distorted sine wave. The circulating-current reactors must limit the current between the two converters.

This AC current within the motor results in an alternating torque, and when backlashes occur in the gear, it also results in noises even when the feed axes are at stand-still. On the other hand, this alternating torque is favorable, when due to the high static friction there is a danger of stick-slip effects. The alternating torque forced by the circulating current keeps the mechanical transmission elements in constant motion, so that no static friction can occur. The losses that will occur in the motor must be taken into account. The circulating current, measured as effective value, should not exceed 10–15% of the rated current of the motor.

This type of connection is the one mostly used on numerically controlled feed drives. Turning and milling machines especially, up to torque requirements of approximately 50 Nm, are with preference equipped with these drives. In the position control loop, a position loop gain of up to $K_v \approx 35 \, \text{s}^{-1}$ can be reached, but the mechanical transmission elements must be carefully dimen-

a) Control angle of about 130°

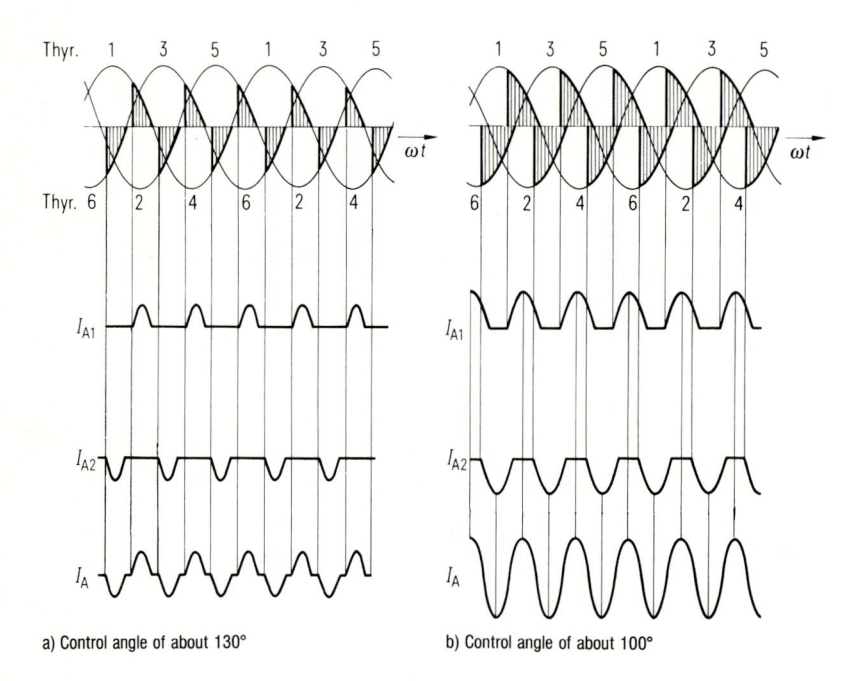

b) Control angle of about 100°

Fig. 3.12
Circulating current build-up for the 3-pulse circulating current-conducting anti-parallel circuit

sioned. 3 current pulses per line period are equivalent to 150 Hz on a 50 Hz line (180 Hz on a 60 Hz line). Typically, mechanical resonance frequencies lie in this range. This can lead to design changes that may become necessary if a machine develops an intolerable noise level. With this type of current converter, it is preferable to use a servo motor that is directly coupled to the feed screw, or a timing belt reduction.

3.3.7 6-Pulse Circulating Current-free Anti-parallel Connection

The 6-pulse circulating current-free anti-parallel circuit requires 12 converter semiconductors, of which two are always connected directly anti-parallel. Compared to the circulating current-conducting versions, the protective circuits and the fusing are here simpler to accomplish; circulating-current reactors are not necessary. With the thyristor modules with insulated construction available nowadays, it is possible to assemble very compact and cost-efficient feed drives. This type of circuit is shown in fig. 3.13.

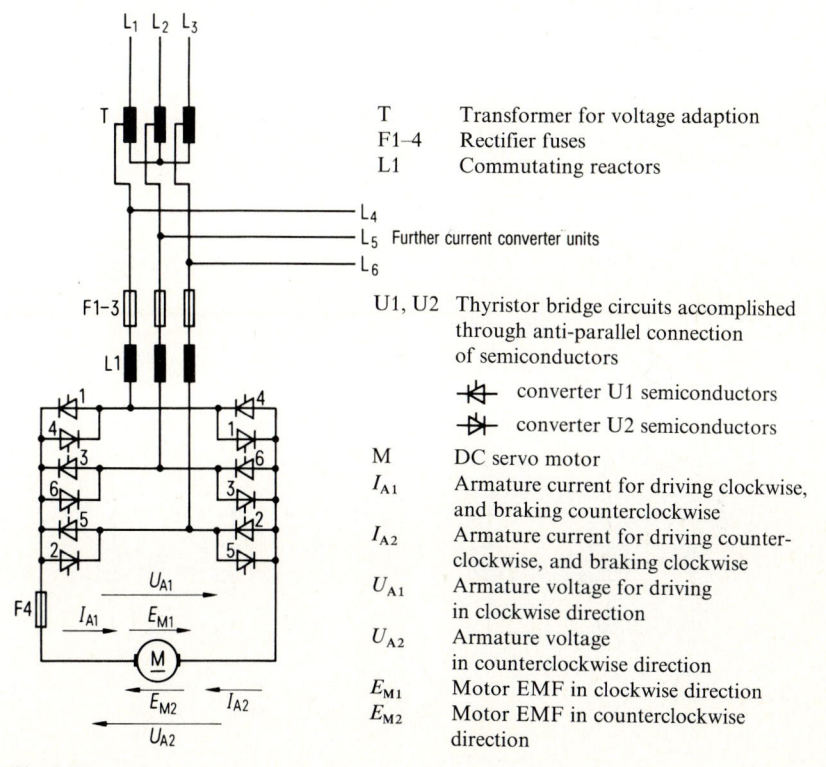

T	Transformer for voltage adaption
F1–4	Rectifier fuses
L1	Commutating reactors
L_4, L_5, L_6	Further current converter units
U1, U2	Thyristor bridge circuits accomplished through anti-parallel connection of semiconductors
	converter U1 semiconductors
	converter U2 semiconductors
M	DC servo motor
I_{A1}	Armature current for driving clockwise, and braking counterclockwise
I_{A2}	Armature current for driving counterclockwise, and braking clockwise
U_{A1}	Armature voltage for driving in clockwise direction
U_{A2}	Armature voltage in counterclockwise direction
E_{M1}	Motor EMF in clockwise direction
E_{M2}	Motor EMF in counterclockwise direction

Fig. 3.13
6-pulse circulating current-conducting anti-parallel connection (B6C)A(B6C)

The converter U1 delivers the motor current I_{A1}, for instance for clockwise driving. In the mean time, the pulses of converter U2 are blocked. If we now brake from this direction, the current converter U1 will be controlled back to the inverter limit, and thus I_{A1} will return to zero. The command stage switches converter U1 completely off, and after a safety period it switches converter U2 on, at the inverter limit. Now the motor EMF E_{M1} drives the braking current I_{A2} against the AC voltage U_{A2}. In the inverter mode, U_{A2} has a negative sign, and thus opposes the motor EMF E_{M1}. The mechanical energy is fed back into the line by the motor as a generator, through the current converter U2. The situation is repeated when driving counterclockwise, with converter U2 as the rectifier, and converter U1 as the inverter. A 4-quadrant operation is possible. Figure 3.14 shows the current and voltage vectors in the speed-torque diagram.

The auto-transformer T, shown in fig. 3.13, is used for the voltage adaption for permanent magnet-excited DC servo motors. A direct line connection is possible, if the motor has an appropriate connection voltage.

Circulating current-free 6-pulse anti-parallel circuits are used nowadays with many numerically controlled feed drives. The dynamic values that can be

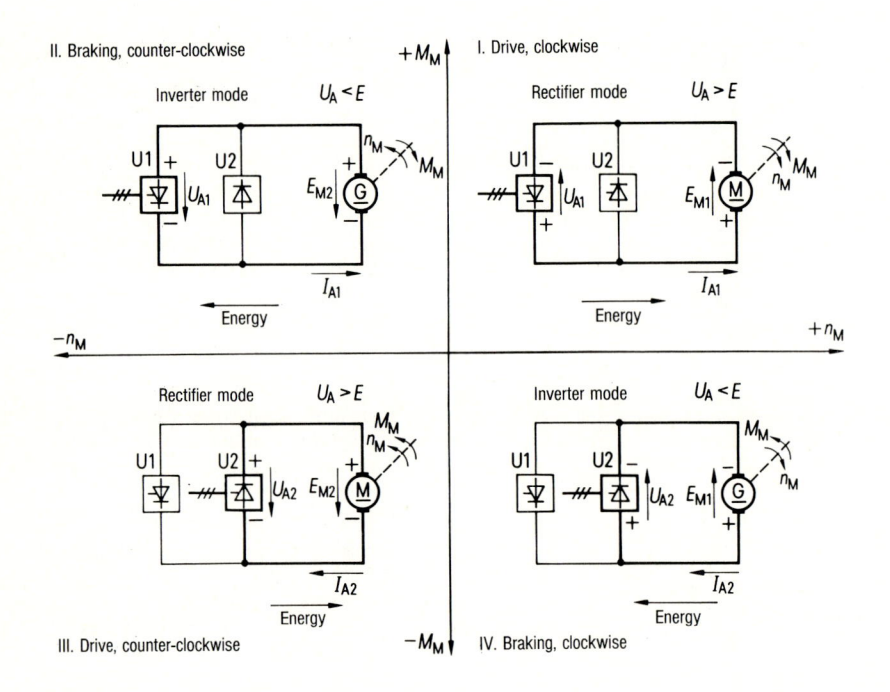

Fig. 3.14
4-quadrant operation of the 6-pulse circulating current-free anti-parallel connection

achieved (response and settling times, delay time), allow the position loop gain that can be reached in the position control loop to be up to $25\ \text{s}^{-1}$. The main applications are for production machines for turning, milling and drilling, preferably with indirect positioning control measuring systems.

When direct position control measuring systems are used, and mechanical reversing errors are present, the reaction times resulting from the current-free pause during the switch-over of the two converters can be disturbing. The hunting effects occurring at positioning must be damped within the control loop, through lower gains and by limiting the integral components in the speed regulator. Taking these measures however, will worsen the control behavior at small position increments.

3.3.8 2-Pulse Circulating Current-conducting Anti-parallel Connection

The circulating current-conducting 2-pulse anti-parallel circuit can consist of either two bridge connections, or of two middle-point connections. The former version requires a total of eight controlled rectifiers, and large circulating current reactors. The latter necessitates only four controllable rectifiers, but also a loadable neutral, which can be achieved with an auto-transformer. Since

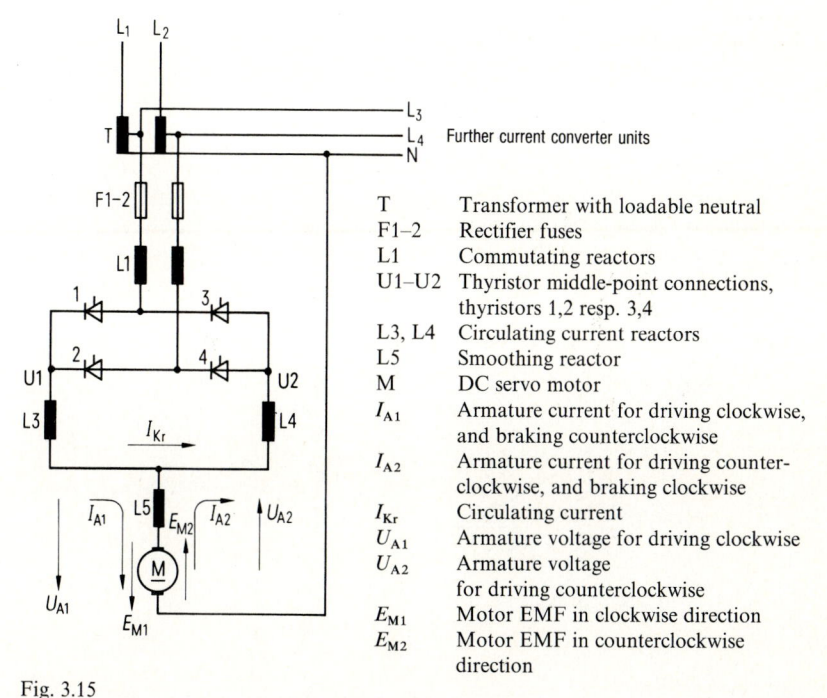

T	Transformer with loadable neutral
F1–2	Rectifier fuses
L1	Commutating reactors
U1–U2	Thyristor middle-point connections, thyristors 1,2 resp. 3,4
L3, L4	Circulating current reactors
L5	Smoothing reactor
M	DC servo motor
I_{A1}	Armature current for driving clockwise, and braking counterclockwise
I_{A2}	Armature current for driving counter-clockwise, and braking clockwise
I_{Kr}	Circulating current
U_{A1}	Armature voltage for driving clockwise
U_{A2}	Armature voltage for driving counterclockwise
E_{M1}	Motor EMF in clockwise direction
E_{M2}	Motor EMF in counterclockwise direction

Fig. 3.15
2-pulse middle-point circuit in circulating current-conducting anti-parallel connection

such a transformer is required anyway for the voltage adaption, the middle-point connection is less expensive.

Figure 3.15 shows the circuit for a middle-point connection. Further feed axes can be connected in parallel to the transformer.

We have again a 4-quadrant operation for the motor, because of the two current direction I_{A1} and I_{A2}, as shown in fig. 3.9. The circulating current I_{Kr} shown in fig. 3.15, is limited through the circulating-current reactors L3 and L4; in this circuit, it does not flow through the motor. The two current converters are controlled accordingly, as for the other circulating current-conducting circuits: starting from a control angle determined by the desired circulating current, converters U1 and U2 are controlled for counter action, i.e. when U1 is in rectifier mode, converter U2 is in inverter mode, and vice-versa. This ensures a delay-free motor current reversal, with the possibility of energy feedback.

Exactly as in fig. 3.11, the output voltage for a 2-pulse converter can be controlled from a maximum positive value at a control angle of $\alpha = 30°$, through zero, to a maximal negative value for control angle $\alpha = 150°$. Rectifier and inverter modes are possible under the conditions described in 3.3.2, i.e. inductivity and energy source are present.

However, since only two half-waves are available per line period, pulses in the current will appear at significantly higher output voltages than those occurring with 3- or 6-pulse current converter circuits. The control caracteristic curves for mixed resistive-inductive and purely inductive loads are shown in fig. 3.16.

For purely resistive loads, the output voltage zero can only be reached at a control angle $\alpha = 180°$. For the common values of armature inductivity and resistance of feed drives, the control characteristic curve lies, however, farther

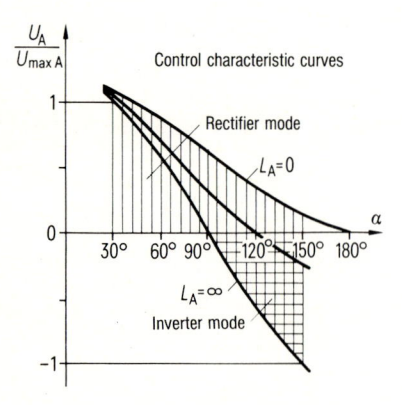

Fig. 3.16 Control characteristic curves of the 2-pulse current converter

down, so that the necessary limitation of the control angle of the 4-quadrant drive can be at $\alpha = 150°$.

The operation with suppressed circulating current can be given as an example for this 2-pulse circuit. Figure 3.17 shows three different control angles. Section a) shows the start position for $\alpha_1 = \alpha_2 = 140°$. The converter U1 and U2 are in series, a 2-pulse circulating current is flowing, and the motor current is zero. In section b), the control angle of U1 has been shifted forward to $\alpha_1 = 120°$, while the control angle of U2 has been shifted back to $\alpha_2 = 160°$. Converter U1 supplies a motor current I_{A1}, and when the control angles overlap, together with converter U2 it delivers the circulating current. The circulating current amplitude is now lower, and the motor current is a direct current with high superimposed AC component.

If converter U1 is shifted even further into the rectifier range, as shown in section c), to $\alpha_1 = 100°$, and the control pulses of thyristors 3 and 4 of converter U2 are allowed to dissappear, then the circulating current will be zero. The circulating current is thus suppressed, and ceases to load the converter. If control angle α is set lower still, e.g. as shown in section d), to $\alpha_1 = 65°$, the armature voltage U_{A1} will then increase, and the motor will rotate faster.

The figure sections b), c) and d) show the induced counter voltage E_{M1}, which is produced by the revolution of the motor. The value relationships between U_{A1} and E_{M1} correspond to the operating mode of the idling motor. For this reason, the motor current pulsates even at high speeds. If the motor is loaded, E_{M1} will be smaller, and the current commutating duration of the individual thyristors will be larger. At still smaller control angles α, the thyristor will conduct current only when the temporary value of the line voltage half-wave is higher than the induced counter voltage. The firing pulse duration must therefore be correspondingly long.

The form factor of the unsmoothed DC voltage at small control angles, lies approximately at 1.5–1.8. The rated torque of a DC servo motor must therefore be reduced to about 50–35%, if the form factor cannot be improved with an additional smoothing reactor in the motor circuit. Out of economical considerations (reactor size), this improvement is limited to a form factor F_i of approximately 1.25, in addition to a reduction of a 30% in the torque.

The feed drives on simpler machines, with lower requirements for positioning time and accuracy, are more and more frequently equipped with this type of current converter circuit. Position loop gains of up to $20\,\text{s}^{-1}$ can be thus reached. The mechanical resonance frequencies must, in any case, be taken into consideration. They cannot be allowed to coincide with doubles of the line frequency, or with any multiples of these. The noise levels that would otherwise result, would be intolerable for the operating personnel. The resonance oscillations of the mechanical transmission elements would also influence negatively the surface quality of the machined work piece.

a) $\alpha_1 = 140°$
   $\alpha_2 = 140°$

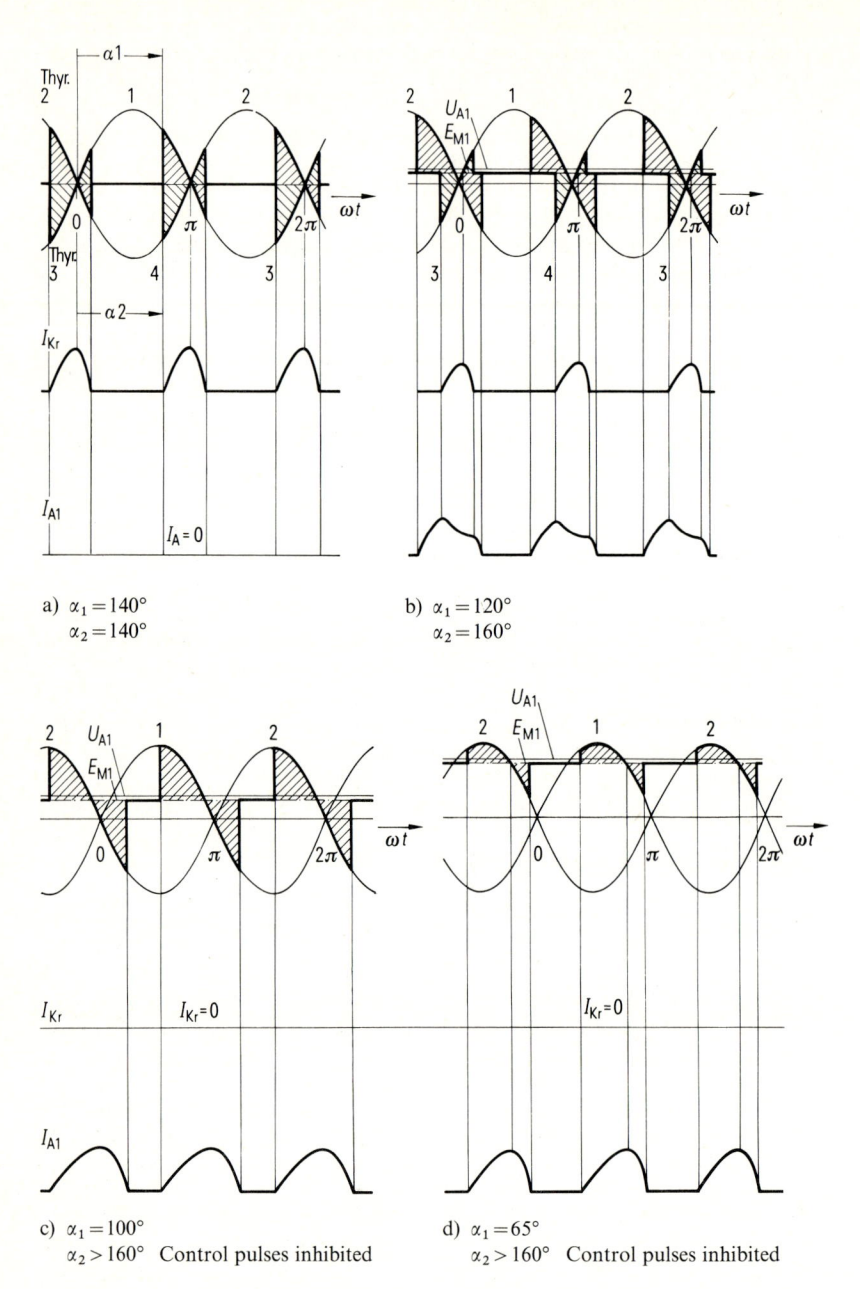

b) $\alpha_1 = 120°$
   $\alpha_2 = 160°$

c) $\alpha_1 = 100°$
   $\alpha_2 > 160°$ Control pulses inhibited

d) $\alpha_1 = 65°$
   $\alpha_2 > 160°$ Control pulses inhibited

Fig. 3.17
Circulating current suppression with 2-pulse reversing current converter

3.4 Controllers

3.4.1 Structure of the Control Loop

The controllers used for feed drives are different in structure as well as design. At present, the analog controllers are used most commonly in practice, but time-discrete digital controllers are already in development. The conventional controllers work with operational amplifiers, and are generally equipped with PI behavior.

The following control structures are in use:

▷ Speed control with subordinated current control (cascaded control)
▷ Speed control with current limitation
▷ Speed control with state monitor and, if necessary, disturbance variable monitor.

Besides those control structures with subordinated control loop commonly used in drive engineering, the speed control with parallel working current limitation has also found a wide range of applications for speed controlled feed drives.

A control concept which will become more meaningful in the future, due to the use of microprocessors, is that of state variable control with monitor. Preliminary practical results show that such systems offer advantages even for feed drives.

The selecting criteria for one or the other system, are:

▷ The achivable nominal angular frequency ω_{0A}
▷ The adjustment demands for the start-up and optimization
▷ Control board costs
▷ Unitary control structure for a particular drive spectrum.

The achievable nominal angular frequency ω_{0A} is affected by the values of the electrical time constant T_{elA} and mechanical time constant T_{mechA}, and is also determined by the dead time T_T, which in turn depends on the pulse number p_{SR} of the converter.

As demonstrated in [3.5], the substitute time constant of a control loop optimized according to absolute optimun is approximately twice the sum of the small time constants, and that of a control loop adjusted according to the symmetrical optimum is up to four times the sum of the small time constants. This sum is mostly determined by the dead time T_T. It can thus be determined that:

If the substitute time constant of the subordinated control loop is larger than the electrical time constant T_{elA} thus compensated, then the system with parallel current limitation is more favorable for the dynamic properties of the speed control loop than the system with subordinated current control.

This means: for the small electrical time constants of the DC servo motors, because of the long dead times, at 2- and 3-pulse circuits the parallel current limitation control is more suitable. For transistor choppers and under circumstances, also 6-pulse converter circuits, a better dynamic behavior can be obtained with a subordinated control system. If hereby we compare the examples of table 2.2 (page 90), we can see that on drive ⑤ the achievable nominal angular frequency could be somewhat improved with a subordinated current control.

The parallel working current limitation presents the advantage, that the dead time contained in the control system has only a small effect on the nominal angular frequency attainable. The disturbance transient response of such a loop is also better, i.e. load changes are controlled more rapidly. Neither do the filter circuits for the current actual value, required in the case of low pulse circuits, have any influence on the dynamic characteristical values of the drive. A comparison of the two systems, though involving a certain simplification, can also be derived from the evaluation of the transfer frequency response curves. Explanations about this point can be found in [3.6].

For systems with state control over a monitor, both methods are possible. The same view points as above apply.

3.4.2 Overcurrent Limitation

Overcurrents, which appear at the control of larger command value changes, are permitted in practice only up to a certain level, taking into consideration the electrical and mechanical load limits of the motor, gear, and current converter. A corresponding over-dimensioning is limited because of economic reasons. Transistor choppers with high peak currents especially, would require considerable expenditure increases.

For position controlled feed drives, the course of speed, respectively velocity command values can be controlled, as shown in chapter 5, at acceleration and deceleration, with a command value modifier, with ramp. The acceleration, or braking current can thus be limited. Even for simple position control loops without command value modifiers, the transfer functions occur only with limited acceleration, respectively deceleration. These are:

$$a_{i\,max} = K_v \cdot \Delta v_v \qquad (3.11)$$

The position loop gain K_v thus determines the necessary acceleration, respectively deceleration current.

At emergency shut-downs or collisions of the moving masses, and during optimization of the speed control loop, speed command value changes in the form of a step must be possible. For this purpose, a fast acting current limitation in the converter must ensure that no critical currents can flow. The commutating limit of the DC servo motor must hereby be taken into account. The current limit must be controlled dependent on the speed. In order to protect the semiconductor components of the power section, an additional time limitation of the overcurrent is generally used. In the example in section 2.3.3, the ramp, respectively braking time for a feed drive is calculated, taking into account the current limitation. The result for this process was found to be approximately 100 ms. Based on this, the required overcurrent load of current converter, motor, and machine can be determined to be about 200 ms. This time period is so designed, that a stepwise speed change can, with certainty, not trigger the supervision, and so that during this time, the current converter elements will not overheat. The dynamic limit range for the permanent magnet-excited DC servo motor shown in fig. 2.2, is also based on this 200 ms period as time admissible limitation.

The demands for overcurrent limitation are met either through the subordinated current control following the speed regulator, or through the parallel current limitation control. The motor current is limited statically through the command value, and dynamically through the time dependent limit stages.

3.4.3 Subordinated Current Control

A subordinated control loop is introduced in order to compensate for a time constant within the control loop, and thus to easier accomplish the optimization of the superceeding control loop. This however, only applies as long as the combined time constant of the subordinated control loop is smaller than the time constant of the control loop itself (3.7).

Figure 3.18 shows the block diagram of the control structure. The power section of the current converter circuit is represented by the dead time T_t. The DC

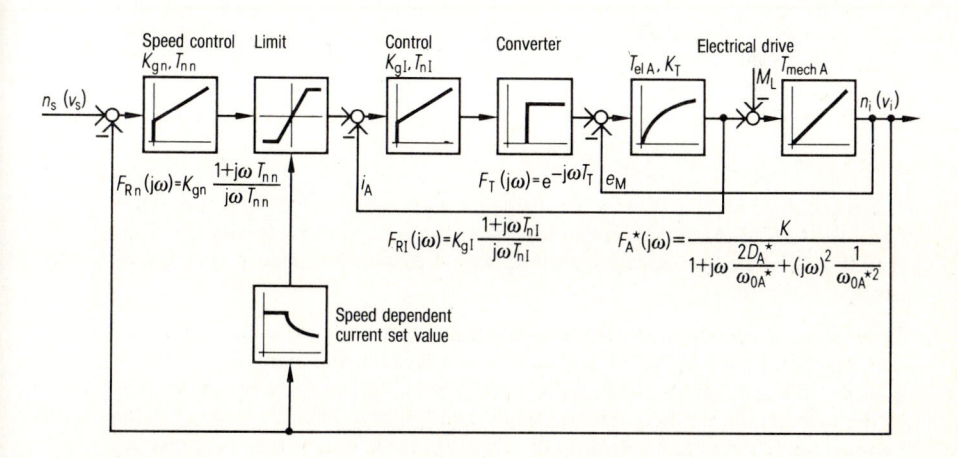

Fig. 3.18
Drive control structure with subordinated current control

servo motor, including the additional internal resistances, inductivities, and inertias of the drive, are shown as a delay of first order with T_{elA}, and through the integrator with T_{mechA}.

The armature time constant T_{elA} can be compensated for in the servo motor drive discussed, by means of the subordinated current control loop. Short current control times in the range of 3–5 ms can be reached in the case of actuators with negligeable dead time. The disadvantage of having to optimize an additional control loop at start-up time, usually makes the user request standard adjustments with fixed feedback circuits.

For very dynamic transistor converters this is easily feasable, therefore the majority of these have a control structure with subordinated current control. There are no difficulties in adapting the circuitry of the current converter. Due to the high switching frequencies, there are no pulses in the current, and the control loop parameters do not change.

In the case of thyristor converters, the subordinated current converter reacts with a prolonged control response in the range of pulsating current. In these cases, adaptively controlled interference into the control structure for correction and amplification, or interference through an appropriately high circulating current, is necessary to avoid the pulsing range (see section 3.3.3).

Drives with thyristor converters have larger dead times, and longer current control response times. If they use a subordinated control loop, this must be carefully optimized. The subordinated current control loop is not suitable for fixed standard feedbacks, because in these cases, dynamic reductions must be traded.

The above mentioned overcurrent limit is very easily accomplished by limiting the speed regulator output signal. The speed dependent interference is derived from the tacho voltage, and through this limitation, it controls the set value of the current.

3.4.4 Speed Control with Current Limit

The parallel current limitation is a protection circuit. It intervenes only upon transgressions of the preset command value. In order to comply with the commutating limit of the motor, the command value is controlled speed dependently.

Different practical solutions are here possible. First, the current limitation can be accomplished through a limit stage, inhibiting the firing pulses in the power section. This results in a switching frequency which depends on the impedance of the power loop, and on the switching delays of the limitation circuit. This two-position control has been found to be applicable only for small transistor choppers, since the drive runs unsteadily at the current limit, due to flip-flopping, and noises on the machine cannot be avoided.

On the other hand, we have the parallel current limitation control. When the preset current command value is exceeded, the speed regulation is suppressed, and the drive accelerates or brakes at the corresponding current limit. Through the appropriate circuitry it is ensured that, after the ramp-up or braking process has ended, the speed regulator takes over the command again, without overshoots. This type of current limitation has been proven effective for thyristor current converters. Figure 3.19 shows the block diagram.

The circuit is supplemented with a firing angle slope limitation. A thyristor acts as a line-triggered switch, i.e. it is activated by a line-synchronized firing pulse to conduct current. It is switched off again, when the current commutates to the next thyristor, respectively also at zero current transit; no control influence is possible in between. For this reason, rapid changes of the speed regulator output voltage must be prevented from affecting the firing pulse control.

The parallel current limitation is used mostly for thyristor converters. This simplifies the optimization, and for the 2- and 3-pulse circuits, the required short current response times can be attained. For 6-pulse circuits, a subordinated current control would serve as well, but since the adjustment demands and the component costs are lower for the parallel current control, here too, the parallel limitation is used with preference. In addition, this allows a unitary concept of control for thyristor converters [3.8].

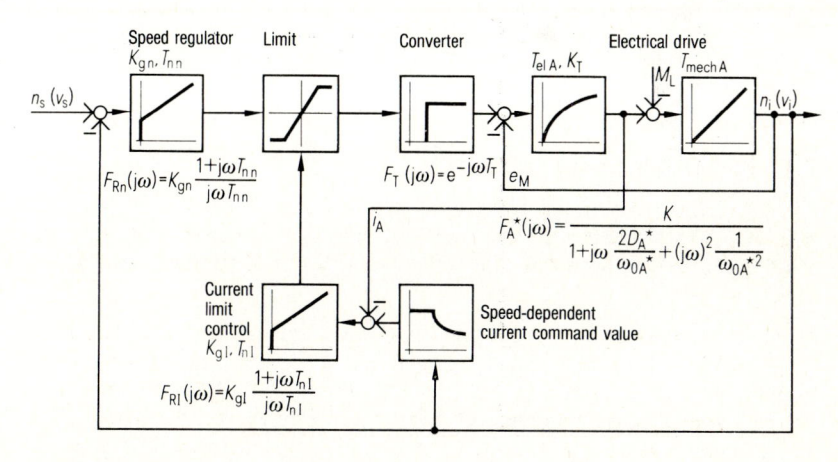

Fig. 3.19
Control structure for drive with parallel current limitation control

3.4.5 Status Monitor

A monitor is a piece of circuitry which serves to determine difficult to measure variables of a control system. A control system, simulating the real control system, is also controlled by the control output signal. The output signal of this model is compared to the real output signal, and the difference between the two signals interferes control-like in the simulated system, in the sense that the difference is decreased. The status values derived from the monitor circuitry are presented to the controller [3.9], [3.10].

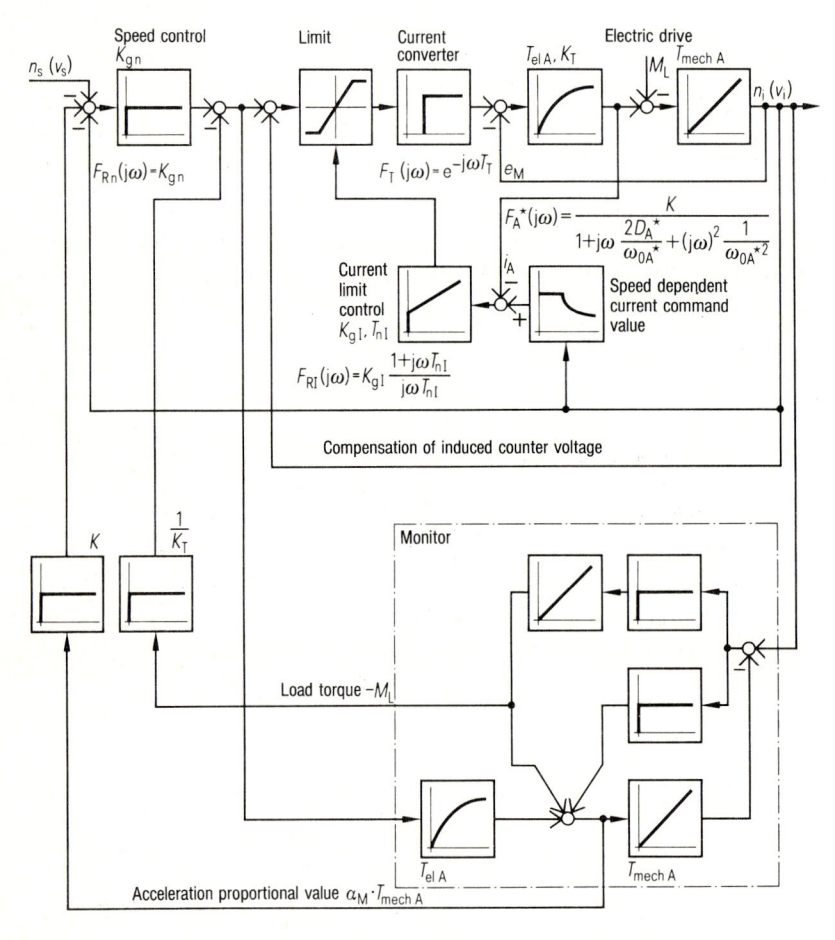

Fig. 3.20
Control structure with monitor, in the example of the feed drive

Figure 3.20 shows the structure of a control with monitor. The values determined are those of the acceleration proportional variable $\alpha_{M\,mech\,A}$, and the one corresponding to the load torque M_L. A damping effect is obtained by feeding the variable proportional to the acceleration, to the input of the speed regulator. A higher gain K_{gn} is thus made possible. The variable proportional to the load torque, through feedback to the summation point after the speed regulator, improves significantly the transient response for load changes control. A simple to adjust proportional control can be used as speed regulator, which will make controlling without overshoots possible. Such an arrangement was tested in practice on a 3-pulse circulating current-conducting drive; the dynamic behavior of the drive could thus be considerably improved.

The compensation of the induced counter voltage (EMF-feedback) additionally present, is used to dampen the control system. With this, higher band width drive requirements (high ω_{0A}) and good damping ($D_A \geqq 0.5$) can be ensured, especially for DC servo motors with low damping gradients ($T_{el} > T_{mech}$).

When dimensioning and adjusting the monitor, we start with the known structure and characteristic values of the control system. The feedback gains in the monitor must be set so, that the difference between the model and the real control variables is reduced as rapidly as possible. Distortions in the speed actual value signal of the tacho generator should, hereby, not have too large an effect; test results have shown a relatively low parameter sensitivity. Despite the somewhat higher component requirement, such monitor circuits can be used on machine tools, and be advantageous for the dynamic behavior [3.11].

3.4.6 Adaptive Speed Control Influence

A characteristic variable of the machine tool feed drive, is the more or less constant static friction of the moving machine slide. This requires a certain torque from the DC drive. For this, the motor draws a basic load current I_{MR}, which results in a voltage drop $I_{MR} \cdot R_A$ across the resistances in the armature circuit. The current converter must thus first be controlled to this voltage value, before the motor begins to turn.

Figure 3.21 shows the course of the referenced EMF E_M for low speeds, and the referenced armature voltage U_A dependent on the speed. Presented next to it is the idealized control characteristic curve of the current converter, and the relationship between the output voltage of the proportional part in the speed regulator and the speed command value at different gains K_{gn1} and K_{gn2}. If the subordinated current regulator is missing, the speed control output voltage must generate a control angle change $\Delta \alpha$ directly, necessary for the desired speed.

Fig. 3.21
Voltage characteristic curves of the DC servo motor, and control characteristic curves of a thyristor current converter

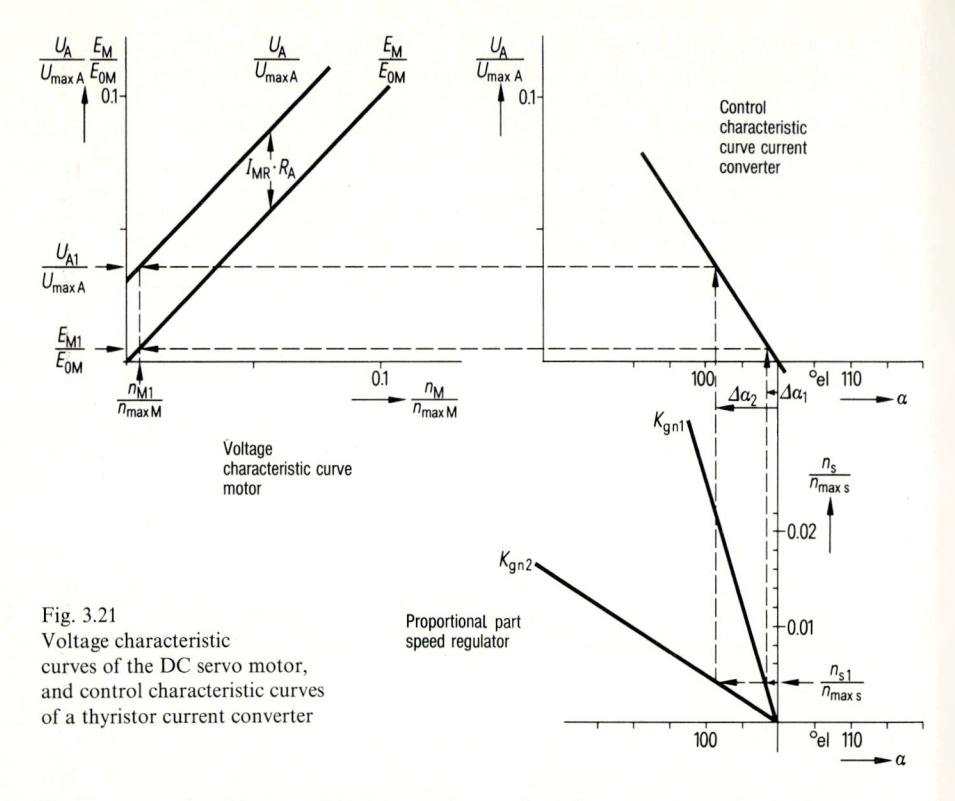

If the motor should turn with the speed n_{M1}, then the armature voltage U_{A1} must be applied. If the command value n_{s1} is applied to the speed control for this purpose, which corresponds to the motor EMF E_{M1} for n_{M1}, then at the set gain of K_{gn1} only the control angle change $\Delta\alpha_1$ can result, and the output voltage will not be sufficient to turn the motor. The integral part of the control must shift the control angle further by the value $\Delta\alpha_2-\Delta\alpha_1$, until the output voltage U_{A1} and the desired speed n_{M1} are reached. This reaction time prolongs the response time.

If however, gain K_{gn2} is set in the speed regulator, the control angle change by $\Delta\alpha_2$ can already be achieved with the proportional jump of the control. The drive reacts significantly faster, because no time is required in this case for controlling the control output voltage through the integral part.

The linearly presented curves of U_A and E_M, and the relationship between control angle α and the current converter output voltage, are not linear in reality. For one, the static friction is higher than the moving friction, and also, the control characteristic curve is flattened because of the pulsating current. In addition, the armature time constant changes according to the pulsating or non-pulsating of the direct current.

These effects result in a significantly changed gain of the control system. The upper section of fig. 3.22 a) shows the measured control system gain of a drive with 3-pulse current converter unit. Since the feed drive works in a control range of $>1:10^4$, the resulting non-linearity must be compensated for through an adaptive control of the control parameters. According to section b) of fig. 3.22, the control gain K_{gn} is increased at low command values, and the integral time of control T_{nn} is reduced, dependent on the command values. The operating points for this change can, in this way, be adapted to the type

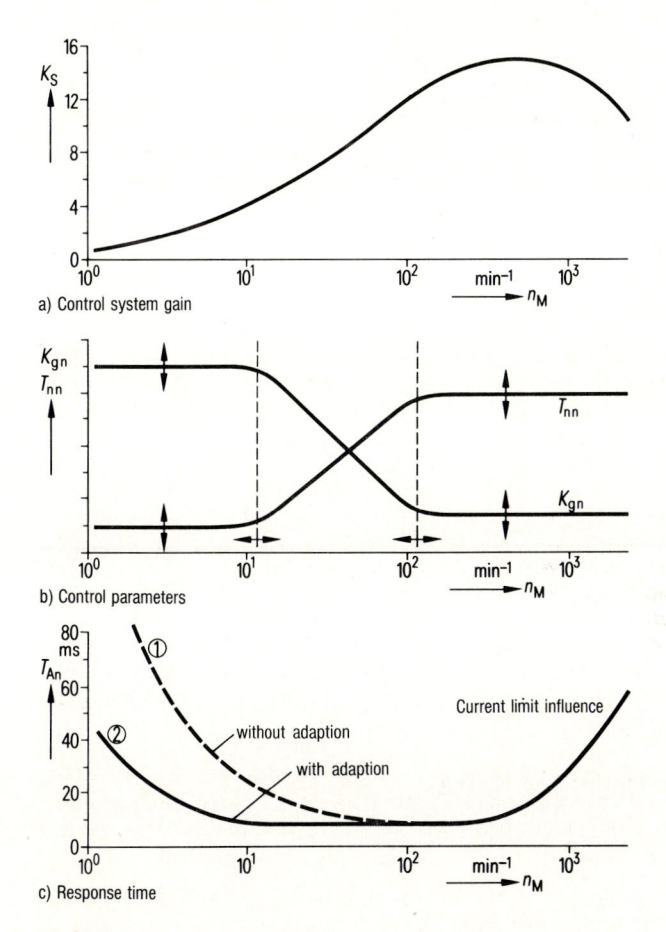

a) Control system gain

b) Control parameters

c) Response time

Fig. 3.22
a) Gain of the control system.
b) Adaptive control of gain and integral time in the speed regulator.
c) Measured response times of a feed drive with 1HU3 076-0AC01 and 3-pulse current converter unit

of drive concerned, and can be factory preset. The adjustment of the gain and integral time for low speeds can be determined during the optimization of the feed drive.

The effects of the non-linearities can be observed in the size of the response time. If measured values of a drive are drawn over the speed range at stepwise command value settings, at low speed values we obtain significant increases in the response times, and as a consequence, reduced nominal angular frequencies ω_{0A} (curve 1 in section c of fig. 3.22).

With the adapted control, significantly shorter response times can be achieved in this range, and thus almost constant nominal angular frequencies ω_{0A} (curve 2 in section c of fig. 3.22).

The response time increases at higher speeds result due to the current limit. This range should be so adjusted during component dimensioning, that it lies outside the working range, because otherwise contour distortions can be expected in the position control loop (see section 2.3).

3.5 Selection Criteria

3.5.1 Characteristic Values of Converter Circuits

As a summary of the discussion about current converter circuits used on feed drives, table 3.1 presents the characteristic values and the results that can be achieved in practice. The dead time T_T and the nominal angular frequency ω_{0A}, determine the position loop gain K_v. (The relationship is derived in table 1.2, and is also presented in fig. 2.12.)

However, the achievable nominal angular frequency is reduced, and the position loop gain K_{verr} becomes lower, due to additional external moments of inertia, and the effects of non-optimal mechanical transmission elements. The values measured in practice are shown in table 3.1.

The last column shows the preferred application ranges of the individual circuits. These data give indications about the ranges in which the current converter circuits should be used. Technical changes of the available components can alter the indicated limits.

3.5.2 Applications for Converter Circuits

For the selection of a current converter, often the deciding reasons are not technical, but for instance:

▷ the unitary concept of the drive within a particular application range, due to spare part and service problems,
▷ experience with an existing type series, because changes require retraining of shop and service personnel,
▷ fixed design requirements, because changing to new drive systems requires extensive and costly new design.

Table 3.1 Characteristic values of the different current converter circuits

Circuit	Mean dead time T_T ms	Form factor F_i	Current rise time T_Anl ms	Nominal angular frequency ω_0A s^{-1}	Position loop gain $K_\mathrm{v\,opt}$ s^{-1}	$K_\mathrm{v\,err}$ s^{-1}	Torque range, preferred application
Transistor chopper	0.25	ca. 1	3–5	400	100	80	Torque approx. 1–30 Nm. Dynamically demanding turning automats, milling automats, small machining centers
6-pulse circulating current-conducting cross connection	1.67	1.05	6–10	300	60	50	Torque approx. 30–120 Nm. Drilling and milling machines of high accuracy, large machining centers, large tracer mills
3-pulse circulating current-conducting anti-parallel connection	3.33	1.15	12–18	250	45	35	Torques approx. 5–50 Nm. Turning and milling machines, accurate production machines, universal use for NC-feed drives
6-pulse circulating current-free anti-parallel circuit	1.67 + ca. 6*)	1.05	6–10	200	30	25	Torques approx. 5–120 Nm. Production lathes and milling machines, boring mills, transfer units, general use
2-pulse circulating current-conducting anti-parallel circuit	5	1.25 (with add. reactors)	20–30	125	23	20	Torques approx. 2–15 Nm. Simple milling machines, machining centers and simple transfer units, simple drives

*) Additional reaction time at torque reversal due to switch-over logic

Requirements	Measuring system		Work piece accuracy	Type of machine	Transistor	Current converter circuits			
	direct	indirect				6-pulse with circulating current	3-pulse with circulating current	6-pulse w/o circulating current	2-pulse with circulating current
high ↓↑ less high			1 μm	Electrode discharge machines, Grinders					
			5 μm	Gauge boring machines Milling machines in the aircraft industry Milling machines in the machine tool industry High accuracy machining centers					
			10 μm	Precision lathes Large lathes Punch and nibbling machines Milling machines for batch production					
			20 μm	Tracer mills Boring and milling machines Production lathes Production machining centers Simple milling machines Transfer units Roller feeds					
			100 μm	Press feeders Other positioning tasks					
					Torques mostly too high	Demand too high	Demand too high	Insufficient dynamic	Insufficient dynamic

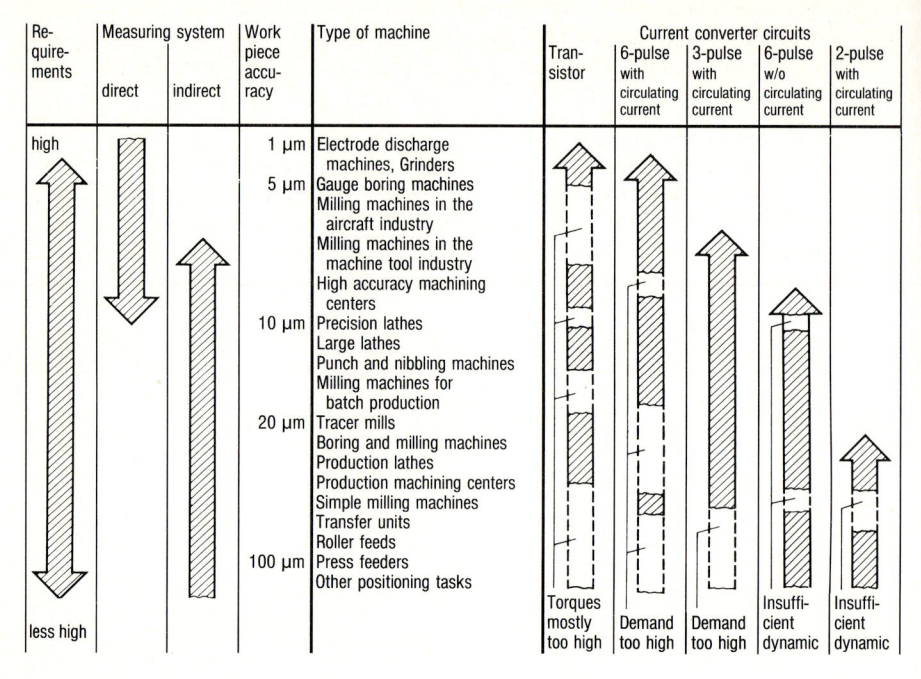

Fig. 3.23 Selection of current converter circuits

In general, a particular machining task requires a corresponding accuracy. This determines the design measures, the numerical control selection, and the measuring system to be used, with its required resolution. These are all factors to be taken into consideration when deciding what type of unit is best suited for which task. Figure 3.23 shows the measuring processes used most frequently with the individual machine types, and the accuracies required on the work piece. It also indicates which converter circuits meet these requirements and what type of unit is commonly used.

High quality current converter circuits, like transistor choppers and 6-pulse thyristor converters with circulating current, can meet almost any requirements. With transistor converters, an exception are the ranges in which higher torques are necessary. For the 6-pulse circulating current-conducting thyristor circuit, at certain applications the cost demand may be too high.

For the 6-pulse circulating current-free and the 2-pulse circulating current-conducting circuits, the dynamic results in certain applications are insufficient. Although these are not critical for the work piece accuracy to be achieved, the positioning time they would determine would be too long.

Due to the inherent reaction time with 6-pulse circulating current-free circuits during torque reversal, there are certain limitations in application. For direct measuring systems, oscillations can occur about the position reached.

The effects of the reaction time is small for indirect measuring systems. The reversing error present in the mechanical transmission system, limits the oscillations of the drive elements to the measuring system, so that they cannot be measured any longer on the feed longitudinal motion. Experience with machines installed in the field shows that satisfactory results can be achieved with a circulating current-free circuit, up to resolution of 2 µm in the position control loop.

3.6 Bibliography

[3.1] Möltgen, G.: Netzgeführte Stromrichter mit Thyristoren 3. Auflage. Berlin und München, Siemens AG, 1974

[3.2] Volkrodt, W.: (siehe Literaturhinweis [2.1])

[3.3] Derksen, J.: Antriebe für Arbeitsspindeln in Werkzeugmaschinen. Siemens-Zeitschrift 49 (1975), Heft 6, Seite 375 bis 380

[3.4] Buxbaum, A.: Regelung von Stromrichterantrieben bei lückendem und nichtlückendem Ankerstrom. Technische Mitteilungen der AEG 59 (1969), Heft 6, Seite 348 bis 352

[3.5] Fröhr, F.; Ortenburger, F.: (siehe Literaturhinweis [1.2])

[3.6] Grotstollen, H.: Comparison of speed controlled DC-drives with and without subordinate current loop. 2. IFAC-Symposium: Control in power electronics and electrical drives, Düsseldorf 1977

[3.7] Keßler, C.: Ein Beitrag zur Theorie mehrschleifiger Regelungen. Regelungstechnik 8 (1960), Heft 8, Seite 261 bis 266

[3.8] Gierse, H.; Engel, D.: Stand der Technik bei Positionierantrieben. Ingenieur Digest (1980), Heft 11, Seite 49 bis 51

[3.9] Pavlik, E.: Anschauliche Darstellung des Beobachters nach Luenberger. Regelungstechnik 26–27 (1978), Hefte 2 und 3, Seiten A5–A11

[3.10] Pavlik, E.: Aspekte des praktischen Einsatzes von Beobachtern für die Prozeßautomatisierung. Regelungstechnische Praxis 38 (1979), Heft 2, Seite 37 bis 44

[3.11] Weihrich, G.: Drehzahlregelung von Gleichstromantrieben unter Verwendung eines Zustands- und Störgrößenbeobachters. Reglungstechnik 26 (1978), Heft 11, Seite 349 bis 354 und Heft 12, Seite 392 bis 397

Ernst, D.; Ströhle, D.: Industrie Elektronik. Berlin-Heidelberg-New York, Springer Verlag 1973

Schröder, D.: Grenzen der Regeldynamik von Regelkreisen mit Stromrichter-Stellgliedern. Regelungstechnik und Prozeß-Datenverarbeitung 21 (1973), Heft 10, Seite 322 bis 329

DIN 41761, Mai 1975, (Vornorm): Stromrichterschaltungen, Benennungen und Kennzeichen

DIN 57558, VDE 0558, Teil 1: Bezeichnungen für Halbleiter-Stromrichter, Netzgeführte Stromrichter

4 Design Versions of the Mechanical Transmission Elements

4.1 Generalities

It is expected of modern machine tools that they exhibit high power, accuracy, and safety of operation. The design of the mechanical transmission elements significantly contributes to the realization of these qualities. This design presents not only a machine construction problem but must, simultaneously, also coordinate the dynamic behavior of the speed controlled DC servo motor with the transfer behavior of the mechanical parts. These mechanical transmission elements are the machine parts that lie between the motor and the work piece, in the power train. The tuning of the transfer characteristics of both components is accomplished with the help of control methods.

The mechanical transmission elements, frequently do not receive enough attention during the design, and especially the manufacturing stage. For this reason, occurring difficulties, which first become obvious during start-up of the position control loop, often have their origin in transmission elements badly dimensioned, or manufactured with insufficient accuracy.

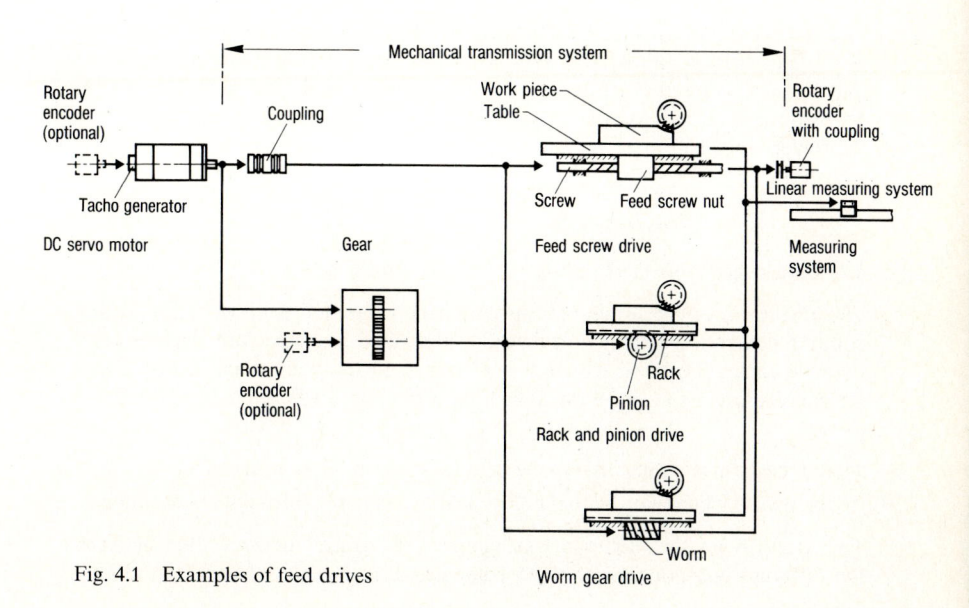

Fig. 4.1 Examples of feed drives

177

In the following, we shall first analyze the transfer behavior of the mechanical parts of the feed drive system with speed controlled DC servo motor. From this analysis, we shall draw conclusions about the dimensioning of mechanical transmission elements. Finally, we shall summarize the basic relationships and diagrams necessary for the calculation, respectively dimensioning, of the mechanical transmission elements.

The task of a feed drive is, to move a work piece or tool with a velocity corresponding to the machining mode, along the desired work piece contour. For this, the most common translational motion of two or more axes acting independently of each other, is accomplished by electrical drive components, e.g. with the linear motor. Space, weight, and heating problems of this type of motor demonstrate its unsuitability as a feed drive, with the present state of the art. For this reason, besides the hydraulic systems, the most commonly used electro-mechanical feed drive today is the DC shunt motor (see chapter 2).

According to fig. 4.1, the following possibilities exist for converting rotational into translational motion:

▷ Drive with feed screw
▷ Drive with rack and pinion
▷ Drive with rack and worm gear

A scheme as the one in fig. 4.2 can be established for the construction elements belonging to the mechanical transmission system. The feed drive can be used, according to this scheme, with or without gear, on the machine tool.

The feed screw drive is normally used for travel ranges of up to about 4 m [4.1]. The stiffness of the screw is often too low for larger travel ranges, so that in these cases it is preferable to switch to the rack and pinion or worm gear principle. For these drive systems, the stiffness is almost independent of the travel range length.

4.2 Requirements for the Mechanical Transmission System

The deviations of the actual contour from the command one, occurring during contour generation with a numerically controlled machine tool, have several possible causes, some of which can be traced to the mechanical transmission system. According to [4.2], the following aspects contribute decisively to such causes:

▷ the mechanical transmission elements possess mass and elasticity;
▷ the transfer behavior of the mechanical transmission elements is not linear.

This stems from the fact, that reversing errors cannot be prevented between the different components of the mechanical transmission elements, and that

Drive w/o gear

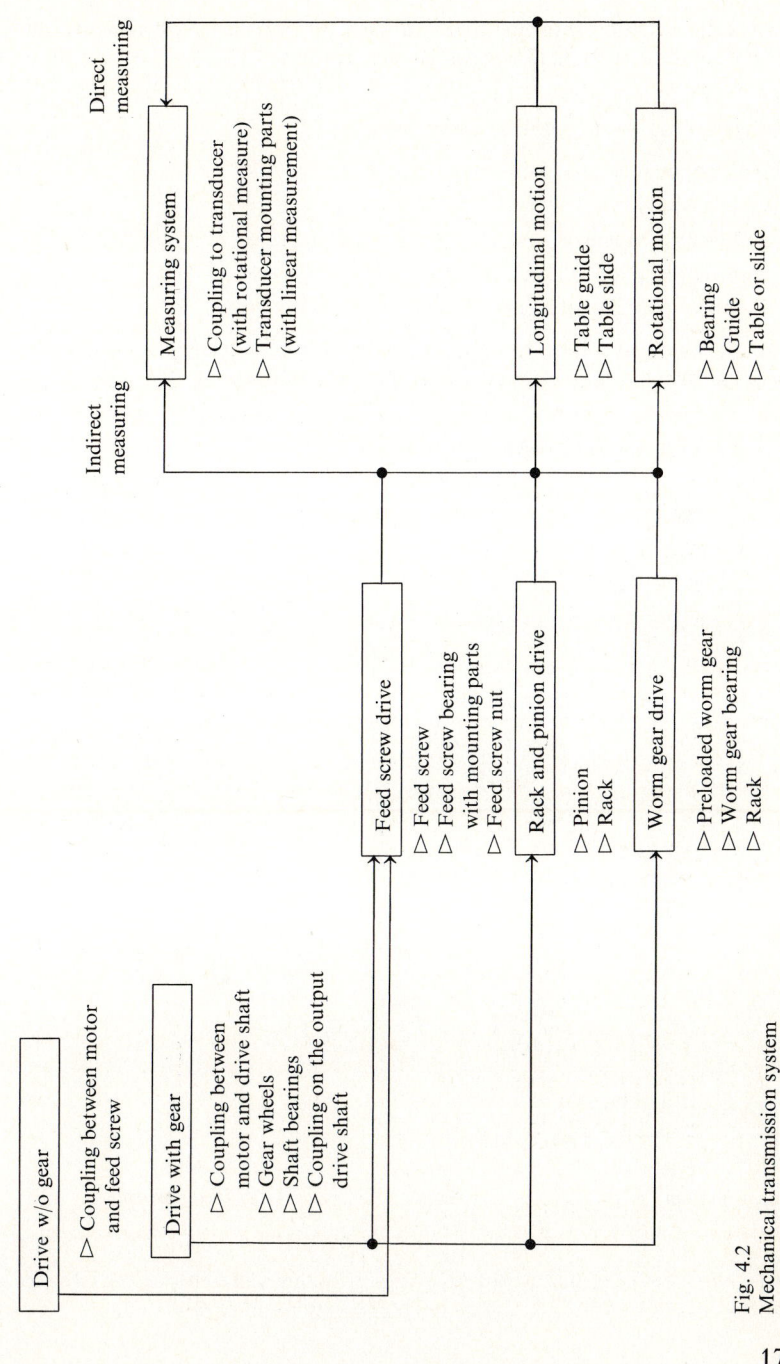

△ Coupling between motor and feed screw

Drive with gear

△ Coupling between motor and drive shaft
△ Gear wheels
△ Shaft bearings
△ Coupling on the output drive shaft

Feed screw drive

△ Feed screw
△ Feed screw bearing with mounting parts
△ Feed screw nut

Rack and pinion drive

△ Pinion
△ Rack

Worm gear drive

△ Preloaded worm gear
△ Worm gear bearing
△ Rack

Measuring system

△ Coupling to transducer (with rotational measure)
△ Transducer mounting parts (with linear measurement)

Longitudinal motion

△ Table guide
△ Table slide

Rotational motion

△ Bearing
△ Guide
△ Table or slide

Direct measuring

Indirect measuring

Fig. 4.2
Mechanical transmission system of a feed drive

a backlash, resulting from the finite stiffness of the mechanical transmission elements and the frictional forces is always present.

The requirements for the construction components of the mechanical transmission system can be summarized under five points:

▷ high nominal angular frequency $\omega_{0\,mech}$
▷ high stiffness
▷ sufficient damping
▷ transfer behavior as linear as possible
▷ as low as possible moment of inertia of the moving parts.

In the following, it will be demonstrated why these requirements are necessary, and what they mean in terms of the design of a feed drive.

4.2.1 Nominal Angular Frequency

The mechanical transmission system components listed in fig. 4.2 possess mass, and have only finite stiffness. They can be treated as one-mass oscillators, and their dynamic behavior is characterized through the nominal angular frequency and the damping gradient. The behavior of the entire mechanical trans-

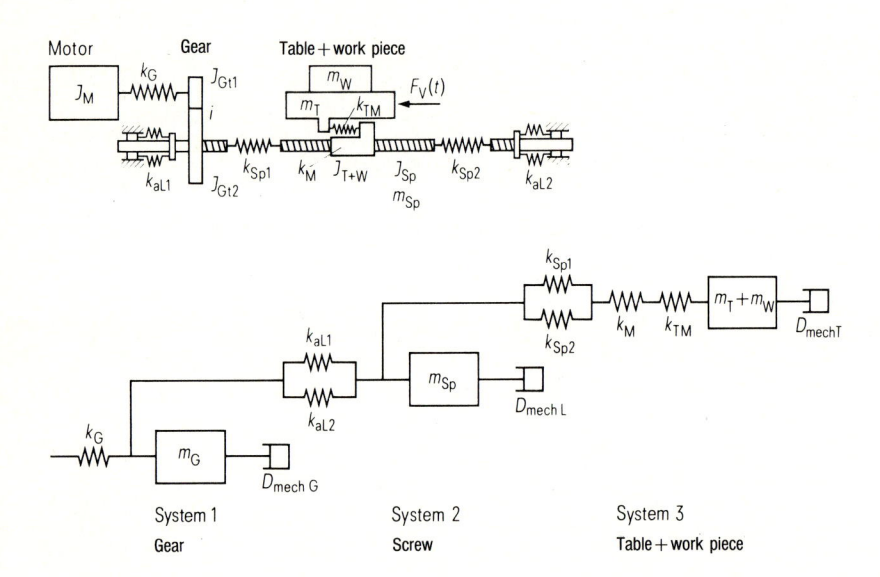

Fig. 4.3
Division of a feed drive into coupled single-mass oscillators with corresponding substitute diagram

mission system can thus be simulated with a model of coupled single-mass oscillators. In fig. 4.3, under the example of the feed screw drive with gears, it is shown how the substitute diagram of the coupled oscillators can be assembled.

The nominal angular frequency of a single-mass oscillator is

$$\omega_{0\,mech} = \sqrt{\frac{k}{m}} \qquad (4.1)$$

k spring constant
m mass

The number of single-mass oscillators determines the number of resonance positions of the mechanical transmission system. The resulting frequency response curve is, however, mainly affected by the nominal angular frequencies lying in the vicinity of the nominal angular frequency of the speed controlled feed drive. Measurements on machines in operation have shown that, as a good approximation, the mechanical transmission elements can be treated as 2nd order systems, whereby for instance for a feed screw drive, the interesting lowest nominal angular frequency is demonstrated generally by the feed screw bearing and feed screw nut system, with the feed screw, table, and work piece mass (system 3 in fig. 4.3). For a rack and pinion drive, the lowest nominal angular frequency is shown as a rule, by the gear with the mass to be moved. For worm gear drives, often built as hydrostatic versions for larger machines, the gear and the hydrostatic preload are weak points, which must therefore be carefully dimensioned. In order not to affect the properties of the highly dynamic DC servo motor, the nominal angular frequencies of the mechanical construction elements must be higher than the drive nominal angular frequency $\omega_{0\,A}$.

As extensive research according to the literature [4.3], [4.4], [4.5] has shown, the factors presented in table 4.1 should be complied with for the positions of the nominal angular frequencies relative to each other.

Table 4.1
Minimal requirements for the relative positions of nominal angular frequencies within the feed drive system

Nominal angular frequency in the position control loop	$\omega_{0\,L}$	40–120 s^{-1}
Cut-off angular frequency in the position control loop	ω_{EL}	0.6–0.7 $\omega_{0\,A}$
Nominal angular frequency of the drive	$\omega_{0\,A}$	2–3 $\omega_{0\,L}$
1st mechanical nominal angular frequency	$\omega_{0\,mech\,1}$	2–3 $\omega_{0\,A}$
Further mechanical nominal angular frequencies	$\omega_{0\,mech\,1+n}$	2–3 $\omega_{0\,mech\,1}$

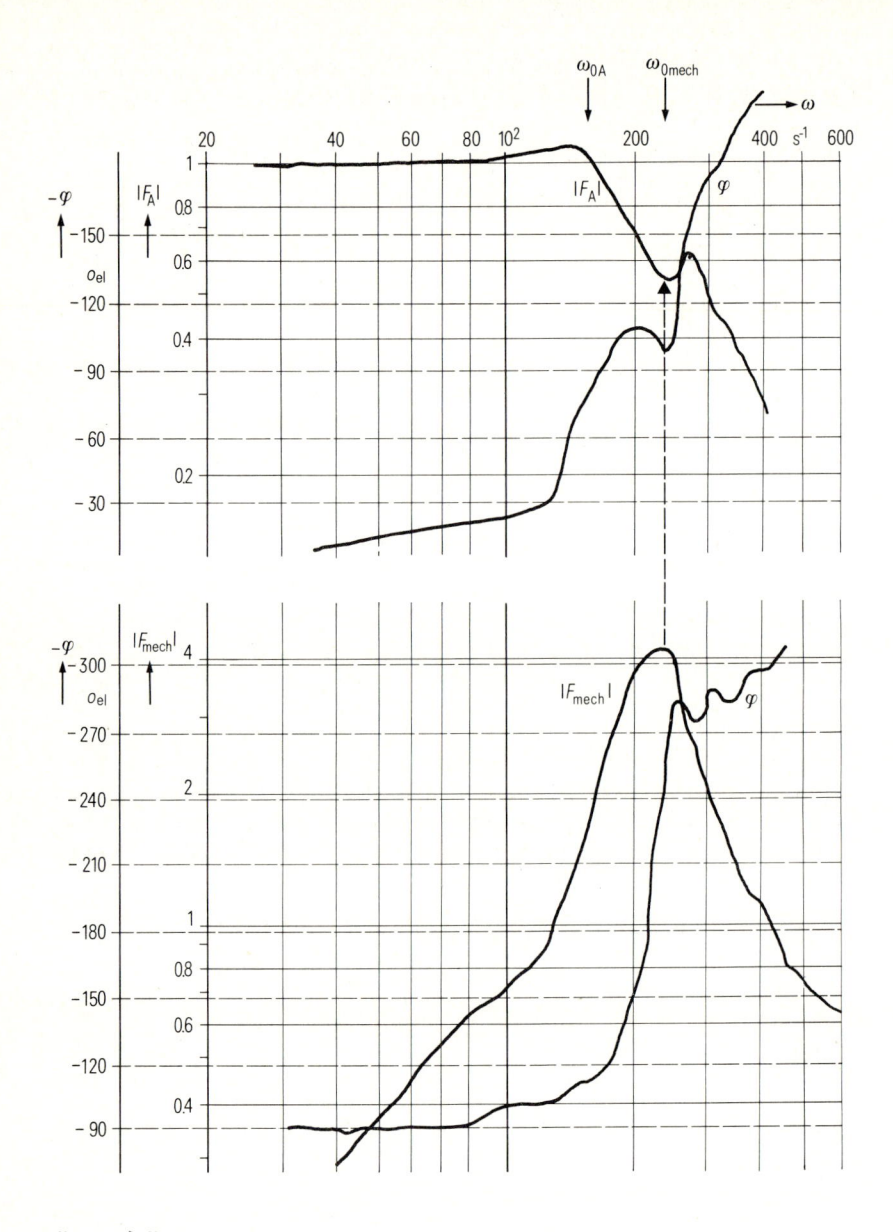

$\omega_{0\,mech\,1} > \omega_{0\,A}$

Fig. 4.4
Frequency response curves of the total feed drive (above), and of the mechanical transmission elements (below)

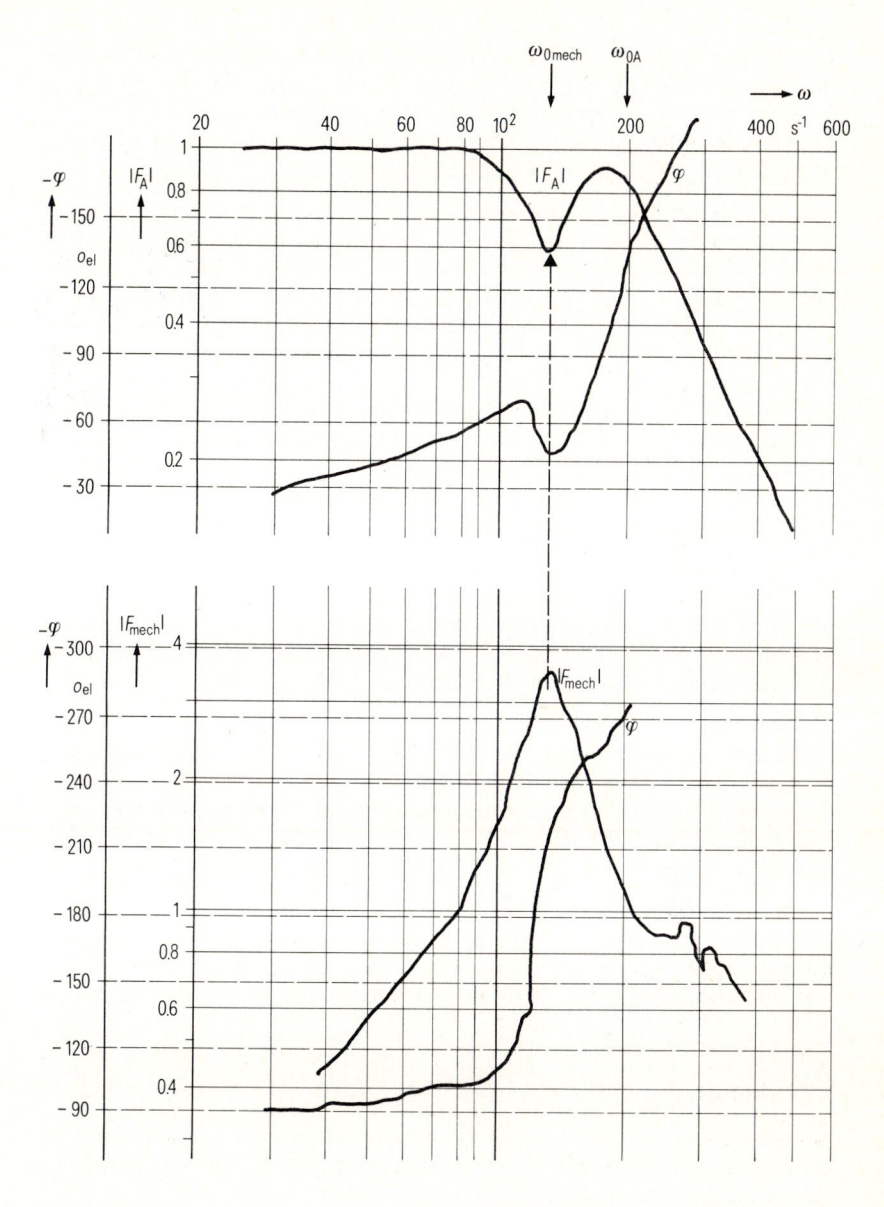

$\omega_{0\,\mathrm{mech}\,1} < \omega_{0A}$

183

It follows from table 4.1 that the speed control loop, subordinated to the position control loop, should have a nominal angular frequency higher by at least a factor of 2 than that of the position control loop. The corresponding applies to the relationship of the 1st. mechanical nominal angular frequency to the nominal angular frequency of the speed control loop.

The magnitude of the dominating mechanical nominal angular frequency $\omega_{0\,mech\,1}$ in relation to the nominal angular frequency of the speed controlled drive $\omega_{0\,A}$, and the relation of the external moment of inertia J_{ext} to J_M, determine the reaction of the mechanical transmission system on the behavior of the speed control loop. This is obvious in view of the frequency response curve of the total feed drive. Figure 4.4 shows two examples; left $\omega_{0\,mech\,1} > \omega_{0\,A}$, and to the right $\omega_{0\,mech\,1} < \omega_{0\,A}$. In both cases, below the frequency response curve of the feed drive F_A, is shown the frequency response curve of the mechanical transmission system F_{mech}, as determined with the help of an acceleration recorder. (Unlike the figures presented in section 1.1.3.4, the diagram shows here the phase angle φ included in the amplitude response curve. These figures present experimentally determined frequency response curves which have been recorded with an X/Y plotter.)

At the mechanical nominal angular frequency, a dip can be observed in the amplitude response curve of the feed drive. The sizing of the mechanical transmission elements on the left side of the figure is good, but the mechanical nominal angular frequency on the right side is too low.

Starting with the nominal angular frequencies calculated in table 2.2 for speed controlled DC servo motors of series 1HU.. and 1GS.., which, depending on the control system, can lie between 60 and 180 s^{-1}, the values for the dominating 1st. mechanical nominal angular frequency must, according to table 4.1, lie at least in the range 120 s$^{-1} < \omega_{0\,mech\,1} < 540$ s^{-1}. This corresponds to nominal angular frequencies $f_{0\,mech}$ in range of approximately 20–90 Hz.

For thyristor fed feed drives, attention must be paid that the mechanical nominal angular frequencies do not approach the pulse frequency determined through the control device. Should they coincide, undesirable machine noises and accelerated wear of the mechanical components would result (see section 3.1.3).

4.2.2 Stiffness

A minimal total stiffness k_{Ges} for the mechanical transmission system can be determined for a particular nominal angular frequency with equation (4.1). The spring constant of a single-mass oscillator is

$$k = 4\pi^2 \cdot m \cdot f_{0\,mech}^2 \qquad (4.2)$$

k spring constant
m mass
$f_{0\,mech}$ mechanical nominal angular frequency

If we take, e.g. for a feed screw drive, the system 3 of fig. 4.2 as the system with the lowest mechanical nominal angular frequency, then the necessary total stiffness of screw, table, and work piece system will be:

$$\frac{k_{Ges}}{N/\mu m} = 3.95 \cdot 10^{-5} \cdot \frac{m_T + m_W}{kg} \left(\frac{f_{0\,mech}}{Hz}\right)^2 \tag{4.3}$$

k_{Ges} total stiffness ($k_{Ges} = \Delta F_{a\,Sp}/\Delta s_a$, according to equation (6.6))
m_T slide mass
m_W work piece mass
$f_{0\,mech}$ mechanical nominal angular frequency

In a rough calculation, the value of the total stiffness k_{Ges} for a sufficiently high nominal angular frequency at a mass of 1,000 kg, must be higher than 100 N/µm. The mass and the necessary stiffness are directly proportional.

The required total stiffness can be also correspondingly determined for the rack and pinion or the worm gear drive, from the required nominal frequency. Calculation methods for the individual stiffness of the mechanical components are presented in section 4.3.1 and 4.3.2.

Besides the demands for sufficiently high mechanical nominal frequency, the stiffness of the mechanical construction components must also satisfy requirements for the desired accuracy, and the stability in the position control loop.

According to fig. 1.18, the elasticity for indirect measuring systems lies outside, and for direct measuring systems inside the position control loop. In indirect measuring systems, resulting reversing errors affect the accuracy of the work piece directly as measurement errors. Depending on the magnitude of the cutting forces, the stiffness requirements in these cases are about ten times higher than those presented above.

In direct measuring systems, the value which is necessary for a sufficiently high nominal frequency, is generally adequate. The low stiffness acts here negatively on the stability of the position control loop. In combination with the speed-independent friction of the machine slide, a reversing error is produced, which leads to oscillations about the position to be controlled (see sections 3.3.7, 4.2.4.2, and 4.2.4.3.2).

4.2.3 Damping Gradient

The damping gradient D_{mech}, together with the nominal angular frequency of the mechanical transmission system $\omega_{0\,mech}$, also determines the behavior of the position controlled feed drive. If, again, we take the behavior of the single-mass oscillator as an approximation for the mechanical transmission system, we obtain:

$$D_{mech} = \frac{1}{2} c_v \sqrt{\frac{1}{k \cdot m}} \qquad (4.4)$$

c_v speed-dependent damping quotient (force/velocity)
k spring constant
m mass

With only one dominating mechanical oscillating element, the total behavior of the feed drive can be approximated through a serial connection of two oscillating elements. We can thus obtain a block diagram as in fig. 4.5, for the position control loop.

In the frequency response curve for the electrical drive F_A, ω_{0A} and D_A stand for the values of the nominal angular frequency and the damping gradient of the speed controlled drive motor, including external moment of inertia and converter-dependent inductivities. $\omega_{0\,mech}$ and D_{mech} refer to the nominal angular frequency and the damping gradient of the particular mechanical component (e.g. screw-table and work piece system) which primarily determines the behavior of the mechanical transmission system. A direct measuring system is presumed.

The linear model of the position control loop presented in fig. 4.5, is the basis for a calculation with the optimizing criterion, of the squared comparison control area [4.6]. Depending on the nominal angular frequency ratio $\omega_{0\,mech}/\omega_{0A}$, on the damping gradient D_A of the speed control loop, and on the derived curves for the minimal squared comparison area, the required damping gradient D_{mech} can be read out of fig. 4.6. It shows the curves for the minimum of the referenced squared comparison control area, at damping gradients $D_A = 0.5$ and $D_A = 0.7$.

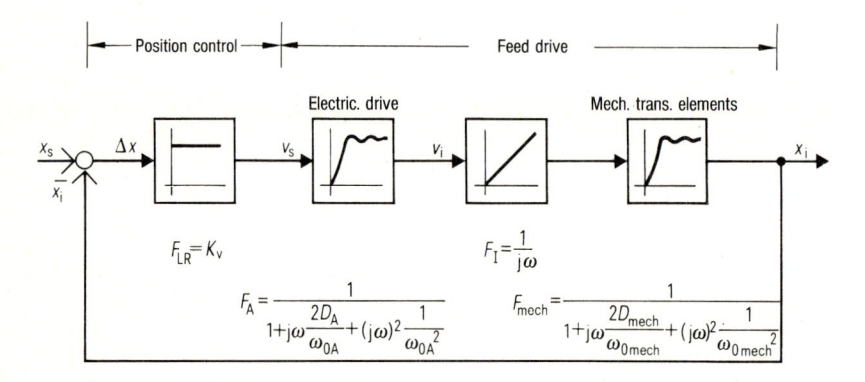

Fig. 4.5
Block diagram of a position control loop with a mechanical oscillating element

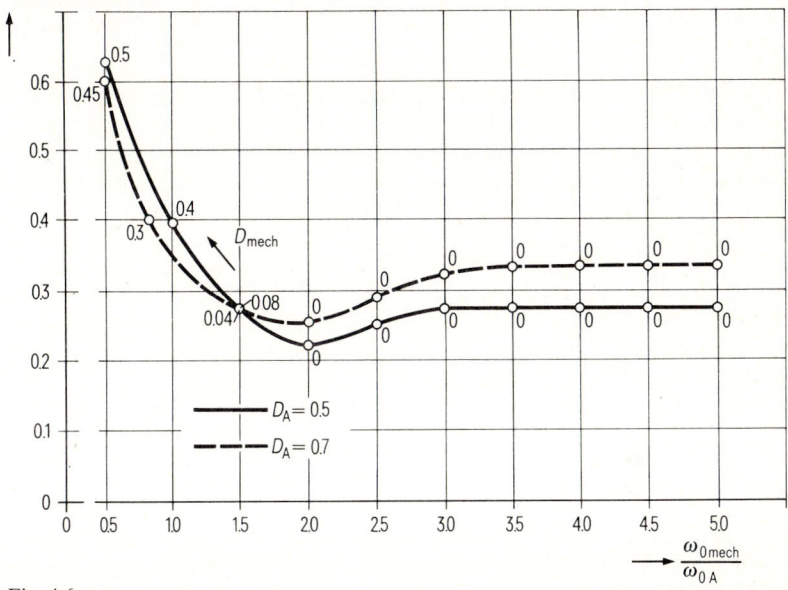

Fig. 4.6
Minimum of the referenced squared comparison control area as function of the nominal angular frequency ratio $\omega_{0\,mech}/\omega_{0\,A}$, for damping gradients $D_A=0.5$ and $D_A=0.7$

The referenced squared comparison control area plotted on the ordinate axis, is a measure for the contour deviation at a stepwise position command value change. From the shape of the curves we can derive, that when the ratio of the nominal angular frequencies decreases, the minimum of the squared comparison control area increases significantly. Both curves have their lowest value at a nominal angular frequency ratio $\omega_{0\,mech}/\omega_{0\,A} \approx 2$.

For nominal angular frequency ratios that lie above this value, there are no special requirements for the damping gradient D_{mech}; the resulting contour deviations are independent of it. If $\omega_{0\,mech}/\omega_{0\,A} < 2$ however, the mechanical damping gradient D_{mech}, dependent on the damping gradient D_A determined in the speed control loop, must have definite values.

For machine tools with roller guides or for hydrostatic bearing, the damping gradient of the table system without any additional damping devices, is generally very low: $D_{mech} \leq 0.1$. According to the above statements, at this damping value it would be sufficient to require a value of ≥ 1.5 for the nominal angular frequency ratio $\omega_{0\,mech}/\omega_{0\,A}$. But since an actual feed drive always has reversing error which counteracts the damping in the total system, a mechanical nominal angular frequency $\omega_{0\,mech} = (2 \cdots 3)\,\omega_{0\,A}$ will be required here.

On the other hand, because of the inherent reversing error, even at $\omega_{0\,\text{mech}} = (2\cdots3)\,\omega_{0\,\text{A}}$, values of at least $0.15..0.2$ ought to be reached for the damping gradient D_{mech}. This is proven by the distance intervals of the nominal angular frequencies presented in table 4.1.

The mechanical damping gradient D_{mech} of a feed table system of a machine tool is determined by the speed dependent damping quotient c_v. According to equation (4.4), D_{mech} is directly proportional to c_v. This damping quotient characterizes the friction proportional to the speed, and determines the feed force neccesary for overcoming the friction, dependent on the velocity. An increase in the damping gradient D_{mech}, e.g. through the tightening of the gibs on the table guides, leads simultaneously to an increase of the speed-independent friction. This friction however, in combination with the elasticity of the mechanical transmission elements, generates the so-called frictional reversing error which affects the properties of the position control loop negatively (see section 4.2.4.2).

A sufficient mechanical damping is necessary not only to prevent oscillations in the position control loop, but even more, because of the disturbance variables occurring on the machine table, which have their origins in the actual run of the machining processes. During milling especially, due to the number of teeth on the tool, oscillations are produced which may range in the vicinity of resonance frequencies of the mechanical transmission elements. If the machine table were undamped, because of the amplitude increase of the resonance point the table would be brought to oscillations known as chatter oscillations. The negative effects on the work piece are known as chatter marks. In order to keep these undesirable chatter manifestations to a minimum, it is necessary to have a sufficient damping of the screw-, table-system [4.7], [4.8], [4.9], [4.10]. In addition, in the case of friction guides, these oscillations can break the lubricating film on the guide rails, which can bring about the distruction of the friction guides through "freezing".

We derive from all this the difficulty to fulfil requirement for:

▷ low, speed-independent friction for accurate positioning, and
▷ high, speed-dependent friction for damping of oscillations and for preventing chatter phenomena during heavy machining processes.

Because of the contradiction in these requirements, empirically, a damping gradient of

$$0.1 \leqq D_{\text{mech}} \leqq 0.2 \tag{4.5}$$

has been found satisfactory. The major portion of the damping should hereby be generated through speed proportional friction in the table guide. This can be best accomplished with a combination of friction and roller guides, or with a hydrostatic guide (see section 4.3.3.3).

4.2.4 Non-linearities

4.2.4.1 Reversing Errors

The reversing error, also known as play or lost motion, can appear in the mechanical transmission system as:

> flank clearance in the gear, and
> torsional play in couplings.

In the feed screw drive we also have the

> axial play in the feed screw bearing, and
> backlash in the transition feed screw/feed screw nut,

and for the rack and pinion and worm gear drives, we have the

> play between pinion, respectively worm, and rack.

The effect of the reversing error is described in the signal flow of the mechanical transmission elements by the behavior of a hysteresis element. The coupled parts are, during traveling through the hysteresis, completely decoupled. The stability of the position control loop is affected, and the position loop gain must be reduced to ensure that the positioning will be free of overshoots. In addition to this reversing error, we have the reversing error produced due to friction and elasticity of the mechanical transmission elements (see section 4.2.4.2). Table 4.2 on page 190 shows how the sum of reversing error and frictional reversing error affects the total reversing error $2\varepsilon_u$, depending on their location.

We can state: the individual reversing errors in the mechanical transmission elements influence the total reversing error more, the closer the corresponding construction elements are to the feed table. The feed screw nut and bearing of the feed screw, respectively the pinion or worm, must therefore always be preloaded. (The effect of the reversing error on the position control loop is described in section 4.2.4.3.)

Reversing errors can be prevented by:

▷ preloading the gear wheels
▷ preloading the feed screw nut against the feed screw
▷ preloading the feed screw bearing
▷ preloading the pinion, respectively worm
▷ aspects of design, e.g. by stiff dimensioning of screw connections and bearing construction elements, selecting larger shaft diameter.

The theoretical possibility of removing reversing errors by preloading involves however, an increase in the speed-independent friction, and a diminished life time of the corresponding machine elements. When the moving construction

Table 4.2 Influence of reversing errors on the total reversing error

Originating location of the reversing error	Increase in the total reversing error by:
Reversing error between feed screw nut and feed screw, resp. between rack and pinion	A full reversing error
Reversing error in the feed screw bearing	A full reversing error
Flank play on the circumference of gear wheel 2 coupled to the feed screw drive	Flank play $\cdot \dfrac{h_{Sp}}{\pi \cdot d_{Gt2}}$
Torsional angle in the coupling, between the motor and the gear of the feed screw drive,	Torsional angle $\cdot \dfrac{h_{Sp}}{360° \cdot i}$
and of the rack and pinion drive	Torsional angle $\cdot \dfrac{2\pi \cdot r_{Ri}}{360° \cdot i}$

d_{Gt2} diameter gear wheel 2
h_{Sp} feed screw lead
i gear ratio
r_{Ri} pinion radius

components are more heavily dimensioned, the effect of the mass increase must be taken into account, because higher moments of inertia are to be expected (see sections 4.2.5 and 2.2.2.6).

References about the effects of reversing errors on the position control loop can be found in [4.11]. They concern experiments that were conducted on a model with hydraulic feed drive, without subordinated speed control. These tests have proven that in a position control loop with reversing error, the friction has a favorable effect on the transient response behavior. The measurements have shown that the undamping effect of the back lash does not have any effect, as long as it is significantly lower than the back lash caused by elasticity and friction. Further elaborations on the design of the mechanical transmission elements can be found in sections 4.3.1 and 4.3.2.

4.2.4.2 Friction

Frictional losses result between two machine components moving relative to each other. For this, the motor must deliever a torque ΣM_R. These frictional losses reduce the efficiency coefficient.

The following frictional losses occur on a feed drive:

    bearing friction
    friction in the gear
    friction in the table guides.

In addition, the feed screw drive exhibits

    friction between feed screw and feed screw nut,

and the rack and pinion drive, respectively worm gear drive

friction between pinion, respectively worm, and rack.

Depending on the materials pairing, the torque that must be delivered in dependence of the velocity can assume certain characteristic curves. Fig. 4.7 shows six possible friction characteristics, dependent on the construction of the table guide.

These six cases of frictional behavior of table guides are characterized in the following manner:

Case 1   Constant friction, independent of velocity.
Case 2   High static friction and increasing friction proportional to the velocity.
Case 3   High static friction and increasing friction proportional to the velocity.
Case 4   Very high static friction; friction decreasing at low velocities and proportionally increasing again at higher velocities.
Case 5   Low static friction and decreasing friction proportional to the velocity.
Case 6   Friction proportional to velocity, only.

Cases 1 through 4 illustrate slide friction, case 5 represents rolling friction (roller guides), and case 6 viscous friction (hydrostatic guides).

The effect is determined by the relation of the friction characteristic to the feed rate.

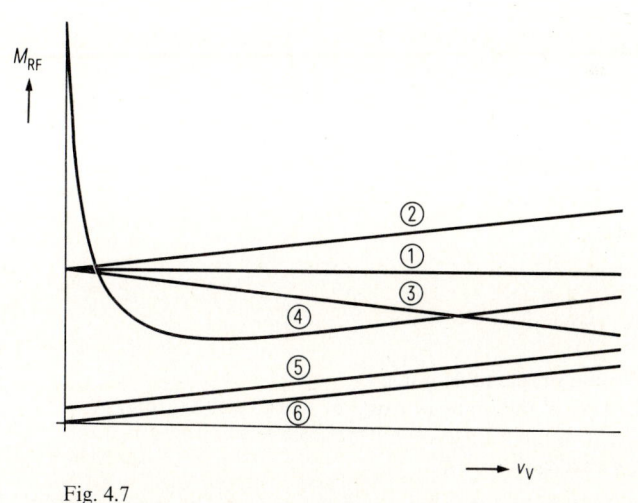

Fig. 4.7
Curves of the torque for the friction in the table guides, basically possible in dependence on the feed rate

Course proportional to the velocity:

▷ The damping gradient is increased, due to the frictional force increasing with the velocity (see equation (4.4)). The damping quotient c_v is the characteristic.
▷ The damping gradient is reduced, due to a frictional force decreasing with the velocity. (When linearizing the frictional characteristic for an operating point, D_{mech} can become negative.)
▷ A strongly negative slope in the range of low velocities, in combination with the elasticity of the mechanical transmission elements, causes the stick-slip effect. This is a jerking sliding motion which results from the periodic alternations of sticking and slipping. The elastic elements are preloaded, due to the high initial break-away torque. During the subsequent motion in the range of decreasing frictional characteristics, they relax, and cause the table to move more than it is desirable. The table comes to a standstill, and the elastic elements get preloaded again. This cycle repeats periodically [4.7].

Course independent of the velocity (also mentioned in the literature as dry friction):

▷ Results in connection with the finite stiffness of the mechanical transmission elements, in the frictional reversing error.

The magnitude of the frictional reversing error $2\varepsilon_{uR}$, is determined by the magnitude of speed-independent friction and the spring constant k. (Equations (2.26), (2.27), respectively (2.30 and (4.19).) For the feed slide, we can derive:

for feed screw and worm gear drive

$$\frac{2\varepsilon_{uR}}{\mu m} = 1.26 \cdot 10^4 \frac{\dfrac{M_{RF}}{Nm} \cdot \dfrac{1}{\eta_{SM}} + \dfrac{M_{RSL}}{Nm}}{\dfrac{h_{Sp}}{mm} \cdot \dfrac{k_{Ges}}{N/\mu m}} \qquad (4.6)$$

and for rack and pinion drives

$$\frac{2\varepsilon_{uR}}{\mu m} = 2 \cdot 10^3 \frac{\dfrac{M_{RF}}{Nm}}{\dfrac{r_{Ri}}{mm} \cdot \dfrac{k_{Ges}}{N/\mu m}} \qquad (4.7)$$

M_{RF} torque for the friction on the table guide
M_{RSL} torque for the friction of the screw bearing (or worm gear bearing)
η_{SM} efficiency coefficient of the feed screw nut, respectively worm
h_{Sp} feed screw lead, worm lead
r_{Ri} pinion radius
k_{Ges} total stiffness of screw, table and work piece system, respectively of the pinion/rack, table and work piece system.

These relationships do not include the frictional reversing error of the gear and of the motor coupling. In that case, the absolute value of the reversing error would generally be larger. Chapter 6 presents directions for the measurement of the total reversing error, which affect the position control loop according to section 4.2.4.3.

The total stiffness of a drive cannot be determined only from the necessary nominal angular frequency $\omega_{0\,mech}$; the resulting frictional reversing error must also be taken into consideration.

Example:

For the feed drive calculated in section 2.3.2.2 on page 101, a total stiffness of 500 N/μm is given. From equation (4.6) results a frictional reversing error of

$$2\varepsilon_{uR} = 1.26 \cdot 10^4 \, \frac{2.13 \cdot \frac{1}{0.926} + 1.584}{10 \cdot 500} \, \mu m = 9.79 \, \mu m$$

The mechanical nominal frequency is derived from the modified equation (4.3)

$$f_{0\,mech} = \sqrt{\frac{500}{3.95 \cdot 10^{-5}(500 + 1{,}000)}} \, Hz = 92 \, Hz$$

These are values that would offer sufficient work piece accuracy even at indirect position measurement. If we however, take the mean value given in section 4.2.2, of approximately 150 N/μm at 1,500 kg, then the frictional reversing error will increase to about 32.6 μm, while the nominal frequency would be reduced to about 50 Hz.

Very often, the frictional reversing error is a decisive factor for the required stiffness. Sufficient nominal frequencies can generally be achieved with lower stiffnesses.

Exactly as in the case of the reversing error, the signal flow of the mechanical elements can show the frictional reversing error, through a hysteresis element. It acts between the motor shaft signal "rotation angle" and the signal "position of the machine slide". The hysteresis behavior determined by the frictional reversing error, by comparison to the hysteresis resulting from normal reversing error, has different properties. With the frictional reversing error, there is no complete decoupling of the connected construction components. The effects of the reversing error in the position control loop is described in section 4.2.4.3.

The requirements for the frictional behavior of the mechanical transmission system can be summarized as follows:

▷ In order to achieve the least contour deviations and the best work piece surface possible, it is desirable to have a velocity proportional friction.

▷ The torque ΣM_R on the motor shaft, generated by the friction, should be in the empirically determined range (M_{0M} = motor torque):

$$0.2 M_{0M} < \Sigma M_R < 0.3 M_{0M} \tag{4.8}$$

▷ Negative slopes in the speed-dependent frictional characteristic should be avoided, because the resulting stick-slip effects cause poor work piece surfaces, lower the positioning accuracy, and cause higher table guide wear.

▷ To avoid frictional reversing errors, the speed-independent (dry) friction must generally be kept low.

Further references to the effects of friction in the mechanical transmission system can be found in the literature [4.2], [4.7], [4.9], [4.12], [4.13].

4.2.4.3 Effects of Reversing Errors and Frictional Reversing Errors

The elasticity, in combination with the friction not proportional to velocity, and/or the reversing error, generally cause the same types of contour errors. For this reason, the total reversing error of the mechanical transmission elements $2\varepsilon_u$ is represented in the substituted diagram of the position control loop, through a common hysteresis element. The values for $2\varepsilon_u$ always refer to the table travel, independently of location or cause, reversing error or frictional reversing error.

At the transfer of a sinusoidal input signal $u(t)$ for instance, a hysteresis element presents the following disadvantages:

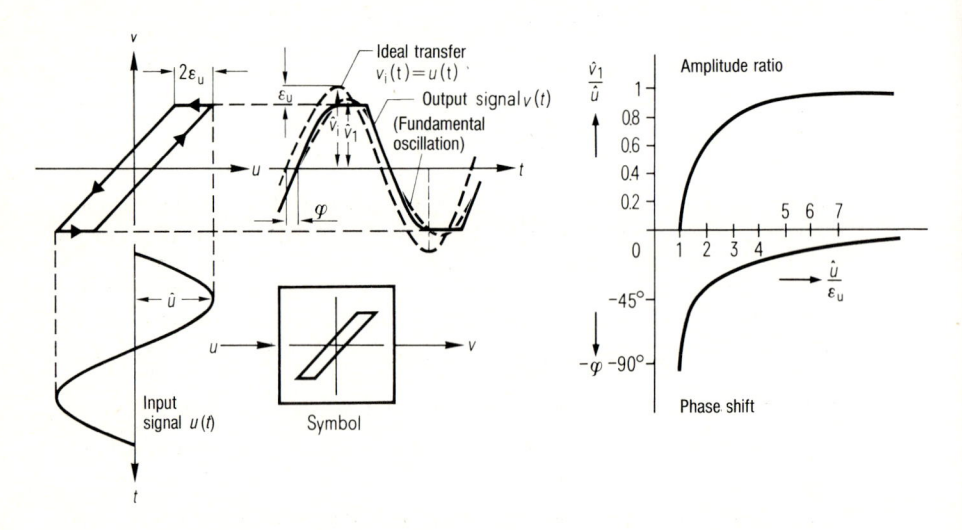

Fig. 4.8 Signal transfer through a hysteresis element

▷ The output signal $v(t)$ is distorted. This distortion increases with diminishing input amplitude $\hat{u}$.
▷ The amplitude of the fundamental oscillation of output signal $\hat{v}$, is reduced by the reversing error. The maximal value of output signal $v(t)$ is lower by half the reversing error $(1 \cdot \varepsilon_u)$ than the amplitude of the ideal output signal $\hat{v}_i$.
▷ The fundamental oscillation of the output signal is shifted by the angle φ, in reference to the input signal.

Figure 4.8 shows the input and output signals of a hysteresis element, as well as the describing function. From it, we can derive the amplitude reduction and the phase shift.

The effects in the position control loop are:

▷ Because of the additional phase shift there is a potential hazard of instability in the loop, especially at low input amplitudes and high position loop gains.
▷ Deviations between the command and the actual contour result, due to the signal distortion.

The location of the position measuring system (direct or indirect), determines differences in the effects of the reversing error and the frictional reversing error on the position loop behavior and the contouring.

4.2.4.3.1 Indirect Position Control Measurement

Figure 4.9 presents the structure of a simplified position control loop. It is apparent that with indirect position measurement, the hysteresis acts outside the position control loop.

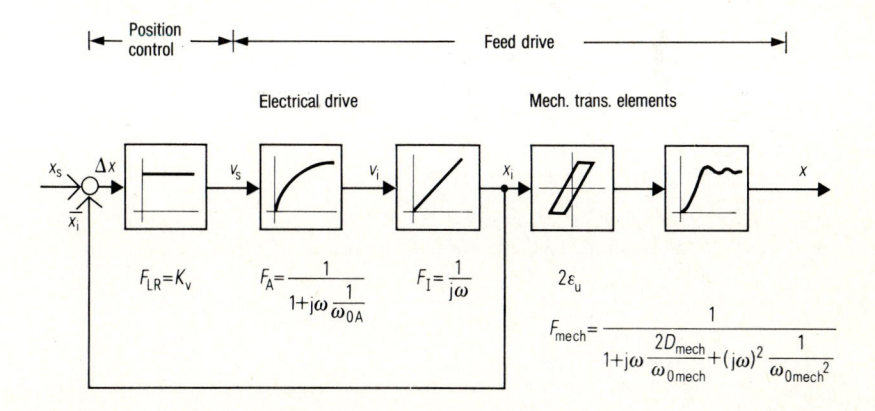

Fig. 4.9
Block diagram of a position control loop with indirect position measurement. Hysteresis outside the position control loop

A deviation between the real position of the table and the actual position as reported by the position measuring system, will not be recognized. During positioning of an axis, the size of the total hysteresis $2\varepsilon_u$ will be entered as an error, up to its full magnitude.

The results of a simulation with a control loop structure like the one in fig. 4.9 on an analog computer, are presented in [4.2]. There is assumed that the nominal angular frequency is $\omega_{0\,\text{mech}} = 220\text{ s}^{-1}$; the damping gradient D_{mech} is set, according to the possible frictional characteristics, like in fig. 4.7. As frictional reversing error, is taken the relatively large value of $2 \cdot 40$ μm; additional reversing errors are not taken into account. The drive is treated as a 1st order delay element, with $\omega_{0\,A} = 80\text{ s}^{-1}$. The position loop gain in the position control loop is set to $K_v = 40\text{ s}^{-1}$.

Figure 4.10 shows the motion of this feed axis over time. The course of the position actual value $x_{i1}-x_{i6}$ depends largely on the given frictional characteristic. If we start with curve 1 with speed-independent friction, in curve 2 we can observe an improvement, due to the speed-dependent friction. By comparison, curve 3 is clearly worse because of the negative slope in the course of the speed-dependent friction.

Curve 4 shows, that it is impossible to execute an exact positioning at high break-away torque and negative slope of the frictional characteristic, with low speeds. We can also observe the jerking sliding aspect due to the stick-slip effect.

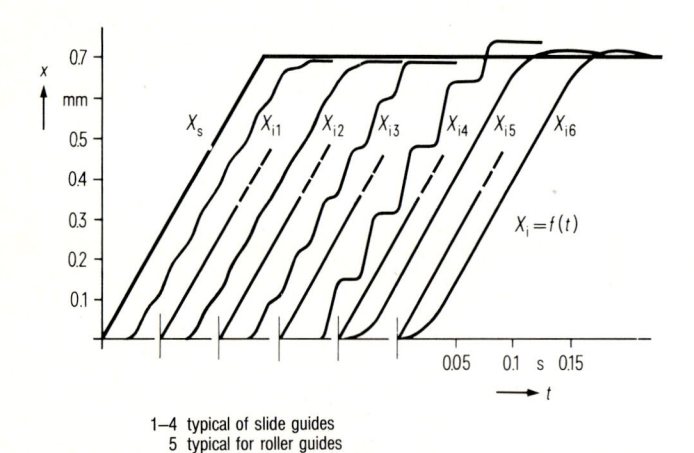

1–4 typical of slide guides
5 typical for roller guides
6 typical for hydrostatic guides

Fig. 4.10
Travel of a machine slide by $x = 0.7$ mm, in a presentation over time with the different friction torque curves according to fig. 4.7 (indirect position measurement)

The curves 5 and 6 with low, respectively no speed-independent friction, are practically undistinguishable. They show an undisturbed course of the actual position value.

Although large frictional forces are presumed for the curves 1–4, they still exhibit significant oscillations. The decisive factor for the mechanical damping gradient D_{mech} is not the absolute value of the friction, but the slope of the characteristic curve $\Delta M_R / \Delta v_v$.

The behavior at the traveling of inclines, in which two feed axes construct the actual contour, was also analyzed for the given data with the analog computer. Fig. 4.11 shows the course, first for a 30° incline, and also for a 10° incline. A frictional characteristic according to curve 1 in fig. 4.7, is presumed.

Three phenomena can be recognized:
▷ Waviness of the actual contour.
  It is a consequence of the oscillating behavior of the position control loop due to the hysteresis, and has nothing to do with the stick-slip effect.
▷ Parallel offset of the actual contour from the command contour.
  It is a direct consequence of the indirect position control measurement and hysteresis. The parallel offset can be calculated from

$$s = |2\varepsilon_{uy} \cdot \cos \beta - 2\varepsilon_{ux} \cdot \sin \beta| \tag{4.9}$$

where $2\varepsilon_{uy}$ and $2\varepsilon_{ux}$ are abbreviations for the total hysteresis in the two axes.

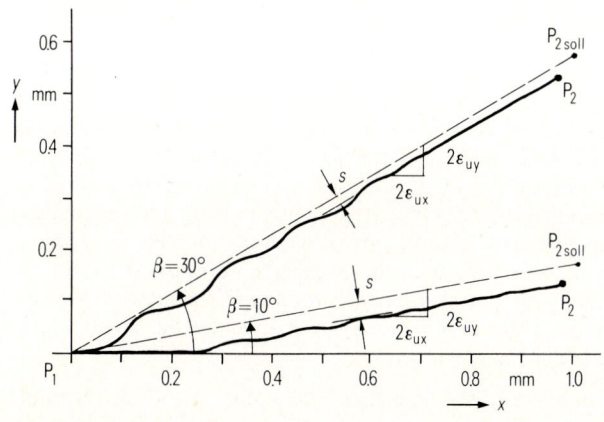

Fig. 4.11
Travel of a 10° and 30° incline with indirect position control measurement

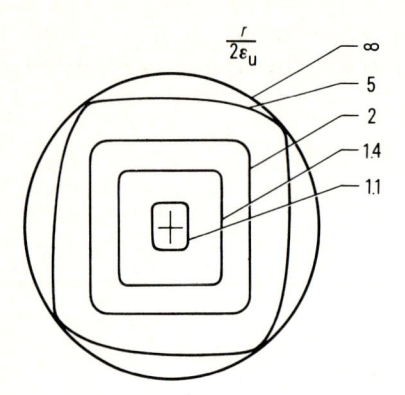

Fig. 4.12
Contour distortions on the circle because of hysteresis (indirect position control measurement)

$\dfrac{r}{2\varepsilon_u}$ — ∞, 5, 2, 1.4, 1.1

▷ Position deviation of the point to be approached.

According to equation (6.5), the distance between the command and actual position amounts to

$$|x_s - x_i| \approx \sqrt{(2\varepsilon_{ux})^2 + (2\varepsilon_{uy})^2}$$

A further simulation shows the contour course at traveling of circles with indirect position control measurement. In fig. 4.12, the parameter given is the relationship between the circle radius r, and the reversing error $2\varepsilon_u$. It is presumed that the value $2\varepsilon_u$ is the same for both, X and Y-axes. It is apparent that the actual contour is always smaller than the command contour. As the radius of the circle becomes smaller, the actual contour at constant hysteresis becomes increasingly distorted.

4.2.4.3.2 Direct Position Control Measurement

At direct position measurement, the hysteresis acts within the position control loop. The block diagram of a simplified position control loop is shown in fig. 4.13. In this case, by comparison to the indirect position control measurement, the hysteresis element affects the stability of the position control loop, and therefore position deviations of the position measuring system are recognized and controlled. Since the reversing error causes a complete decoupling, but the frictional reversing error is to be treated without complete decoupling, when evaluating the effects of the two non-linearities we must make certain distinctions. The stability in the position control loop is primarily affected by the reversing error, while contour errors are caused by the combination of reversing and frictional reversing errors.

In order to avoid overshoots of the actual position value at a given reversing error, it is necessary to reduce the position loop gain. The achievable value of the position loop gain K_{verr} of a position control loop with reversing error, lies below the optimal value of the error-free position control loop. Figure 4.14 from [4.11] shows this relationship. In this case, the achievable position loop gain was measured on a reconstructed feed table without subordinated speed control. It is interesting that in this example, from a reversing error of more than about 20 µm, the achievable position loop gain remains at a constant value. Independently of the reversing error size, the achievable position loop gain becomes maximally 40% smaller than in the linear position control loop. The additional frictional reversing error in this example was approximately 12 µm.

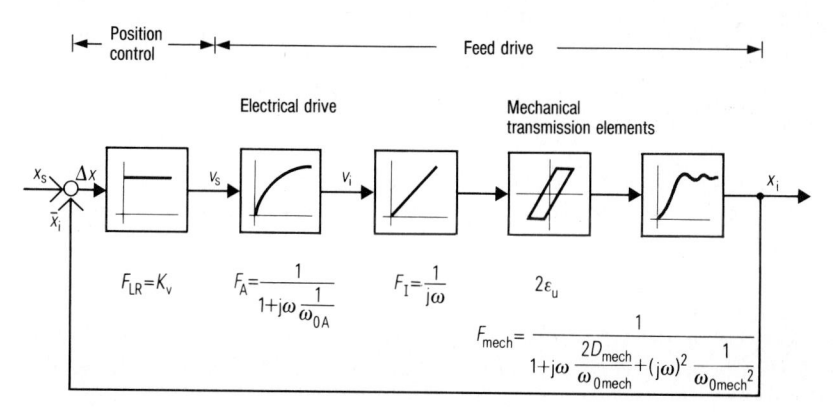

Fig. 4.13
Block diagram of a position control loop with direct position measurement. Hysteresis inside the position control loop

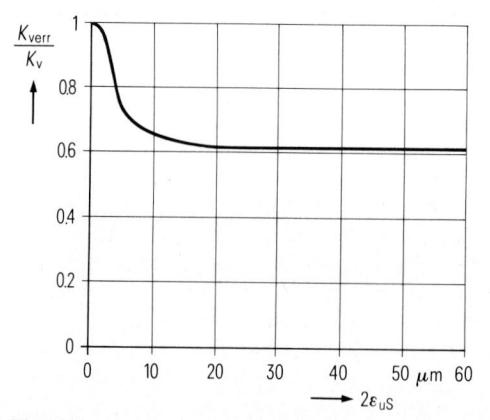

Fig. 4.14
Attainable position loop gain K_{verr} of the position control loop with reversing error, referenced to the gain of the position control loop free of reversing error, dependent on the value of the reversing error

According to this test, the optimal position loop gain of $0.2\,\omega_{0A} \leq K_v \leq 0.3\,\omega_{0A}$ derived in equation (1.52) must be reduced because of the reversing error, to $0.12\,\omega_{0A} \leq K_v \leq 0.18\,\omega_{0A}$. This shows that even with direct position measuring systems, it is necessary to get the mechanical transmission elements as free of reversing error as possible.

Example:

Drive 1 of table 2.2 on page 90, with the circulating current-free anti-parallel circuit, is allowed to have a position loop gain of

$$K_v \approx (0.065 \ldots 0.1)\,\omega_{0A} = 9.5 \ldots 14\;\text{s}^{-1}$$

even with a reversing error of 10 μm and taking into account the dead time. This value could be improved to

$$K_v \approx (0.1 \ldots 0.16)\,\omega_{0A} = 14 \ldots 23\;\text{s}^{-1},$$

by using a circulating current-conducting connection with lower dead time.

When evaluating contour deviations, the total reversing error from backlash and frictional reversing error must be taken into consideration. If the course of the motion of a feed axis over time is similar to that seen in fig. 4.10, then, according to fig. 4.15, in the presence of reversing errors we will have overshoots. The command position will be reached, but under circumstances, only after several over and undershoots. Another possibility is that, depending on the size of the reversing error and the settings of the position and speed

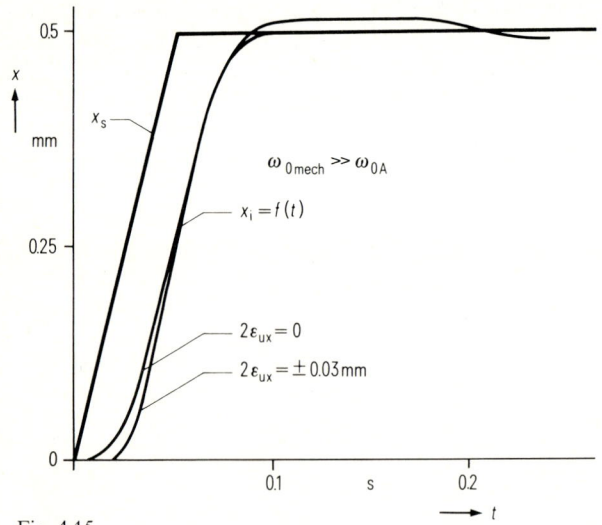

Fig. 4.15
Driving a coordinate axis by $x = 0.5$ mm, in the presence of reversing error (direct position control measurement)

controls, continuous oscillations will occur. These oscillations about the position lead, when driving axis-parallel, to visible markings on the work piece. This can be reduced only by using a less sensitive speed regulator in the current converter unit.

The resolution of the position measuring also plays a role in the determination of the behavior of the position control. As shown in fig. 1.23, the insensitivity range $2\,\varepsilon_{u2}$ acts in the position control loop as a hysteresis element. This results in oscillations whose amplitude is larger or equal to ε_{u2}. For incremental position control systems, this insensitivity range is equal to ± 1 increment. In order to reduce the effects on the work piece, the resolution must be at least 2–3 times higher than the desired measuring accuracy.

A contour curve at traveling of inclines according to fig. 4.16, can also be determined for the direct position measurement with a computer simulation, under the same assumptions as those for fig. 4.11.

By comparison to the indirect position control measurement, at direct position measurement the parallel offset disappears completely, and the waviness in the actual contour is reduced. Deviations from the command contour occur only at the start of the motion, until the hysteresis is overcome. This however, happens relatively fast since the drives, in the initial absence of the position feedback, are accelerated accordingly. The end positions will be overshot by about 15 µm, but this error will be corrected subsequently. The overshoot width depends on the position loop gain, which is set especially high in our example.

The signal distortions determined by the reversing error, cause an enlargement of the actual contour at direct measuring systems. The contour course at traveling of circles, and the deviations that resulted in the process, were measured on a milling machine. The results are presented in fig. 4.17.

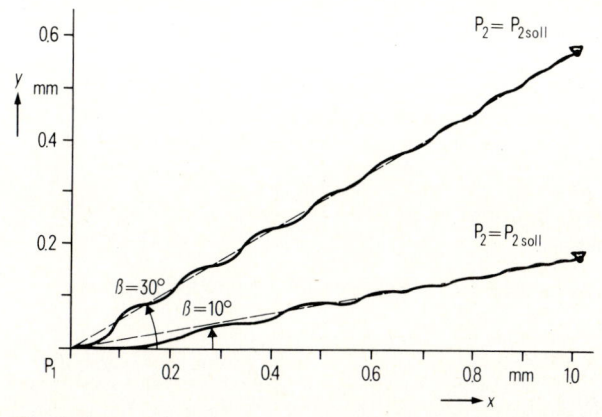

Fig. 4.16
Traveling a 10° and a 30° incline, with direct position measurement

Circle diameter	Circle diameter	Circle diameter
9 mm	0.9 mm	0.09 mm
Reversing error	Reversing error	Reversing error
0.2 mm	0.2 mm	0.2 mm

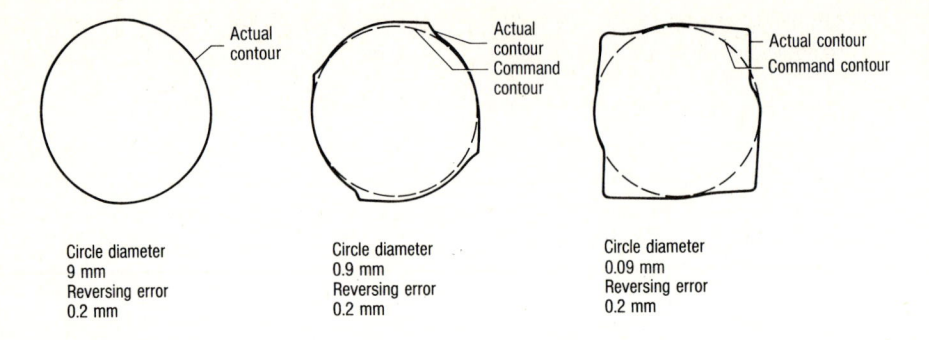

Fig. 4.17
Experimentally measured contour distortions due to reversing error, with direct position measuring system

The deviations from the command circle increase as the relationship $r/2\,\varepsilon_u$ decreases. They are usually lower than the reversing error itself, but depend on the contour velocity and on the dynamic properties of the drive.

4.2.5 Moment of Inertia

All mechanical transmission elements posses mass. The moving parts as a unit determine the external moment of inertia, which has a considerable effect on the dynamic behavior of the entire feed drive. In the mechanical time constants of a drive, this is considered as the external moment of inertia J_{ext} reflected onto the motor shaft (see section 2.2.2.1).

Section 2.2.2.6 presents the effect of the external moment of inertia on the attainable nominal angular frequency ω_{0A} of the drive. Also presented are the standard values for the proper size of J_{ext} in reference to the motor moment of inertia J_M. In the following sections, values are given for the moments of inertia of the individual mechanical transmission elements. Also presented in these sections are formulas for the geometric forms used on feed drives, and the conversion formulas for the external moment of inertia reflected on the motor.

4.2.5.1 Cylindrical Bodies

The moment of inertia of a cylinder, reflected on the longitudinal axis, is

$$J = \frac{1}{8} m d^2 \qquad (4.10)$$

J moment of inertia
m mass
d diameter

For a steel cylinder with a density of $\rho = 7.85 \cdot 10^3$ kg/m³, we obtain as adjusted dimensional equation

$$\frac{J}{\text{kg m}^2} = 0.77 \cdot 10^{-12} \cdot \left(\frac{d}{\text{mm}}\right)^4 \cdot \frac{l}{\text{mm}} \tag{4.10.1}$$

l length of the cylinder.

Table 4.3 shows the moments of inertia and masses referenced to 1 m. For other materials this can be calculated proportional to the relationship of the densities.

Table 4.3
Referenced moments of inertia and masses of cylindrical steel bodies
($\rho = 7.85 \cdot 10^3$ kg/m³)

Cylinder diameter	Cylinder cross section	to 1 m length		Cylinder diameter	Cylinder cross section	to 1 m length	
		Referenced moment of inertia	Referenced mass			Referenced moment of inertia	Referenced mass
d mm	A cm²	J' kg m²/m	m' kg/m	d mm	A cm²	J' kg m²/m	m' kg/m
10	0.785	0.0000077	0.617	110	95.3	0.1128350	74.6
15	1.767	0.0000390	1.39	120	113.10	0.1598075	88.8
20	3.142	0.0001233	2.47	130	132.73	0.2201125	104
25	4.909	0.0003010	3.85	140	153.94	0.2960750	121
30	7.069	0.0006242	5.55	150	176.71	0.3901500	139
35	9.621	0.0011565	7.55	160	201.1	0.5050750	158
40	12.57	0.0019730	9.86	170	227.0	0.6436750	168
45	15.90	0.0031602	12.5	180	254.5	0.8090250	200
50	19.64	0.0048165	15.4	190	283.5	1.0043500	223
55	23.76	0.0070520	18.7	200	314.2	1.2330750	247
60	28.27	0.0099877	22.2	210	346.4	1.4988250	272
65	33.18	0.0137572	26.0	220	380.1	1.8053500	298
70	38.48	0.0185037	30.2	230	415.5	2.1566500	326
75	44.18	0.0243845	34.7	240	452.4	2.5570000	355
80	50.27	0.0315675	39.5	250	490.9	3.0105000	385
85	56.74	0.0402300	44.5	260	530.9	3.5217000	417
90	63.62	0.0505650	49.9	270	572.6	4.0832000	449
95	70.88	0.0627725	55.6	280	615.8	4.7372000	483
100	78.54	0.0770675	61.7	300	706.9	6.2425000	555

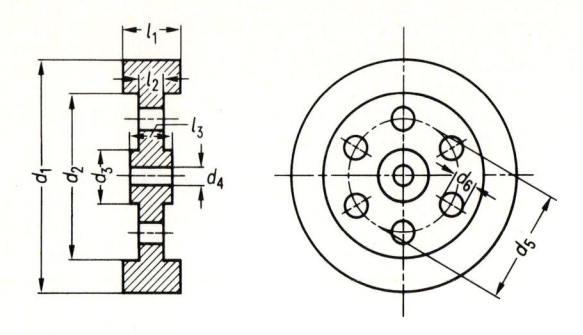

Fig. 4.18
Diameter and length data for the calculation of moments of inertia for gear wheels

In order to decrease the moment of inertia, gear wheels are often hollowed; this results in cylindrical bodies of different lengths (see fig. 4.18). The mass can be additionally reduced through holes in the wheel body.

The moment of inertia of such a wheel made out of steel, is

$$\frac{J}{\text{kg m}^2} = 0.77 \cdot 10^{-12} \left[\frac{l_1}{\text{mm}} \cdot \left(\frac{d_1}{\text{mm}}\right)^4 - \frac{l_1 - l_2}{\text{mm}} \left(\frac{d_2}{\text{mm}}\right)^4 + \frac{l_3 - l_2}{\text{mm}} \left(\frac{d_3}{\text{mm}}\right)^4 \right.$$
$$\left. - \frac{l_3}{\text{mm}} \left(\frac{d_4}{\text{mm}}\right)^4 - k \frac{l_2}{\text{mm}} \left[\left(\frac{d_6}{\text{mm}}\right)^4 + 2 \left(\frac{d_6}{\text{mm}}\right)^2 \left(\frac{d_5}{\text{mm}}\right)^2 \right] \right] \quad (4.11)$$

The lengths and diameters are to be interpreted according to fig. 4.18.
k = number of holes of diameter d_6.

In case of other materials, this must be recalculated according to the relationship of the densities.

4.2.5.2 Linearly Moved Masses

The moment of inertia of the linearly moved mass $m_W + m_T$ of a feed screw drive is, according to fig. 2.13, reflected onto the feed screw

$$J_{T+W} = (m_W + m_T) \left(\frac{h_{Sp}}{2\pi}\right)^2 \quad (4.12)$$

According to fig. 2.14, for the rack and pinion drive it is reflected onto the pinion

$$J_{T+W} = (m_W + m_T) r_{Ri}^2 \quad (4.13)$$

$m_W + m_T$ linearly moved mass
h_{Sp} feed screw lead
r_{Ri} pinion radius

4.2.5.3 Gear Ratio

The moment of inertia is transmitted from one shaft onto the other, through a gear ratio i, according to

$$J_1 = \frac{J_2}{i^2} \tag{4.14}$$

J_1 moment of inertia reflected on shaft 1 with speed n_1
J_2 moment of inertia reflected on shaft 2 with speed n_2
i gear ratio $\frac{n_1}{n_2}$

(Indices according to figures 2.13, 2.14)

Through the gear, we can achieve a moment of inertia reduction proportional to $1/i^2$, while the speed and the torque are converted proportional to $1/i$, respectively to i (see equations (2.29), (2.31), (2.32), (2.33), and (2.48). The linear acceleration at the feed slide is proportional to $1/i$ (equation (2.43), respectively (2.44)). It follows that:

▷ A gear ratio is advantageous for large linearly moved masses as well as for rotating masses. The feed motor should be driven at high speeds.
▷ Acceleration drives should be equipped with gear ratios adapted according to equation (2.45), and should have moments of inertia as low as possible.
▷ Rotating elements situated in the proximity of the motor should be constructed with masses as low as possible.

For feed screw drives with or without gear and small gear ratio, the clutches and the feed screw should be designed with moments of inertia as low as possible. (According to section 4.2.2, the stiffness should be taken into account.) The pinion diameter of rack and pinion drives should be kept small.

4.3 Calculating the Slide Elements of the Drive

4.3.1 Feed Screw Drives

4.3.1.1 Construction and Requirements

The mechanical transmission system discussed here, consists of feed screw bearing, feed screw, feed screw nut, and feed slides. Backlash-free, adjustable ball screw drives are used almost exclusively. The feed screw is driven either through a gear, or directly by a feed motor. Depending on the type of machine, the feed slide carries the work piece or the tool; it is also called feed table.

A decisive characteristic for the dynamic behavior is the total stiffness k_{Ges}. As already stated in section 4.2.2, it represents the relationship of the force on the machine table to the table's change of position. It determines, in combination with the mass of the feed screw, machine table, and work piece, primarily the dominant first mechanical nominal angular frequency $\omega_{0\,mech\,1}$.

This total stiffness is composed of the individual stiffnesses of the screw bearing, feed screw itself, feed screw nut unit and its mounting. It is always smaller than one of these individual stiffnesses.

For very long feed screws the moment of inertia becomes too large. In these cases, resolutions at which the ball nut is driven are, under circumstances, more advantageous. The stiffness of the screw, which contributes in determining the achievable nominal angular frequency, must be taken into account. For long, horizontally mounted feed screws, also the bending cannot be disregarded any more.

The substituted diagrams and the inverted values of the smallest total stiffness are shown in fig. 4.19, for four of the most common feed screw bearings [4.1].

Here, the concepts are:

K_{aL} axial spring constant of an axial bearing (bearing elements and bearing)
K_{Sp} spring constant of the feed screw of length l_{Sp}
k_{Sp1} k_{Sp2} spring constant of the feed screw parts.

The total stiffness of the feed screw can be derived from

$$\frac{1}{k_{Sp}} = \frac{1}{k_{Sp1}} + \frac{1}{k_{Sp2}} \tag{4.15}$$

k_M spring constant of the feed screw nut unit, i.e. stiffness of the parts within the feed screw nut
k_{TM} spring constant of the feed screw nut mounting to the table
k_{Ges} resulting total stiffness
l_{sp} feed screw length

The following four possibilities are compared for the axial mounting of the feed screw bearing:

Version 1   The feed screw is supported simply on both sides axially, and is heavily preloaded to pull.
Version 2   The feed screw is doubly supported on both sides axially, with preloaded double bearings, and is lightly preloaded to pull.
Version 3   The feed screw is fixed only on one side, through an axially preloaded double bearing. The other side is supported only radially.
Version 4   The feed screw is fixed only on one side through an axially preloaded double bearing. The other side is not supported.

The advantages and disadvantages of the individual versions are summarized in table 4.4.

Fig. 4.19
Different versions of feed screw drive, with substituted and total stiffness

Version	Feed screw bearing	Substituted diagram		Total stiffness	Correction factors for	
					critical speed b_{Krit}	buckling strength b_{Kn}
1	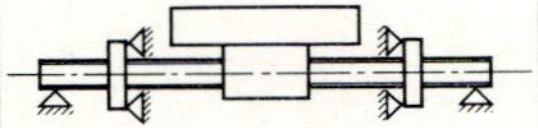	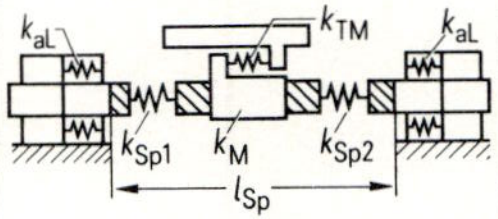	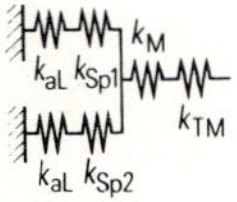	$\dfrac{1}{k_{\text{Ges}}} = \dfrac{1}{2k_{\text{aL}}} + \dfrac{1}{4k_{\text{Sp}}} + \dfrac{1}{k_{\text{M}}} + \dfrac{1}{k_{\text{TM}}}$	0.65	1
2	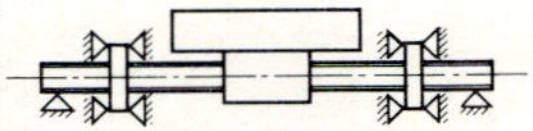	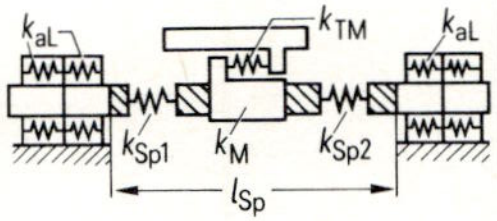	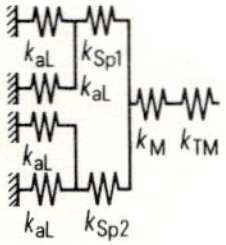	$\dfrac{1}{k_{\text{Ges}}} = \dfrac{1}{4k_{\text{aL}}} + \dfrac{1}{4k_{\text{Sp}}} + \dfrac{1}{k_{\text{M}}} + \dfrac{1}{k_{\text{TM}}}$	1.44	4
3	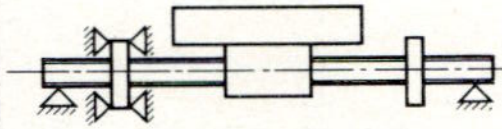	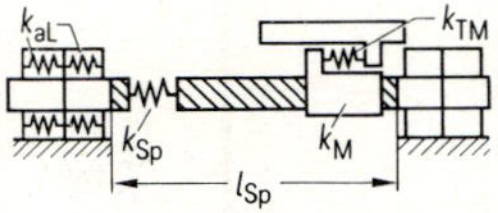	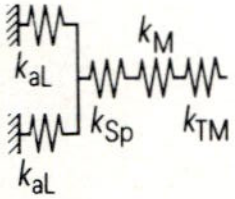	$\dfrac{1}{k_{\text{Ges}}} = \dfrac{1}{2k_{\text{aL}}} + \dfrac{1}{k_{\text{Sp}}} + \dfrac{1}{k_{\text{M}}} + \dfrac{1}{k_{\text{TM}}}$	1	2
4	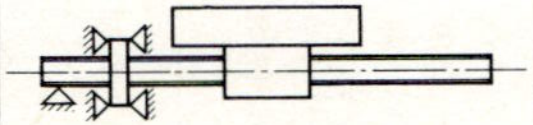	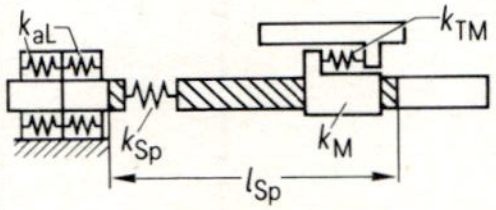	like 3	like 3	0.21	0.25

Table 4.4 Advantages and disadvantages of individual bearing versions

Version	Advantage	Disadvantage
1	High total stiffness	A high feed screw preload is necessary, bearing life lower than in ②.
2		Feed screw preload cannot be exactly predetermined.
3	Cost efficiency higher	The stiffness is considerably altered with the position of the slide. Lower total stiffness.
4		

When heated, the feed screw elongates. Without preload of the feed screw, the bearings in version 1 would be unloaded, and axial play would result. In version 2, the heat expansion causes an upset in the feed screw, admissible only to a certain extent; in addition, the bearing preload becomes asymmetrical. Both can be reduced through a slight preload of the feed screw to pull.

Version 2 is preferable for precision feed axes. Simple machines however, are generally designed for version 4 or, for long feed screws, for version 3. Indications about the calculations and improvement of stiffness of the individual elements are given in the following sections.

4.3.1.2 Feed Screw

Besides the ball screw, as feed screw in the feed drive, the hydrostatic trapezoidal thread screw can also be used. A normal trapezoidal thread screw however, cannot be used because of too high friction and high reversing error. The hydrostatic trapezoidal thread screw is used specifically for large machine tools. Its design and manufacturing are complex, and therefore it is expensive. [4.7] shows an example of design.

The following points must be given special attention when using a ball screw drive in a feed drive:

stiffness	thread lead accuracy
moment of inertia	critical speed
efficiency coefficient	buckling strength (bending)
friction and heating	lifetime
thread lead	lubrication.

Design references to these points are made in the following sections. Further data can be derived from the manufacturing literature for threaded drives.

4.3.1.2.1 Stiffness of the Feed Screw

The first design criterion for the dimensioning of the feed screw is its tensile, respectively its compression strength. For this:

$$k_{Sp} \approx \frac{\pi}{4} \cdot \frac{E \cdot d_{KSp}^2}{l_{Sp}} \qquad (4.16)$$

or as adjusted dimensional equation for steel

$$\frac{k_{Sp}}{N/\mu m} \approx 1.65 \cdot 10^{-1} \frac{\left(\frac{d_{KSp}}{mm}\right)^2}{\frac{l_{Sp}}{m}} \qquad (4.16.1)$$

d_{KSp} core diameter of the feed screw (for exact calculations, the mean value from d_{KSp} and d_{Sp} can be used)
E elasticity modulus (steel $2.1 \cdot 10^{11}$ N/m²)
l_{Sp} total free length of the feed screw.

Depending on the feed screw bearing type, we obtain different values, according to fig. 4.19, for the feed screw stiffness. In the versions 1 and 2 with bearings on both ends, in the worst case only half of the feed screw length acts at the middle position of the feed screw nut. The two springs in the substituted diagram are switched in parallel. Thus, for the feed screw stiffness, we obtain four times the value of k_{Sp}. For one sided bearing as in versions 3 and 4, by comparison, the lowest stiffness is equal to the value k_{Sp}, as calculated with equation (4.16) (see the equations in fig. 4.19).

From fig. 4.20, the tensile, respectively compression strengths of the feed screws can be derived in dependence on the screw core diameter d_{KSp}, for the different bearing versions.

For small screw diameters, under circumstances, the torsion of the feed screw converted into the longitudinal movement of the feed slide, can also influence the tensile, respectively compression strength. The torsional stiffness of the feed screw is

$$k_{ToSp} \approx \frac{\pi}{32} \frac{G \cdot d_{KSp}^4}{l_{Sp}} \qquad (4.17)$$

or for steel

$$\frac{k_{ToSp}}{Nm} \approx 7.84 \cdot 10^{-3} \frac{\left(\frac{d_{KSp}}{mm}\right)^4}{\frac{l_{Sp}}{m}} \qquad (4.17.1)$$

G shearing modulus (steel $8 \cdot 10^{10}$ N/m²)
d_{KSp} screw core diameter (for exact calculations, the mean value from d_{KSp} and d_{Sp} can be used)
l_{Sp} total free length of the feed screw.

The position change resulting from torsion of the feed screw, measured at the feed slide, is:

$$\Delta s_{\text{ToSp}} = \frac{1}{2\pi} \cdot \frac{M_{\text{Sp}} \cdot h_{\text{Sp}}}{k_{\text{ToSp}}} = \frac{1}{4\pi^2} \cdot \frac{F_{\text{aSp}} \cdot h_{\text{Sp}}^2}{k_{\text{ToSp}}} \qquad (4.18)$$

M_{Sp} torque on the feed screw
F_{aSp} feed force in the axial direction of the feed screw
h_{Sp} feed screw lead.

The position change resulting from tensile or compression force on the feed screw, measured on the feed slide, is:

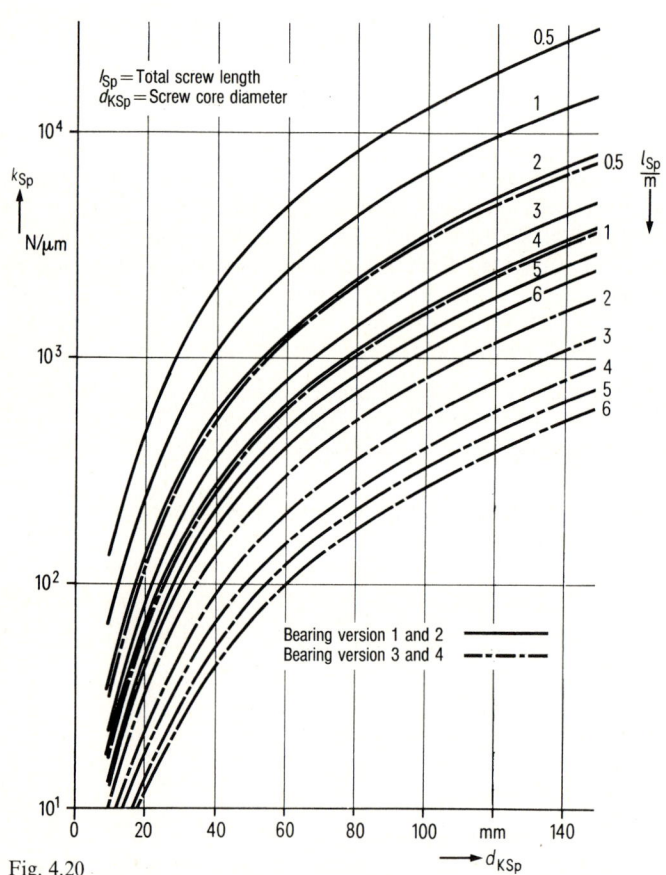

$$\Delta s_{\text{ZSp}} = \frac{F_{\text{aSp}}}{k_{\text{Sp}}} \qquad (4.19)$$

Fig. 4.20
Tensile, respectively compression strength of the feed screw; bearing versions as in fig. 4.19

If we place both in a relationship to each other, we obtain for steel

$$\frac{\Delta s_{\text{ToSp}}}{\Delta s_{\text{ZSp}}} = 0.53 \left(\frac{h_{\text{Sp}}}{d_{\text{KSp}}}\right)^2 \quad (4.20)$$

This means that for a ratio $d_{\text{KSp}}/h_{\text{Sp}} > 4$, the longitudinal movement due to torsion reflected on the table movement is now only $<3.3\%$ of the position change due to tensile or compression strength.

The torsion must be considered in respect to possible torsion oscillation caused, e.g. by the line or pulse frequency of the current converter. Especially for stiff couplings of the motor, disturbing beat phenomena can occur in the speed control loop.

The torsion resonance frequency of a cylinder fixed on one side is

$$f_{0\,\text{To}} = \frac{1}{2\pi}\sqrt{\frac{k_{\text{To}}}{J}} \quad (4.21)$$

k_{To} torsion spring constant
J moment of inertia.

If two masses are coupled through an elastic connection (e.g. a turn elastic clutch), and the two masses can rotate freely, the torsion resonance frequency is

$$f_{0\,\text{To}} = \frac{1}{2\pi}\sqrt{k_{\text{To}}\left(\frac{1}{J_1} + \frac{1}{J_2}\right)} \quad (4.22)$$

where J_1 and J_2 are the moments of inertia of the two masses.

In practice, masses are often coupled through shafts with staggered diameters. This makes the calculation of one or several resonance positions more difficult. More detailed explanations can be found in the corresponding literature [4.14], [4.15].

For a given table mass, the feed screw stiffness k_{Sp} must be so selected, that together with the other individual stiffnesses of the feed screw system, it should amount to a sufficient total stiffness, and thus a high mechanical nominal frequency can be achieved. Also, because of the requirement for a small frictional reversing error, the total stiffness ought not to become too small (see sections 4.2.2 and 4.2.4.2).

For a given bearing type and feed screw length, the only remaining variable which can affect the feed screw stiffness is its diameter d_{Sp}. It should be noted that the related moment of inertia increases with the fourth power of d_{Sp}. This reduces the drive nominal angular frequency $\omega_{0\,\text{A}}$ (see section 2.2.2.6).

One way to estimate the necessary feed screw diameter is offered in the diagram of figure 4.21, according to [4.1]. The following values achievable in practice are presumed:

k_{TM} 1,000 N/μm, stiffness of the feed screw nut mounting to the table
k_M 700 N/μm, feed screw nut unit stiffness
k_{aL} 800 N/μm, axial bearing stiffness

The additional values are:

$f_{0\,mech}$ Mechanical nominal frequency of the feed screw table and work piece system $\left(f_{0\,mech} = \dfrac{1}{2\pi} \cdot \omega_{0\,mech}\right)$

$m_T + m_W$ table and work piece mass (m_W can be ommited, depending on the type of machine)
l_{Sp} total free length of the feed screw
d_{KSp} feed screw core diameter.

It can be seen, that for double sided bearings only about half the feed screw diameter is necessary for the same nominal frequency. This only applies however, under the given presumptions for the stiffness of the other parts.

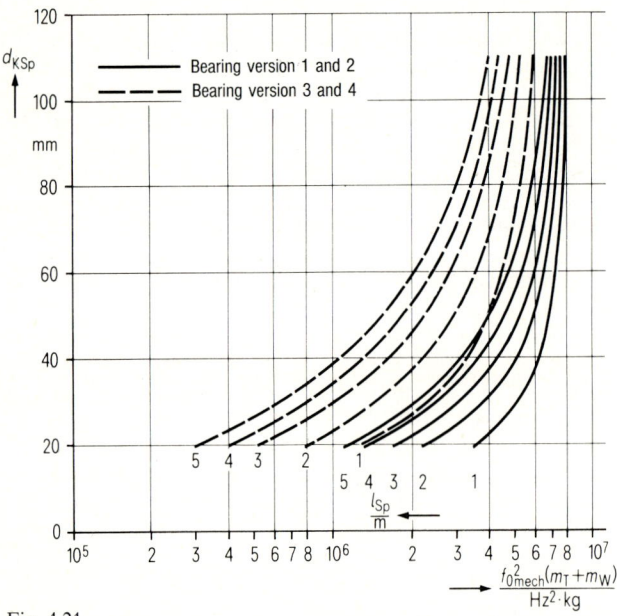

Fig. 4.21
Diagram for the estimation of the feed screw diameter d_{KSp} in relationship to mechanical nominal frequency, table and work piece masses, bearing type according to fig. 4.19, and feed screw length

For larger $f_{0\,\text{mech}}^2(m_T+m_W)$ products, it can be seen from the steep slopes of the curves, that only a disproportionately high diameter increase would result in the desired nominal frequency. A rack and pinion drive is more suitable for these ranges.

Example:

A mechanical nominal frequency of 40 Hz is to be obtained on a feed screw drive. The table and maximal work piece mass is 3,000 kg; the travel range, same as the feed screw length, is 2 m. The feed screw diameter should not be over 50 mm, since otherwise the moment of inertia of the mechanical transmission elements would be too large. How should this feed screw be supported, and what diameter d_{KSp} is required?

The values for the individual stiffness of the other parts are the same as those used for fig. 4.21. We obtain

$$f_{0\,\text{mech}}^2 \cdot (m_T+m_W) = (40\text{ Hz})^2 \cdot 3{,}000\text{ kg} = 4.8 \cdot 10^6\text{ Hz}^2\text{ kg}.$$

It becomes obvious from fig. 4.21, that the given problem cannot be solved using a single sided axially supported feed screw (versions 3 or 4 according to fig. 4.19 would give $d_{KSp} \approx 87$ mm).

With double sided axial bearing of the feed screw (version 1 or 2), the feed screw core diameter of $d_{KSp} = 37$ mm would be sufficient, and a feed screw with a rated diameter of 40 mm can be selected.

For longer, horizontally mounted feed screws, the tensile and compression forces must be taken into account, because a compressive force would increase the sagging due to the feed screw's own weight. If only one load direction is necessary during machining (e.g. as for lathes and horizontal milling machines), the arrangement of the axial bearings must be selected in such a manner, for bearing types 3 and 4 of fig. 4.19, that the feed screw only be exposed to tensile forces.

The feed screw diameter is determined generally by the stiffness requirements for a high nominal frequency, and for a low frictional reversing error. The requirements concerning buckling safety and sufficient leeway from critical speeds usually necessitate only smaller diameters.

4.3.1.2.2 Moment of Inertia of the Feed Screw

Fast reacting drives require total moments of inertia as low as possible. For the feed screw, this means that the diameter for a given length should be selected as small as possible. This contradicts the requirement for a large diameter, necessary for high feed screw stiffness. The two demands are to be balanced according to the economical point of view that applies. Hereby, the stiffness of the feed screw can also be influenced by the type of bearing used, according to fig. 4.19 (see the example given under section 4.3.1.2.1). Other criteria for the diameter of the feed screw are the critical speed, as in section 4.3.1.2.6, and the buckling strength, according to section 4.3.1.2.7.

The basic relationships necessary for the calculation and conversion of moments of inertia are given in section 4.2.5. The effects of moments of inertia that are too high on the dynamic of the feed drive, can be offset with a gear ratio. The optimizing procedures for this are given in [4.16] and [4.17].

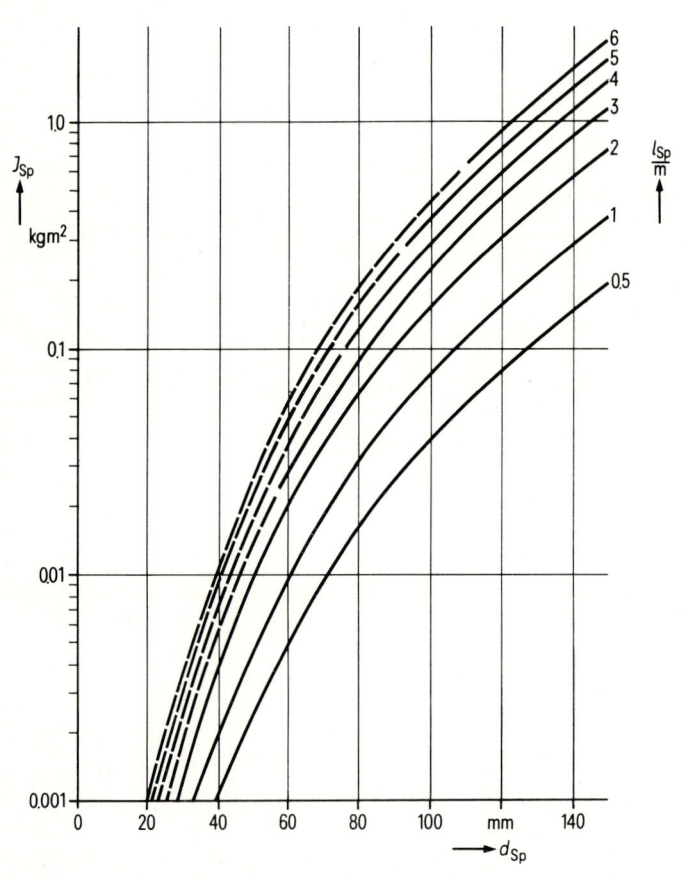

Fig. 4.22
Feed screw moment of inertia, as function of the screw length l_{Sp} and diameter d_{Sp}

The feed screw moment of inertia J_{Sp} can be determined, for the most common diameters and lengths, from the diagram in fig. 4.22, where

d_{Sp} feed screw diameter
l_{Sp} total feed screw length.

The curves for the ratio $l_{Sp}/d_{Sp} > 50$ are shown in dotted lines, since this range should be avoided because of the danger of buckling.

When calculating according to fig. 4.22, the thread leads are disregarded; the moment of inertia derived is larger than the actual moment of inertia. As it is demonstrated in [4.16], depending on the lead and feed screw diameter, these deviations can amount to about 20%. However, as long as the external moment of inertia J_{ext} reflected on the motor shaft remains within the limits given in section 2.2.2.6, these deviations can be allowed. This is especially true in view of the fact, that when calculating the feed screw moment of inertia, the coupling hubs or bearing parts often cannot be determined exactly.

4.3.1.2.3 Efficiency

The efficiency of typical ball screw drives lies, as shown in fig. 4.23, between 0.8 and 0.95 (also see equation (2.28)).

Hereby,
h_{Sp} feed screw lead
d_{Sp} feed screw diameter

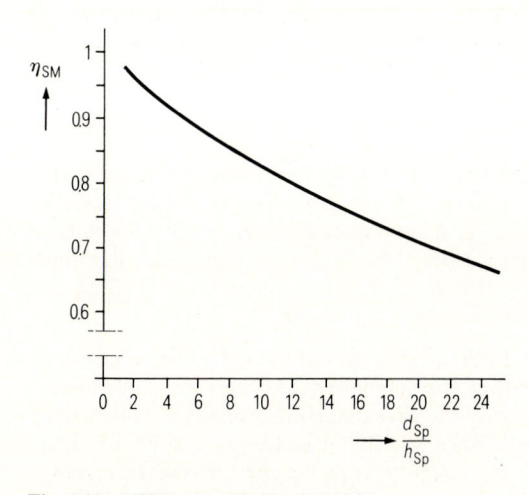

Fig. 4.23   Efficiency of ball screw drives

When ball screws are used, because of the high efficiency coefficient, the heating of the feed screw is low in the range of the feed screw nut. This is of significance in the case of indirect position control measurement. The longitudinal changes of the feed screw are not recognized by the indirect measurement system, and therefore cause position errors (longitudinal expansion of steel ≈ 12 µm/m and per degree temperature). At longitudinal expansions of type 1, according to fig. 4.19, this longitudinal change can eliminate the bearing preload, and lead to reversing errors.

Because of this, for machines with frequent rapid traverse movements, it is advisable to check the occurring losses. In the process, it should be established what amount of heat remains in the feed screw; it causes a temperature rise, and thus longitudinal expansion. The residual heat flows partially to other machine parts, and partially to the surroundings.

4.3.1.2.4 Feed Screw Lead

The feed screw lead h_{Sp}, in combination with the gear ratio i possibly used and the maximal motor speed n_{maxM}, determines the rapid traverse rate v_{Eil}. The relationships are presented in section 2.3.6, with the equations (2.46), (2.47), and (2.48).

Different optimization techniques, which have as purpose the achievement of the highest table acceleration, can for instance be found in [4.1], [4.16] and [4.17]. In these processes, the feed screw lead and gear ratio are altered. The different variations for a particular feed task are given in section 2.3.5.4.

Ball screws are manufactured, generally with no additional costs, only with certain leads. Optimization according to the stated parameters is therefore only advisable for economical considerents. Because of this, for applied feed drives the leads used are mostly the 5, 10, and 20 mm leads, as stated in the DIN 69051.

4.3.1.2.5 Lead Accuracy

Lead deviations in the thread of ball screws at indirect position measurement, cause a position error (pitch errors are not recognized by an indirect position measurement system). Generally, the pitch errors are given per 300 mm thread length, and refer to a feed screw temperature of 20 °C. The pitch deviation is cumulative, i.e. increasing over the entire length of the feed screw. The standard quality for feed drives is 0.01 mm/300 mm length.

For numerically contour controlled machine tools with direct position measurement, lead deviations that are within this standard quality category, are generally admissable. With indirect position measurement however, it is necessary to narrow this pitch error range, since it is fully effective as positioning error. In these cases, the feed screw manufacturer can be given a tolerance range, and for instance, the tolerances can be pushed into the minus direction. Preload and heating of the feed screw balance each other; the remaining error goes towards zero. (Certification requirements for ball screw drives are listed in DIN proposal 69051, part 3.)

4.3.1.2.6 Critical Speed

At speeds within the critical speed range, the feed screw goes into bending oscillations. The feed screw speeds for the rapid traverse movements in practical operations should therefore remain 20% below the critical speed. The critical speed n_{Krit} can be determined with the aid of fig. 4.24.

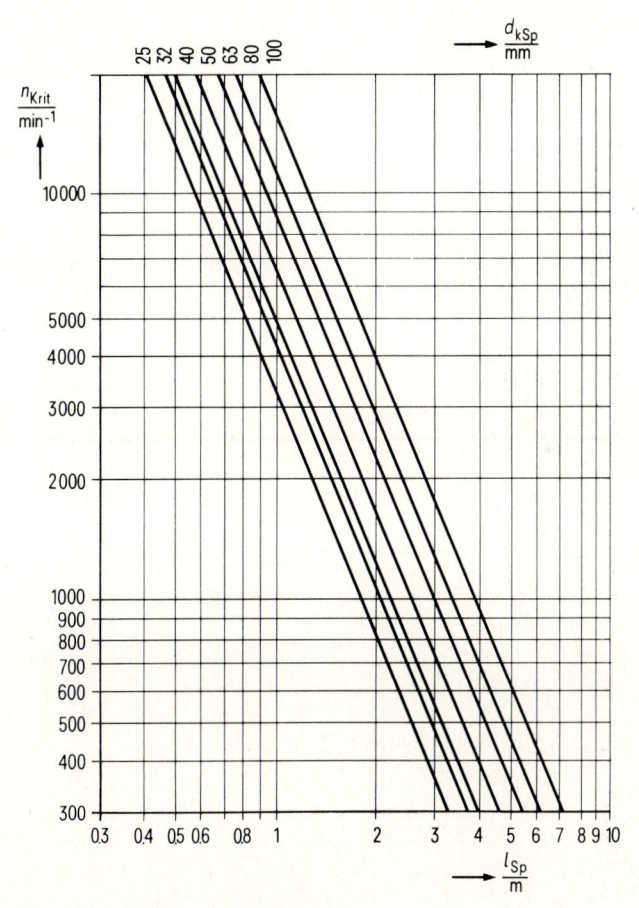

Fig. 4.24 Critical speed of the feed screw

The maximal admissable feed screw speed is:

$$n_{max\,Sp} = 0.8\, n_{Krit} \cdot b_{Krit} \tag{4.23}$$

n_{Krit} critical speed according to fig. 4.24
b_{Krit} correction factor depending on the bearing type, as in fig. 4.19
l_{Sp} feed screw length
d_{KSp} core diameter of the feed screw

For numerically controlled machines, due to the stiffness requirements, the feed screw diameters are generally so big that the critical speeds are not reached under normal circumstances. Only for a free feed screw end ($b_{krit} = 0.21$ in fig. 4.19) is there a danger that the critical speed range may be reached.

4.3.1.2.7 Buckling Strength

In the process of accelerating the machine table and machining, forces in the direction of the feed axis are generated, which must remain smaller than the admissable buckling load. Larger forces would result in buckling (bending) of the feed screw.

The maximal admissable force $F_{Kn\,sp}$ in the direction of the feed axis can be approximated with the following adjusted equation:

b_{Kn} correction factor of the buckling strength for different bearing versions, according to fig. 4.19

$$\frac{F_{KnSp}}{N} = 3.4 \cdot 10^{-2} \cdot b_{Kn} \frac{\left(\frac{d_{KSp}}{mm}\right)^4}{\left(\frac{l_{Sp}}{m}\right)^2} \tag{4.24}$$

l_{Sp} feed screw length between bearings
d_{KSp} core diameter of the feed screw

This maximal admissable force $F_{Kn\,sp}$ must remain larger than:

▷ the machining force F_{Vl} resulting from the machining process, in addition to the force for overcoming the friction, or

▷ the acceleration force F_{Vb} necessary to accelerate the table through the screw, in addition to the friction force.

For the load determined by the machining force, it must be true that

$$F_{Kn\,Sp} > F_{VL} + \frac{2\pi}{h_{Sp}} M_{RF} \tag{4.25}$$

h_{Sp} feed screw lead
F_{Vl} machining force
M_{RF} torque for friction

The acceleration force is proportional to the actual acceleration $a_{i\,max}$. For many CNC-controls, the rated acceleration can be selected through the programming, by means of a command value modifier. The maximal actual acceleration can be calculated according to the explanations in chapter 5.

For the load determined through the acceleration, it must hold that

$$F_{Kn\,Sp} > F_{VB} + \frac{2\pi}{h_{Sp}} M_{RF} = (m_T + m_W) \cdot a_{i\,max} + \frac{2\pi}{h_{Sp}} \cdot M_{RF} \qquad (4.26)$$

m_T feed table mass
m_W work piece mass
$a_{i\,max}$ maximal actual acceleration of the work piece

For velocity controlled command value generation, the set position loop gain K_v determines the occurring acceleration. Here, it must be true that

$$F_{Kn\,Sp} > F_{VB} + \frac{2\pi}{h_{Sp}} M_{RF} = (m_T + m_W) \cdot K_v \cdot v_{Eil} + \frac{2\pi}{h_{Sp}} \cdot M_{RF} \qquad (4.27)$$

K_v position loop gain
v_{Eil} rapid traverse feed rate

The product $K_v \cdot v_{Eil}$ corresponds to the maximal occurring acceleration (see also equation (3.11)).

Example:

The feed drive calculated under section 2.3.2.2 on page 101, is to be equipped with a bearing in version 2 (see fig. 4.19). The correction factor b_{Kn} is thus 4. The maximal force allowed in the direction of the feed axis with a feed screw diameter of 33 mm and a screw length of 2 m, is calculated with equation (4.24) to be:

$$F_{Kn\,Sp} = 3.4 \cdot 10^{-2} \cdot 4 \cdot \frac{33^4}{2^2} \text{ N} = 40.3 \cdot 10^3 \text{ N}$$

The check of the load due to machining, with the machining force of 20,000 N, the feed screw lead of 10 mm, and the friction torque of the table guides of 2.13 Nm, is, with equation (4.25):

$$F_{VL} + \frac{2\pi}{h_{Sp}} M_{RF} = (20{,}000 \text{ N} + \frac{2\pi}{10 \cdot 10^{-3} \text{ m}} \cdot 2.13 \text{ Nm}) = 21.3 \cdot 10^3 \text{ N}$$

The load is only about 50% of the maximum admissible force. A bearing version as in 3 or 4 in fig. 4.19 would, however, not be possible.

According to equation (4.26), the dynamic load for an acceleration of $a_{i\,max} = 1 \text{ m/s}^2$ is:

$$F_{VB} + \frac{2\pi}{h_{Sp}} \cdot M_{RF} = 1{,}500 \text{ kg} \cdot 1 \text{ m/s}^2 + 1{,}330 \text{ N} = 2.83 \cdot 10^3 \text{ N}$$

it is thus way below the admissable force $F_{Kn\,Sp}$.

For the velocity controlled position command value generation, according to equation (4.27), for a setting of the position loop gain of $K_v = 16.6 \text{ s}^{-1}$ $\left(1\frac{\text{mm/min}}{\mu\text{m}}\right)$ and the rapid traverse feed rate of 10 m/min, the load due to acceleration becomes:

$$F_{VB} + \frac{2\pi}{h_{Sp}} \cdot M_{RF} = 1{,}500 \text{ kg} \cdot 16.6 \text{ s}^{-1} \frac{10 \text{ m/min}}{60 \text{ s/1 min}} + 1{,}330 \text{ N} = 5.5 \cdot 10^3 \text{ N}$$

The maximal acceleration occurring is

$$a_{i\,\text{max}} = 16.6 \text{ s}^{-1} \cdot \frac{10 \text{ m}}{60 \text{ s}} = 2.77 \text{ m/s}^2$$

Here also, the difference up to the calculated buckling force is very large, but it should be recognized that the load of the feed drive without command value modifier would be considerably higher.

4.3.1.3 Ball Screw Nut

Three versions are used for ball screw nuts (see fig. 4.25):

▷ the reversing pipe system   (advantage: shock-free run)
▷ the axial reversing system   (advantage: no protruding parts)
▷ the reversing piece system   (advantage: economical)

In the *reversing pipe system,* the balls make several turns around the feed screw. Several of these ball circuits can be assembled side by side. A reversing pipe belongs to each of these circuits, and it returns the balls to the start point.

Reversing pipe system   Axial reversing system   Reversing piece system

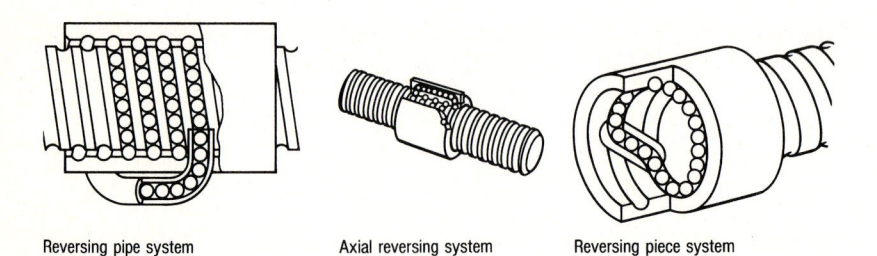

Fig. 4.25
Ball screw nut versions
(figures courtesy of Warner Electric and Kugellager GmbH)

The number of turns and of ball circuits is listed in the data sheets of the ball screw drive. The higher this number is, the higher the loadability and the stiffness of the feed screw nut will be; the friction between feed screw and screw nut will however, also increase in the same proportion. When the number of balls is too high within a circuit, the bearing and separation balls should be arranged alternately; the separation balls have smaller diameters and reduce the friction. If the balls are fed-in cleanly tangentially, a shock-free, uniform run can be achieved. Thus this system allows high feed screw speeds and accuracy.

The *axial reversing system* has the ball feedback within the housing. Because of the necessarily heavy reversing, the uniformity of the ball movement is affected. Space efficiency can be accomplished by designing the ball feedback channel externally as a key.

With the *reversing piece systems,* immediately after a looping, the balls are fed back through the reversing piece to the starting point. For this reason, it is necessary to have at least two offset loops on the circumference, in order that the entire circumference of the feed screw be covered. This screw nut has smaller dimensions and is easier to manufacture.

Because the feed screw, the feed screw nut, and balls cannot be manufactured completely play-free, for the application on a feed drive there must be a possibility for preload. This is accomplished by splitting the nut in two halves, thus creating a double nut. With a preload force as the one used on bearings (section 4.3.1.4), the axial play can be reduced and the stiffness increased. According to fig. 4.26, this can be achieved by either expansion or contraction of two nut halves.

Preload directed outward

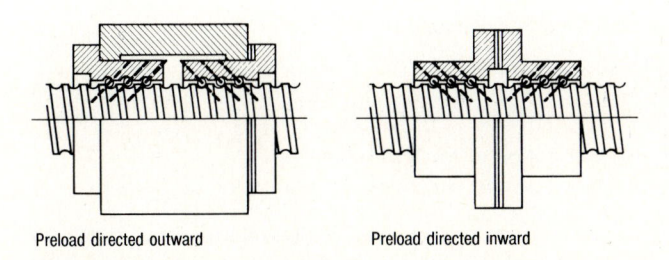

Preload directed inward

Fig. 4.26
Possibilities for generating the preload in the feed screw nut
(figure courtesy of Warner Electric)

When expanding two feed screw nut halves, the preload of the screw nut is achieved by using thicker spacing disks. (This is the so-called O-preload, fig. 4.26, left.) Hereby, the screw is exposed to a pulling tension, since the preload of the double nut is directed outwards. The expansion of the two feed screw nut halves is particularly advisable, since it hereby prevents deformations of the nut.

When contracting two feed screw nut halves, the preload is produced by using thinner disks. (This is known as X-preload, fig. 4.26, right.) The screw is under a contracting force. If, in the machining process, the feed screw nut heats up faster than the feed screw, the preload decreases.

In a different mechanical design, the two feed screw nut halves are twisted together, and are fixed in the preload position either through dowel pins in the frame or with a safety ring.

The stiffness of the feed screw nut can be almost doubled through preloading, as compared with the not preloaded nut and depending on the profile of the ball orbit. The preload force here is approximately $1/3$ of the mean axial load of the screw nut. Lower preload values reduce the stiffness, and higher ones increase the friction; in both cases the reversing error is increased. Too much preload causes, in addition, indentations in the thread leads, and results in flaking of the ball surface; the lifetime is thus reduced. The maximum preload forces as stated by the feed screw manufacturer should therefore be adhered to.

The stiffness values k_M of the preloaded ball screw nut unit move, depending on the feed screw diameter, preload force, and the number of bearing thread leads, within the following limits:

d_{Sp} = 16 mm ... 40 mm ... 63 mm ... 100 mm
$k_M \approx$ 350 N/μm ... 900 N/μm ... 1,400 N/μm ... 2,200 N/μm

With the help of diagrams 1 and 2 in fig. 4.27, the axial shift s_{SpM} between feed screw and ball screw nut can be approximated closely. With its aid, the stiffness k_M of the ball nut unit can be approximated.

The stiffness of the nut units is calculated from:

$$k_M = \frac{F_{aSp}}{\Delta s_{SpM}} \qquad (4.28)$$

F_{aSp} axial force in the direction of the feed screw
Δs_{SpM} shift between feed screw and ball screw nut unit.

From diagram 1 we can derive the factor b_{St}, with which the Δs_{SpM}-scale of diagram 2 can be expanded. Then, we can read out of diagram 2 the axial shift between screw nut and feed screw, dependent on the external axial force F_{aSp} and the axial preload force F_{aVM} of the ball screw nut unit, and we can calculate the stiffness with equation (4.28).

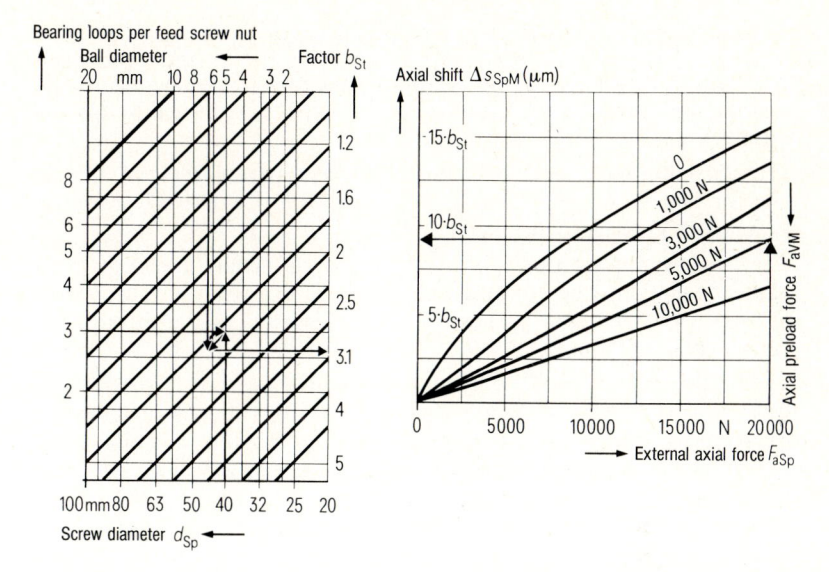

Fig. 4.27
Diagram for determining the axial shift Δs_{SpM} between screw and ball screw nut unit (figure courtesy of Eisenmann and Hommes)

Example:

For a feed screw diameter of 40 mm, a ball diameter of 6.35 mm, and 3 loops of a ball per nut element, the factor $b_{St} \approx 3$ (diagram 1 of fig. 4.27).

With a force of $F_{aSp} = 20{,}000$ N in the axial direction of the screw, and a preload force of $F_{aVM} = 5{,}000$ N, the axial shift becomes

$$\Delta s_{SpM} \approx 9 \cdot 3 \; \mu m = 27 \; \mu m.$$

For the stiffness of the screw nut unit, we obtain with equation (4.28) a value of

$$k_M = \frac{20{,}000 \text{ N}}{27 \; \mu m} = 740 \text{ N}/\mu m.$$

4.3.1.4 Feed Screw Bearings

The bearings of feed screws have to contribute to the accuracy and dependability of a feed drive, by having high load capability, high stiffness, and low axial play, as well as few maintenance demands. Nowadays, completely assembled combined axial radial bearings are available which satisfy all these requirements. It must however, be taken into consideration that a stable support for the rings on the feed screw as well as of the bearing block must be provided.

223

For the bearings of a feed screw, the following points must be checked:

▷ The basic arrangement of the bearings (on both or just one side of the feed screw) affects the total stiffness of the screw system significantly (see fig. 4.19).

▷ The axial stiffness k_{aL} of the individual bearings is part of the total stiffness, and for this reason, should be as large as possible. A bearing preloading is necessary in order to increase the stiffness.

▷ The bearing mounting parts are to be treated as serially mounted spring elements, and must be designed with corresponding stiffness. They should have as few intersections as possible; their elasticity adds to the bearing's deformation.

▷ The bearing friction is part of the total friction of the feed drive. It should be as low as possible, and only slightly tied proportionally to the speed. The bearing friction causes heating of the bearing, and thus also of the feed screw. This results in longitudinal expansions which can eliminate the preload.

4.3.1.4.1 Stiffness and Preload

Cylindrical roller bearings are to be preferred for use as axial bearings, because their stiffness values in the diameter range $20 \cdots 100$ mm are approximately 3 to 6 times higher than the stiffness values of axial ball bearings. The axial bearings for a numerically controlled machine tool, depending on the diameters, should have the following stiffness values:

Feed screw diameter	20 mm	...	100 mm
Axial stiffness	1,500 N/µm	...	7,000 N/µm

Stiffness data are listed in the catalogs of the manufacturer. Figure 4.28 shows two versions of feed screw bearings.

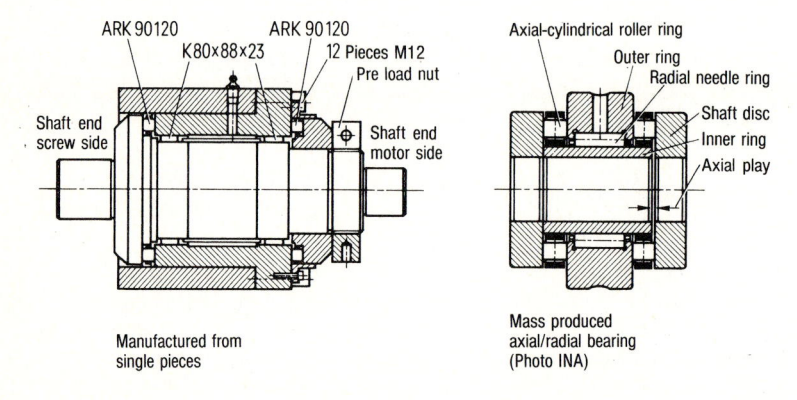

Fig. 4.28 Examples of combined axial-radial bearing (left, see [4.1])

Each bearing has manufacturing tolerances and axial play. The spring characteristic curve in the beginning section is not linear. This range's higher elasticity is avoided through a preload force, and the additional load created achieves a smaller elasticity distance. If two bearings are preloaded against each other, the two spring curves, as it is shown in fig. 4.29, can be presented graphically in a mirror image diagram. The point of intersection is determined by the preload force $F_{a\,VL}$.

The preload force $F_{a\,VL}$ is to be selected so, that the operationally occurring load $F_{a\,Sp}$ added to a safety measure, is just below the point where it would lead to a relaxation of the bearing. This maximal force in the direction of the screw is generally the maximal machining force F_{VL} with the addition of the force necessary to overcome friction. As calculated in the example on page 219, the axial force determined by the acceleration is normally much lower.

In order not to allow the friction to become too large as a result of large preload forces, and thus affect the lifetime of the bearing, at high feed forces one side of the bearing is to be allowed to relax. The thereby produced reduction in bearing stiffness can be afforded.

Figure 4.29 shows the effect of the preload on the spring distance Δs_{aL}. An external force $F_{a\,Sp\,1}$ causes the spring distance $\Delta s_{aL\,1}$. If, by comparison, an unloaded bearing is considered, we obtain for the same force approximately twice the spring distance. As long as the preload is not eliminated ($F_{a\,Sp} < F_{max\,a\,Sp}$), the resulting stiffness of the preloaded double bearing is about twice as great as that of the single bearing.

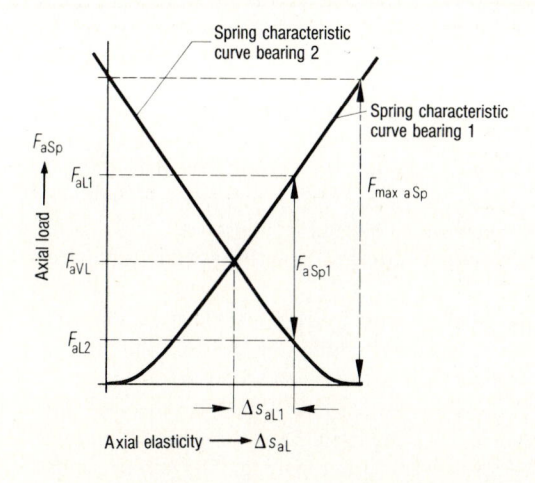

Fig. 4.29 Characteristic curves of two preloaded axial bearings

The catalogs of the manufacturers list the data for bearing stiffnesses. The resulting values for combined axial-radial bearings are given under preload. The axial elasticity Δs_{aL} can be determined from the values for the bearing stiffness k_{aL} and the active force F_{aSp}, to:

$$\Delta s_{aL} = \frac{F_{aSp}}{k_{aL}} \tag{4.29}$$

From the diagram in fig. 4.29, the partial forces in the axial direction $F_{aL\,1}$ and $F_{aL\,2}$ can be determined; these values are necessary for the calculation of the lifetime of the roller bearings.

In addition to the above mentioned doubling of the stiffness of preloaded bearings, fig. 4.19 also shows the parallel effect of two screw bearings in bearing version 2. The individual stiffness of a bearing contributes fourfold value to the total stiffness.

It must be taken into account that for double sided screw bearings of version 1 in fig. 4.19, a bearing preload is only possible through screw preload. The screw preload is simultaneously the bearing preload force, and must be considered when dimensioning the bearings. The screw preload can however, be kept at low values in version 2, so that it does not significantly affect the bearings.

4.3.1.4.2 Influence of the Bearing Mounting Elements

In order to narrow the gap between the theoretical and the practical achievable value for the bearing stiffness, it is necessary to carefully design and manufacture both the bearing and the bearing mountings.

The following guide lines can be given for the design and manufacturing of bearings and bearing mountings:

▷ As few as possible intersection areas should be allowed in the design. Each contact surface area represents an additional spring, because the roughness peaks must be crushed, until the full area bears. For this reason, in fig. 4.28, for instance, the bearings are axial-cylindrical roller bearings without separation roller disks.

▷ The roller areas of the roller bearings must overlap and must exhibit high plane accuracy, since otherwise, when preloading the bearing, the preload force and the load will not be equally distributed between the rolls.

▷ The bearing mounting parts must be constructed so, that they can be preloaded significantly more than the bearing itself. They must be stiff enough to keep the bearing from being skewed during loading. The stiffness must be uniform in both directions, as much as possible.

▷ The mounting areas of the bearing must be parallel to the mounting area of the place of mounting. The areas themselves ought to be as smooth as possible, so that the bearings have a large attaching surface. The shaft disks and the outer rings must reach at least up to the middle circle of the roller.

▷ The screws for mounting the bearing must be calculated not only for sufficient strength, but also for sufficient stiffness. Measurements on the mounting parts of functional machines show that often the screws display insufficient stiffness. When loading, relaxations due to the machining force must be avoided through preloads.

Further references about the subject can be found in the literature [4.1] and [4.18]. The total stiffness of a bearing should be, in the feed screw diameter range of 20 mm to 100 mm, between 500 N/μm and 2,500 N/μm.

4.3.1.4.3 Friction and Temperature Behavior

The total friction torque of axial-cylindrical roller bearings is the sum of the idle and load friction torques. It consists of:

▷ Roll friction, the load being the determining factor.
▷ Friction of the cage and the roller body, for which the determining factor is the load and its direction, as well as the speed, the lubrication, and wear state.
▷ Liquid friction, which depends on the type of bearing, speed, amount and viscosity of the lubricant.

Under circumstances, we have in addition the friction of the bearing seal. Because of the different influencing variables, calculations can be approximated only for one particular state of operation at a time.

The lubricant and its viscosity have a significant influence. A recirculating oil lubrication is most favorable for a feed screw bearing, because the heat generated through friction is removed, and can be radiated without harm to the surrounds through the oil reservoir. This prevents heat expansions of nearby machine parts. An oil mist lubrication is also advantageous for the temperature behavior, since only small amounts of lubricant stay on the bearing, thus reducing the liquid friction. It is suitable for high speed bearings with low loads. If high speeds occur only occasionally (rapid traverse movements), an oil bath lubrication is also sufficient. Lubrication, type as well as amount, should be determined in cooperation with the manufacturer of the bearings.

The friction torque can be roughly determined according to equation (2.27). The power loss thus calculated must be removed from the bearing.

The power loss amounts to:

$$\frac{P_{RSL}}{kW} = 0{,}1047 \cdot 10^{-3} \frac{M_{RSL}}{Nm} \cdot \frac{n_{Sp}}{\min^{-1}} \qquad (4.30)$$

The temperature rise in the bearing is very difficult to estimate in advance, since temperature resistances can be known only inaccurately. The heat is transmitted to the feed screw, the bearing mounting parts, and to the lubrication medium.

Figure 4.30 shows the course of the frictional factor μ_{SL} of an axial-cylindrical roller bearing, in dependence on the load ratio: dynamic load rating to equivalent bearing load. The parameter is the cinematic viscosity of the lubricant.

For an axial roller bearing, the mean value for the frictional factor can be taken between 0.003 and 0.005. At low loads, where liquid friction predominates, the frictional factor increases to 0.006-0.009.

At given loads and dimensions, the bearing friction can only be influenced by the viscosity of the lubricant. According to the studies presented in [4.19], the cinematic viscosity for axial-cylindrical roller bearings should not get below about 20 mm$^2 \cdot$s^{-1}, if mixture friction is to be avoided. In these studies, when the viscosities were too low and the bearing was loaded, a negative slope of the characteristic curve for the friction torque as function of speed, was observed over the entire range of speeds investigated (compare fig. 4.7).

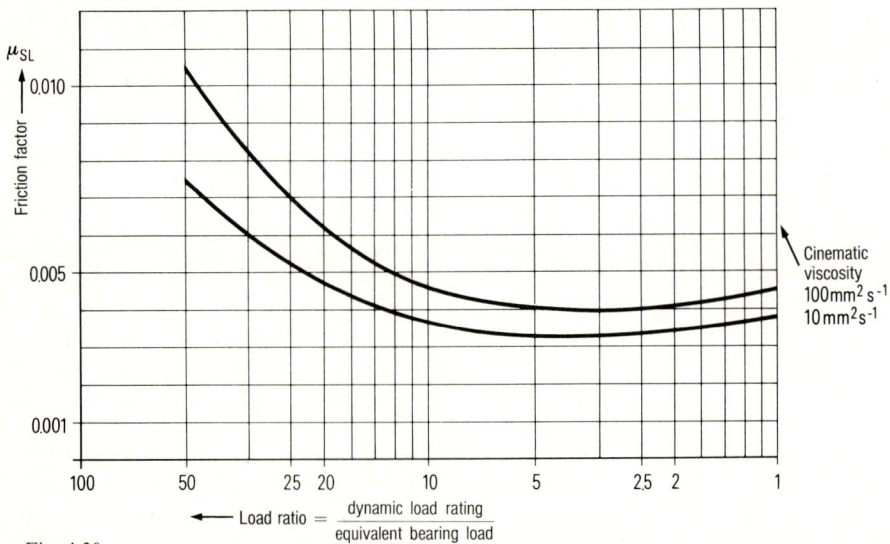

Fig. 4.30
Friction factors for axial cylindrical roller bearings (Figure courtesy of INA)

Lubricant viscosities that are too high increase the bearing friction, because of the high hydrodynamic losses. Data concerning recommended lubricants can be found in the catalogs of bearing manufactures. These give values for the cinematic viscosity according to load ratio, speed, and bearing diameter.

Grease lubrication is also a possibility, but it excludes the removal of heat through the lubricant. The maximal achievable limit speeds can only be 25% of those that can be reached with oil lubricants. The load ratio hereby presumed is 10. It is recommended that information should be obtained from the bearing manufacturers and the distributors of lubricants.

4.3.2 Rack and Pinion Drives

4.3.2.1 Range of Applications, Influencing Variables

For feed ranges of up to approximately four meters, it is possible to obtain higher stiffness and thus a higher mechanical nominal frequency, by using a ball screw drive instead of a rack and pinion drive. For longer ranges, the feed screw diameter must be increased to improve the stiffness. This results in an increase of the external moment of inertia. By comparison, the moment of inertia of a rack and pinion drive at an equal feed length can be reduced more, if the necessary gear between the drive motor and the rack is a multistage version, and large wheel diameters can be avoided in the first stages. (Further explanations about gears are presented in section 4.4.)

The total stiffness of the rack and pinion drive is independent of the travel length. The following individual stiffnesses, however, have an influence on the total stiffness:

▷ Torsional stiffness of the pinion and gear shafts
▷ Bending strength of the pinion and gear shafts
▷ Radial stiffness of the shaft bearings
▷ Bending strength of the gear, pinion, and rack teeth.

These individual stiffnesses are included and converted in a substitute stiffness acting in the feed drive direction of movement.

A generally applicable formula for the total stiffness of a rack and pinion drive, similar to that for a feed screw drive, cannot be devised, because the mechanical design introduces too many variables for such an equation. It can be said, however, that generally the total stiffness of a rack and pinion drive results from the serial connection of springs with the individual stiffnesses stated above. According to fig. 4.19, the inverse value of the total stiffness results from the sum of the inverse individual stiffness values (see the example given in section 4. 3. 2. 2. 5 on page 233). The nominal frequency of the mechanical transmission elements $\omega_{0\,mech}$ can then be calculated by using equation (4.1).

For these reasons, rack and pinion drives are to be found mostly on machines with long travel ranges. Random lengths of feed distances can be generated by adding rack elements. Also small, dynamically demanding feed drives, as for instance in nibbling machines, can be partially equipped with rack and pinion drives, to maintain a small external moment of inertia reflected onto the motor shaft.

4.3.2.2 Stiffness

4.3.2.2.1 Torsional Stiffness of Shafts

Torsional stiffness of a shaft, according to equation (4.17)

$$k_{To} = \frac{\pi}{32} \cdot \frac{G \cdot d^4}{l} \qquad (4.31)$$

G shearing modulus (steel $\approx 8 \cdot 10^{10}$ N/m²)
d shaft diameter
l free length of the shaft (distance between torque attack points)

Converting the torsional stiffness into a substitute stiffness in the direction of the feed drive motion, we obtain:

$$k_{ers\,To} = k_{To} \cdot \frac{i_n^2}{r^2} \qquad (4.32)$$

r radius of the gear wheel on the torsioned shaft
i_n gear ratio between the location of the torsion origin, and the rack and pinion

From this, according to table 4.2 on page 190, one can see that the reversing error due to torsion or gear play has more effect on the length travel, the closer the corresponding component in the gear train is to the rack. It also can be observed that a smaller gear wheel diameter results in more favorable stiffness values.

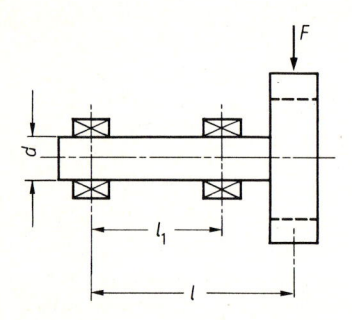

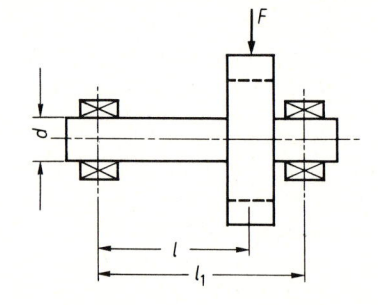

Fig. 4.31 Bending strengths of shafts

4.3.2.2.2 Bending Strength of Shafts

The bending strength of shafts depends on their bearing, and on the point of attack of the force. Figure 4.31 illustrates both possibilities.

The bending strength for a point of attack of the force outside the bearing (fig. 4.32, left) is:

$$k_B = \frac{3 \cdot \pi}{64} \frac{E \cdot d^4}{(l-l_1)^2 \cdot l} \tag{4.33}$$

and for a point of attack of the force between the bearings (fig. 4.31, right):

$$k_B = \frac{3 \cdot \pi}{64} \frac{E \cdot d^4 \cdot l_1}{(l_1-l)^2 \cdot l^2} \tag{4.34}$$

E elasticity modulus (steel $\approx 2.1 \cdot 10^{11}$ N/m²)
d shaft diameter
l, l_1 length measurements on the shaft according to fig. 4.31

These stiffness values, depending on the teeth format, are to be converted into a longitudinal stiffness. The determining factor is the pressure angle α_0, which for involuted gear teeth is 20° (DIN 867). The substitute bending strength in reference to the direction of the table movement is approximately:

$$k_{ersB} \approx k_B \frac{i_n^2}{\tan^2 \alpha_0} \tag{4.35}$$

i_n gear ratio between location of the bending origin, and the rack and pinion
α_0 pressure angle

4.3.2.2.3 Radial Stiffness of Shaft Bearings

The bearing play and the non-linear starting stiffness can be reduced through preloading in the radial direction too. Angular ball bearings or, for heavier loads, tapered roller bearings can be used (e.g. on the pinion of the rack). Single row angular ball bearings or tapered roller bearings can only accept forces in one direction. Because of this, for gear shafts with preloaded bearings, two bearings acting in opposition must be installed either in X-arrangement (the shaft is exposed to pressure), or in O-arrangement (the shaft is exposed to tension).

Such preloadings must be dimensioned exactly, and executed carefully. The heat expansion of the gear shaft changes the preload. If, in the X-arrangement, the shaft heats up and the heat is removed through the outer ring, during the operation this will lead to a narrowing of the play. If the heat flow is inversed, the play increases. For O-arrangements, these relationships are reversed.

According to [4.13], it is necessary that the stiffness of each bearing be determined separately. The following values can be taken (data from factual literature of SKF for play-free adjustment) as approximate values for the radial stiffness:

Diameter range:		30 mm	... 100 mm
Ball bearing:	k_{rL}	80 N/μm	... 150 N/μm
Tapered roller bearing:	k_{rL}	300 N/μm	... 900 N/μm
Cylindrical roller bearing:	k_{rL}	1,500 N/μm	... 6,000 N/μm.

According to fig. 4.31, depending on the point of attack of the force, the resulting radial bearing stiffness of the doubly supported shaft will be:

Force outside the bearing points (fig. 4.31, left)

$$k_{resrL} = k_{rL} \frac{l_1^2}{l_1^2 + 2l \cdot (l - l_1)} \tag{4.36}$$

Force between the bearings (fig. 4.31, right)

$$k_{resrL} = k_{rL} \frac{l_1^2}{l_1^2 - 2l(l_1 - l)} \tag{4.37}$$

This resulting stiffness must be converted, according to the bending strength, into a longitudinal reference stiffness. The radial stiffness of the shaft bearing in reference to the direction of the table motion is, according to equation (4.35):

$$k_{ersresrL} \approx k_{resrL} \cdot \frac{i_n^2}{\tan^2 \alpha_0} \tag{4.38}$$

If $l_1 = 2l$, it can be seen that the resulting stiffness for an attack point of the force in the middle of the bearings is the highest possible, i.e. $2k_{rL}$. For this reason, it may be advantageous, that when the machining forces are high, the rack and pinion be mounted between the bearings. For overhanging arrangements, too large an overhang should be prevented.

Further methods for calculating bearing stiffness can be found, e.g. in [4.19]. The manufacturers' catalogs and [4.20] also offer further information in this direction. Depending on loading forces and preloads, different values can result for the stiffness.

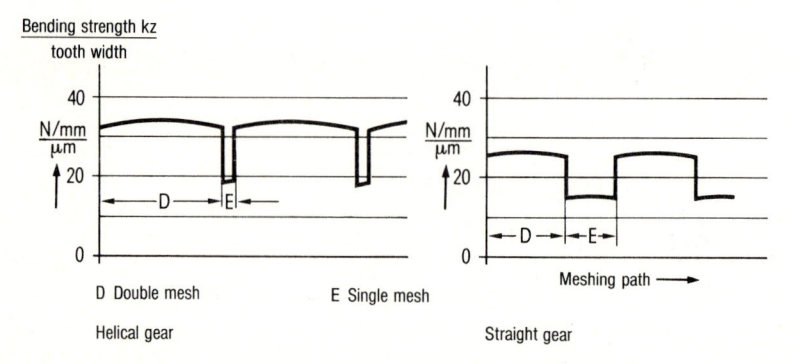

Fig. 4.32 Reference bending strengths of double pinions

4.3.2.2.4 Bending Strength of the Gear Wheel, Pinion, and Rack Teeth

The bending strength of a teeth pair is not constant but changes over the meshing area, and with the number of simultaneously meshing teeth; it is also dependent on the type of gear used. With a meshing larger than two, as it is possible with a helical gear, one can generally prevent an unsteady stiffness curve. The bending strength that changes over the meshing area can, under circumstances, cause self-excited torsional oscillations in the gear, and thus the related noises.

Gears with helical gear wheels have higher bending strengths than those with straight teeth. Figure 4.32 gives approximate values for bending strengths, in reference to the tooth width of a teeth pair [4.21].

The value taken from fig. 4.32 and multiplied by the tooth width, is a stiffness k_z, which is referenced to the pitch circle circumference of the teeth pair. For wheels within the gear, this value is multiplied by the square of the gear ratio i_n^2, in order to maintain the stiffness referenced on the table movement.

If computations regarding the torsion are to be conducted, then the bending strength of the teeth must be converted into a corresponding torsional stiffness. From equation (4.32) we obtain the torsional stiffness:

$$k_{To} = k_z \cdot r^2 \tag{4.39}$$

r gear wheel radius
k_z tooth stiffness on the circumference

4.3.2.2.5 Example

The stiffness and the mechanical nominal frequency are to be evaluated for the gear shown in fig. 4.33. The dimensions are given in the drawing; the first three gear stages and the necessary preload of the pinion should be disregarded (see section 4.3.2.3).

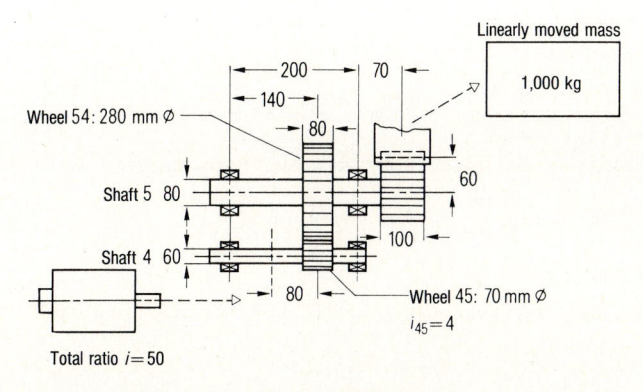

Fig. 4.33   Example of a rack and pinion drive

Torsion of the Pinion Shaft

The length 1 is the distance between the two pinions. According to equation (4.31):

$$k_{\text{To 5}} = \frac{\pi}{32} \frac{8 \cdot 10^{10} \text{ N/m}^2 \cdot (8 \cdot 10^{-2} \text{ m})^4}{1.3 \cdot 10^{-1} \text{ m}} = 2.47 \cdot 10^6 \text{ Nm}$$

Conversion into the stiffness in the direction of the table movement, according to equation (4.32), results in:

$$k_{\text{ers To 5}} = 2.47 \cdot 10^6 \text{ Nm} \frac{1}{(6 \cdot 10^{-2} \text{ m})^2} = 6.86 \cdot 10^8 \text{ N/m} = 6.86 \cdot 10^2 \text{ N/}\mu\text{m}$$

Torsion of Shaft 4:

The length of the shaft is here 80 mm. We obtain:

$$k_{\text{To 4}} = \frac{\pi}{32} \cdot \frac{8 \cdot 10^{10} \text{ N/m}^2 (5 \cdot 10^{-2} \text{ m})^4}{8 \cdot 10^{-2} \text{ m}} = 6.13 \cdot 10^5 \text{ Nm}$$

Converted on the pinion shaft and into the direction of the table motion, we get

$$k_{\text{ers To 4}} = 6.13 \cdot 10^5 \text{ Nm} \frac{16}{(35 \cdot 10^{-3} \text{ m})^2} = 8.0 \cdot 10^9 \text{ N/m} = 80 \cdot 10^2 \text{ N/}\mu\text{m}$$

Bending Strength of the Pinion Shaft:

According to equation (4.33), this will be:

$$k_{\text{B 5}} = \frac{3\pi}{64} 2.1 \cdot 10^{11} \frac{\text{N}}{\text{m}^2} \cdot \frac{(8 \cdot 10^{-2} \text{ m})^4}{(7 \cdot 10^{-2} \text{ m})^2 \cdot 2.7 \cdot 10^{-1} \text{ m}} = 9.56 \cdot 10^8 \frac{\text{N}}{\text{m}} = 9.56 \cdot 10^2 \frac{\text{N}}{\mu\text{m}}$$

Converted into the direction of the table motion with $\alpha_0 = 20°$, according to equation (4.35) we obtain:

$$k_{\text{ers B 5}} \approx 9.56 \cdot 10^{-2} \text{ N/}\mu\text{m} \cdot \frac{1}{0.364^2} = 72.2 \cdot 10^{-2} \text{ N/}\mu\text{m}$$

Bending Strength of Shaft 4:

According to equation (4.34):

$$k_{\text{B 4}} = \frac{3\pi}{64} \frac{2.1 \cdot 10^{11} \text{ N/m}^2 \cdot (6 \cdot 10^{-2} \text{ m})^4 \cdot 2 \cdot 10^{-1} \text{ m}}{(6 \cdot 10^{-2} \text{ m})^2 \cdot (1.4 \cdot 10^{-1} \text{ m})^2} = 11.34 \cdot 10^8 \frac{\text{N}}{\text{m}} = 11.34 \cdot 10^2 \frac{\text{N}}{\mu\text{m}}$$

Converted in the direction of the table motion according to (4.35)

$$k_{\text{ers B 4}} = 11.34 \cdot 10^2 \text{ N/}\mu\text{m} \frac{16}{0.364^2} = 1370 \cdot 10^2 \text{ N/}\mu\text{m}$$

Bearing Stiffness of Shaft 5:

The bearing stiffness with which we calculate for shaft 5 is 900 N/μm. We obtain, with equation (4.36):

$$k_{\text{res r L 5}} = 900 \text{ N/}\mu\text{m} \frac{(2 \cdot 10^{-1} \text{ m})^2}{(2 \cdot 10^{-1} \text{ m})^2 + 2 \cdot 2.7 \cdot 10^{-1} \text{ m} \cdot 7 \cdot 10^{-2} \text{ m}} = 4.63 \cdot 10^2 \frac{\text{N}}{\mu\text{m}}$$

Converted in the direction of the table motion, with a pressure angle $\alpha_0 = 20°$, according to equation (4.38):

$$k_{\text{ers res rL 5}} \approx 4.63 \cdot 10^2 \text{ N/}\mu\text{m} \frac{1}{0.364^2} = 34.9 \cdot 10^2 \text{ N/}\mu\text{m}$$

Bearing Stiffness of Shaft 4:

On shaft 4, we calculate with a bearing stiffness of 400 N/μm. According to equation (4.37) we get:

$$k_{\text{res r L 4}} = 400 \text{ N/}\mu\text{m} \frac{(2 \cdot 10^{-1} \text{ m})^2}{(2 \cdot 10^{-1} \text{ m})^2 - 2 \cdot 1.4 \cdot 10^{-1} \text{ m} \cdot 6 \cdot 10^{-2} \text{ m}} = 6.9 \cdot 10^2 \frac{\text{N}}{\mu\text{m}}$$

and converted in the direction of the table motion, with (4.38):

$$k_{\text{ers res r L 4}} \approx 6.9 \cdot 10^2 \text{ N/}\mu\text{m} \frac{16}{0.364^2} = 832 \cdot 10^2 \text{ N/}\mu\text{m}$$

Stiffness of the Gear of Rack and Pinion:

From fig. 4.32, for the straight teeth we take a stiffness of approximately 20 N/μm per mm tooth width.

For the rack and pinion pairing we obtain:

$$k_{Z\text{Ri}} = 20 \frac{\text{N/mm}}{\mu\text{m}} \cdot 100 \text{ mm} = 20 \cdot 10^2 \text{ N/}\mu\text{m}$$

Stiffness of the Gear of Teeth Wheels 45/54:

The stiffness in reference to the table motion for the wheel pair 45/54 is:

$$k_{\text{ers Z 45}} = 20 \frac{\text{N/mm}}{\mu\text{m}} \cdot 80 \text{ mm} \cdot 16 = 256 \cdot 10^2 \text{ N/}\mu\text{m}$$

Total Stiffness

For the serial connection of the individual elastic elements:

$$\frac{1}{k_{\text{Ges}}} = \frac{1}{k_{\text{ers To 5}}} + \frac{1}{k_{\text{ers To 4}}} + \frac{1}{k_{\text{ers B 5}}} + \frac{1}{k_{\text{ers B 4}}} + \frac{1}{k_{\text{ers res r L 5}}} + \frac{1}{k_{\text{ers res r L 4}}} + \frac{1}{k_{Z\text{Ri}}} + \frac{1}{k_{\text{ers Z 45}}}$$

$$\frac{1}{k_{\text{Ges}}} = \left(\frac{1}{6.86} + \frac{1}{80} + \frac{1}{72.2} + \frac{1}{1370} + \frac{1}{34.9} + \frac{1}{832} + \frac{1}{20} + \frac{1}{256}\right) \cdot 10^{-2} \frac{1}{\text{N/}\mu\text{m}}$$

The total stiffness thus becomes:

$$k_{\text{ges}} = 3.9 \cdot 10^2 \text{ N/}\mu\text{m}$$

Nominal Angular Frequency:

From equation (4.1), the mechanical frequency of a mass to be moved of 1,000 kg will be:

$$\omega_{0\text{ mech}} = \sqrt{\frac{3.9 \cdot 10^2 \text{ N/}\mu\text{m}}{1,000 \text{ kg}}} = 6.23 \cdot 10^2 \text{ s}^{-1}$$

which corresponds to a nominal frequency of

$$f_{0\,mech} = \frac{6{,}23 \cdot 10^2 \text{ s}^{-1}}{2\pi} \approx 100 \text{ Hz}$$

From the example worked out here, it can be seen that the torsion stiffness of the pinion shaft and the tooth stiffness of the rack and pinion pair have the lowest values. It is also apparent that large shaft diameters and small distances between bearings are necessary, in order to obtain mechanical nominal frequencies of approximately 100 Hz. Because of this, on rack and pinion drives, the pinion and the shaft must be dimensioned appropriately.

By selecting the spatial position of the meshing point of wheel 45 on the circumference of wheel 54, one can keep at low levels either the influence of the bearing elasticity, or the bending stiffness of shaft 5 on the reversing error generated additionally between pinions 4 and 5. As shown in the drawing, the radial component of the machining force will bend upwards the shaft 5, between the bearings. Simultaneously however, because of the elasticity of the bearing, shaft 5 will also be moved downwards on the right. Depending on the magnitude of the bending and the bearing stiffness, one would mount wheel 4 to counteract, or on the same side with the force at the pinion.

4.3.2.3 Preload at the Pinion

In the previous example it was shown, that the torsion stiffness of the pinion shaft and the rack and pinion pair have a significant influence on the total stiffness. From the explanations concerning the frictional reversing error, given under section 4.2.4.2, it is clear that the total stiffness is to be given most consideration with respect to the reversing error. For large machines especially, a direct measuring system is frequently used in which the reversing error causes undesirable oscillations.

By installing a second pinion, which is mechanically preloaded to oppose the first pinion through a torque, the torsion stiffness can be raised, and the reversing error can be prevented. In such a mechanical preload, an elastic force acts in axial direction, e.g. on a shaft in the gear. Two opposite helical pinions on this shaft, through axial displacement, give an opposite torque through the two gear branches A and B on the two rack pinions. Fig. 4.34 (from [4.22]) shows a gear designed in this manner. The helical gear wheels on the following shafts bring axial forces onto the bearing positions. The bearings must be designed accordingly.

Another way to preload the rack and pinion system, is to use two motors with the appropriate gears. The motors are hereby given opposite torques at stillstand. In addition to the load torque, when turning in one direction, the driving motor must also overcome the preload torque.

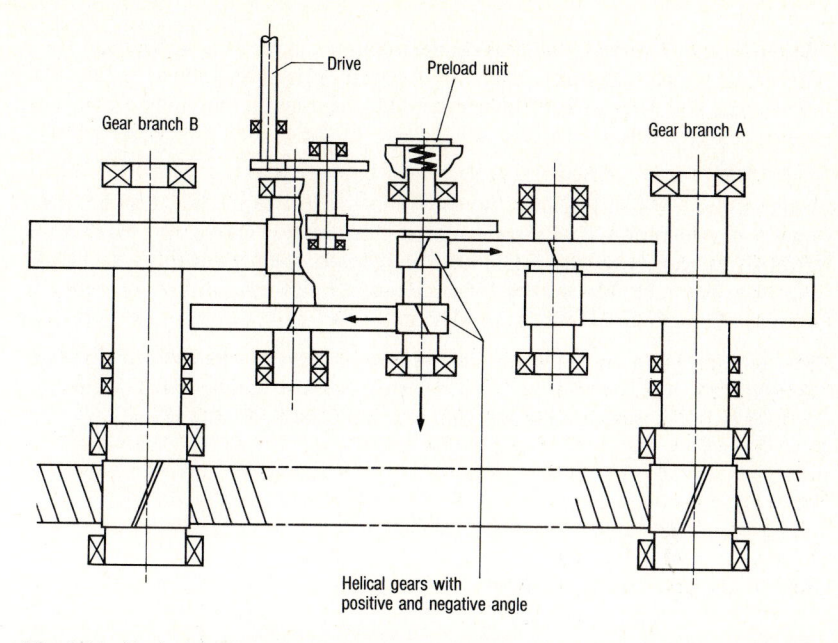

Fig. 4.34 Feed gear for a preloaded rack and pinion drive

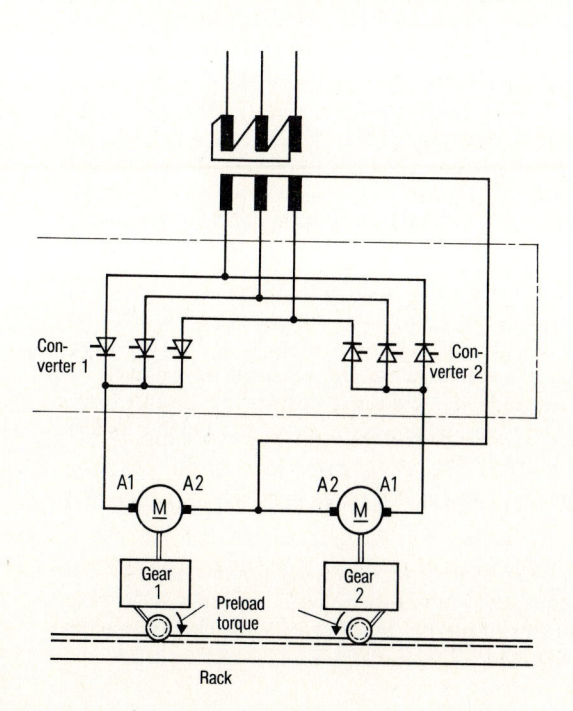

Fig. 4.35 Electrically preloaded rack and pinion drive with two motors

When using a 3-pulse circulating current-conducting anti-parallel connection, there is a simple electrical solution to this problem. As shown in fig. 4.35, the motors are switched into both output lines. The circulating current generates the opposing torques, since the connections of the motors have been selected accordingly.

When using this arrangement, the external moment of inertia must be given special consideration. The second motor must be accelerated and decelerated through its gear by the driving motor. The motors selected for this task should therefore be preferably electrically-excited servo motors with low moments of inertia. The gear should not be self-locking.

For pinions preloaded this way, the stiffness must be determined for each pinion, and when calculating the total stiffness, the pinions must be treated as working in parallel, under the consideration of preload torques.

4.3.3 Machine Slide and Guides

4.3.3.1 Machine Slide

On numerically controlled machines, high accelerations and decelerations occur frequently. The forces necessary for these act on the feed screw axis, respectively the rack. The weight force of the table, however, acts on the guideways, respectively on the vertical slides at the center of gravity. This results in a torque that skews the table. Therefore, when designing the machine table, it must be taken into account that the feed axis and rack should be located as close as possible to the acting point of the weight force. Skewing during motion results in measuring deviations, and at direct measuring systems these can cause oscillations.

For vertical axes, a compensation of the table or slide weight is possible with the use of a counterbalance. According to [4.23], for contour controlled machines the hydraulic counterbalance is the most advantageous because of dynamic reasons. When counterbalance is used, the crossrail, respectively the machine column only has to guide the slide and absorb the cutting forces. This has a positive effect on the permanent accuracy of the guides, and the feed drives can have smaller dimensions.

Further details concerning the design of the slide can be found in the literature [4.7], [4.24].

For a dynamic treatment of the feed drive, the mass of the machine table must be converted into an equivalent moment of inertia, in order to be taken into account as external moment of inertia, together with the clutch, gear and feed screw. The conversion formulas are summarized under section 4.2.5.

4.3.3.2 Table Guides

4.3.3.2.1 Requirements

The table guide has a strong influence on the accuracy of a numerically controlled machine because it determines the direction of motion of the part to be guided, and through the frictional behavior, it affects the positioning accuracy and uniformity at low feed rates. Table guides require high stiffnesses, low wear, and good frictional behavior over the entire feed rate range.

Table 4.5 Properties of table guides on machine tools

	Friction guide	Roller guide	Hydrostatic guide	Aerostatic guide
Friction and wear behavior	unfavorable, modified by material selection	favorable	very favorable	exceptionally favorable
Stick-slip danger	present	inexistent	inexistent	inexistent
Requirement for material and surface quality	very high	high	low	low
Measures for achieving high permanent accuracy	very costly	less costly	not applicable	not applicable
Stiffness	usually very good	good, if guide is preloaded and components stiff enough	variable, depending on oil supply system. High stiffness with membrane pressure valves	not so good
Damping gradient	very high but not constant	low	high, relatively easily modified by design	very low

Four guide types are mainly used in the NC machine construction:
- friction guides
- roller guides
- hydrostatic guides
- aerostatic guides

Table 4.5 (from [4.23]) gives an overview of the properties of these types of guides. Further information and references can be found in [4.7]. [4.25] contains methods of calculation for guide systems.

4.3.3.2.2 Friction Guides

The friction guide is the guide type used most frequently in machine tool construction. Figure 4.36 shows the possible design versions. Small distances between the two guideways are advantageous for the side guide (left in the figure). They make skewing less likely, and the play is reduced through thermal effects. The gibs allow play-free adjustments, and a readjustment of run-in wear through a subordinated tapered gib. The prerequisite for this however, is a uniform wear over the total length of the guideway. Countergribs prevent the slide from tilting or lifting off the bed.

Advantages of Friction Guides

By comparison to the hydrostatic and roller guides, the friction guide presents the advantage of simpler construction, high damping gradient, and high stiffness (also see [4.12]). With careful manufacturing, it can be produced more accurately than the other types of guides.

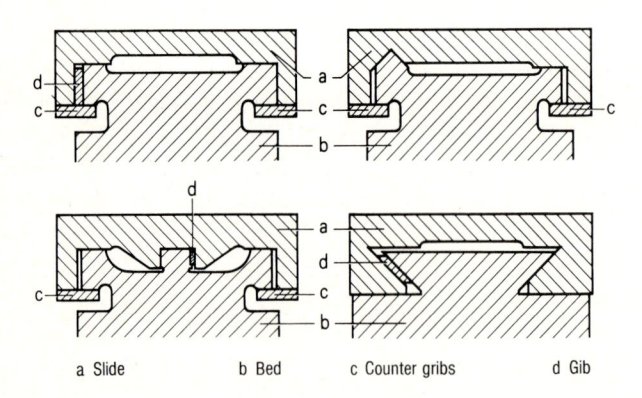

Fig. 4.36 Design versions of friction guides

a Slide   b Bed   c Counter gribs   d Gib

Disadvantages of Friction Guides

At low feed rates, on friction guides there generally exists a limit or mixed friction (which represents a danger of stick-slip effects). The fact that at low feed rates the hydrostatic lubrication film is missing, can cause wear phenomena and surface damage, due to contact between the material of both friction surfaces. Besides the frictional wear, when there is unfavorable material pairing or at extreme loads, freeze wear can occur, and it can lead to loss of the entire guide.

The friction work generated by the shifting of the slide surfaces is converted into heat. The slides lose shape, and thus lead to inaccuracies in the work piece. The guideways must be covered, since intrusion of foreign particles into the guides must be prevented under all circumstances. A sufficient amount of lubrication must also ensure that the drive is completely lifted off the guideway.

The friction, wear, and the tendency towards stick-slip effects can be reduced for the friction guide with the following measures:

> suitable material pairing
> good surface quality of the guides
> addition of molybdenum disulfide or graphite to the lubricant

The cast iron/steel material pairings used in the past have been replaced by plastic coated guide surfaces, which glide on polished steel areas. This combination of materials has very good sliding properties, with no tendency toward stick-slip effects. The plastics used have as a basis polytetrafluorethylene (teflon), and are maintenance-free. They can operate without lubrication, and have very low wear values. One can count on frictional factors μ_F of about $0.05 \cdots 0.1$.

Further references concerning friction guides are found in [4.7] and [4.26].

4.3.3.2.3 Roller Guides

In order to attain high permanent accuracy on a numerically controlled machine tool through uniform, low frictional values, roller guides are frequently used [4.27]. In these cases, tractor rollers with reciprocating roller bodies are generally used. Besides the quality of the guiding surface, determining factors for the accuracy are the dimension and form accuracy, and the guide of the roller body. During motion, due to the varying number of bearing roller bodies, an alternating stiffness and a variation in the height of the support occurs. A soft, shock-free run can be achieved through the special design of the run-in and run-out zones.

In order to increase the stiffness, roller guides are preloaded like the roller bearing. The roller elements can be adjusted and individually preloaded with gibs or excentric adjustment devices. The guideway consists of hardened and polished steel rails which are clamped, glued, or screwed onto the machine bed. Figure 4.37 presents four design versions of roller guides.

Advantages of Roller Guides

By comparison to friction guides, they show only slight differences between stick and slide friction, and therefore no stick-slip effect can occur. The low static friction, by comparison to the friction guide, means a reduction in the frictional reversing error, and thus a reduction in contour deviations. For the same masses, smaller feed motors can be used.

Disadvantages of Roller Guides

The low friction value of roller guides can lead, because of the associated very low damping gradient, to disturbance phenomena (chatter), if the contour controlled machine must, during motion, absorb periodically alternating machining forces, and if the guide or the slide drive have low stiffness or play. The following are ways of increasing the damping gradient:

    Mixed friction-roller guide.
    Heavier preload of the roller guide (limited possibility).
    A stiffer layout of the force transmission elements (ball screw, feed screw bearings, etc.).

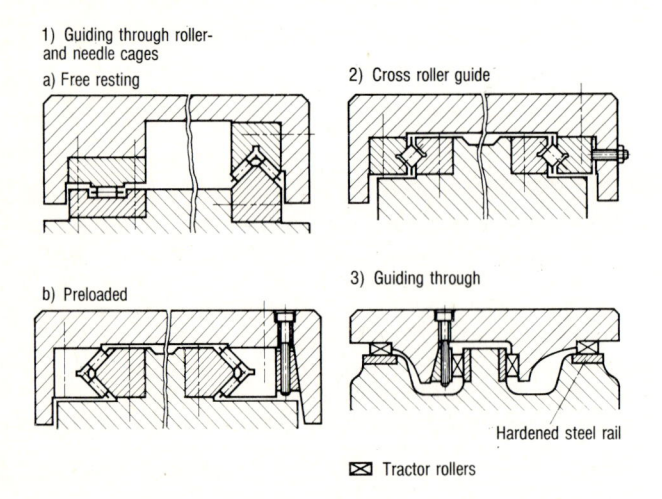

Fig. 4.37 Roller guides design versions

The frictional behavior of roller guides is similar to that of loaded cylindrical roller bearings. The friction coefficient is smaller by at least one power of ten than that of friction guides. One can calculate with about $0.003 \cdots 0.01$ for μ_F.

For roller guides to function perfectly, absolute cleaness is necessary. It is therefore important to pay special attention to the covering of the guide rails.

4.3.3.2.4 Hydrostatic Guides

Because of its high design demands, the hydrostatic guide is utilized only on large machine tools, i.e. starting with slide weights of about $3 \cdot 10^3$ kg. Fig. 4.38 shows two examples of hydrostatic guides.

The two moving surfaces are separated through an externally generated oil pressure. An oil film is maintained during the operation, on which the slide virtually floats. The gap between the two moving areas measures about 20 µm to 80 µm. The countergrip shown in fig. 4.38, is on one hand used to increase the static stiffness through preload, and on the other hand it prevents the table from lifting at alternating load direction.

Advantages of the Hydrostatic Guide

It has the best friction and wearing behavior of all the guides discussed thus far, since no direct contact occurs hereby. The hydrostatic guide is wear-free,

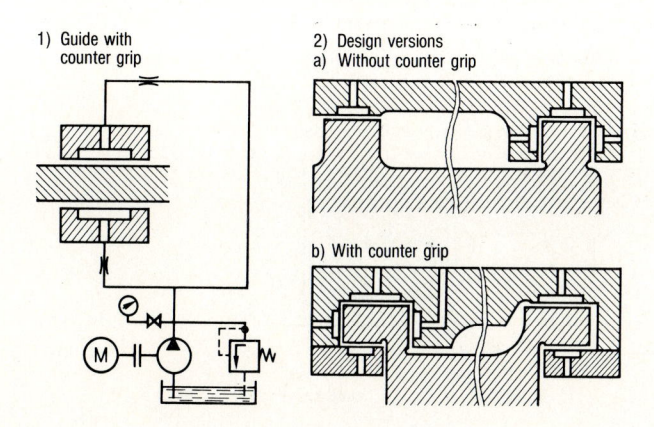

Fig. 4.38 Design versions of hydrostatic guides

243

and has no static friction. In all operation modes, it works with pure fluid friction, i.e. even large loads can be moved almost without friction.

Through these characteristics the motion achieved is highly uniform, so that no stick-slip effects can occur. The low friction, and thus low wear, ensure high permanent accuracy. Because of the so-called squeeze-film effect, the hydrostatic guide has very good damping. Any small faults in the support surfaces are balanced out through the oil film. The necessary manufacturing accuracy is, by comparison to other types of guides, lower.

Disadvantages of Hydrostatic Guides

The design of the hydrostatic guide is complex and demanding, and the control of the oil supply is also relatively complex. For the guide, to also be able to absorb excentric loads, several oil pockets must be always present. The ability to vary the pressure in the individual pockets is necessary in order to accomodate varying loads. Pressure compensation between individual pockets is not allowed. Before the motion is initiated the pressure must build up, and only after the motion has ended is it allowed to fall down again (control demand). High stiffness in the guide can only be achieved by means of countergrips [4.23].

Sealing the system is a difficult problem; contract seals should be avoided, because they are generally characterized by heavily varying friction. The oil return requires large channel crossections and natural slope-flow, or the use of suction pumps.

4.3.3.2.5 Aerostatic Guides

These guides use air as the separating and bearing medium between the slide surfaces. The principle is similar to that of the hydrostatic guides, but the air bearing has its special properties. Fig. 4.39 (from [4.7]) presents an example of aerostatic guide design on a table with ball screw bearings.

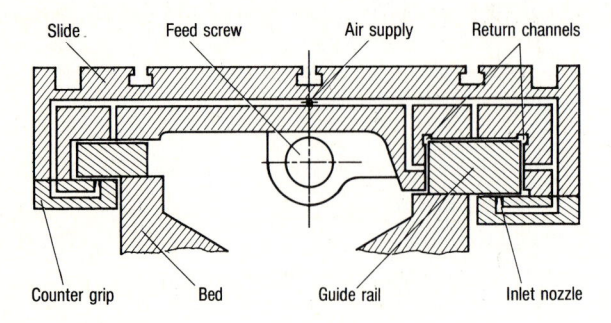

Fig. 4.39 Aerostatic guide design on a feed slide

Advantages of Aerostatic Guides

The exceptionally low friction and the wear-free operation are characteristic of aerostatic guides. As in the case of the hydrostatic guide, stick-slip effects cannot occur. Return pipes and seals are not necessary. The supporting ability and the low friction losses are independent of the velocity of the slide and ambient temperature. Contamination of the bearing air cushion cannot occur.

Disadvantages of Aerostatic Guides

The low viscosity of air must be mentioned for its effect on the support ability and on the damping gradient. Pure air support systems are very poorly damped, the slides are easily brought to oscillations. When calculating the air gap, the air amount or the support capabilities, the compressibility, turbulence, and the form of the nozzle variables must be taken into consideration, which is not a simple task. Air dryers are necessary at the location where the air is compressed, and the guideways may have to be manufactured out of corrosion-resistant materials.

In order to prevent the oscillations excited through the machining process, the aerostatic guide can also be used in combination with a friction guide. With larger pressure on the guide, the air cushion is compressed so far that friction will occur on an area designed for the purpose.

Further references about air support can be found in [4.28] and [4.29].

4.3.3.3 Damping

References about the magnitude of the damping gradient necessary for the stability of the position control loop are made in section 4.2.3. The value stated for D_{mech} between 0.1 and 0.2 must be achieved mainly through the sliding and frictional properties of the guide. The damping thereby results from three different physical processes:

▷ Material damping, due to elastic and plastic deformations of the roughness of the surface.
▷ Frictional damping, resulting from movement around the limit zones of the surface.
▷ Fluid damping (squeeze-film effect), resulting from squeezing and suctioning the lubricant between the surfaces.

Frictional and fluid damping predominate for flat and slide surfaces. The theoretical principles and the results arrived at in practice, are summarized in [4.10], [4.12] and [4.30] of the additional literature.

The variables that have been determined to have a significant influence on the damping gradients of friction and roller guides, are:

    the magnitude of pressure on the area,
    the type of lubricant used,

type of friction,
the magnitude of the nominal angular frequency $\omega_{0\,mech}$,
and the size of the total stiffness.

The damping capability of the friction guide is determined primarily by the area pressure occurring in the slide pairing. By comparison, for roller guides the force with which the roller bodies are pressed between the guide rails, has only marginal influence on the damping gradient of the guide. Figure 4.40 presents the experimentally determined damping gradients for three different guide types, in dependence on the area pressures [4.10].

The low damping that can be seen in fig. 4.40 for the roller guide can be improved by using a combined friction and roller guide. The forces on the guide are hereby divided: one portion will be absorbed by the roller guide and the other by the friction guide. If this ratio is properly chosen, the system will retain the advantages of the low friction of roller guides, but with sufficient damping.

For friction, and combinations of friction and roller guides, an increase in the total stiffness of the feed screw-table system results in a significant improve-

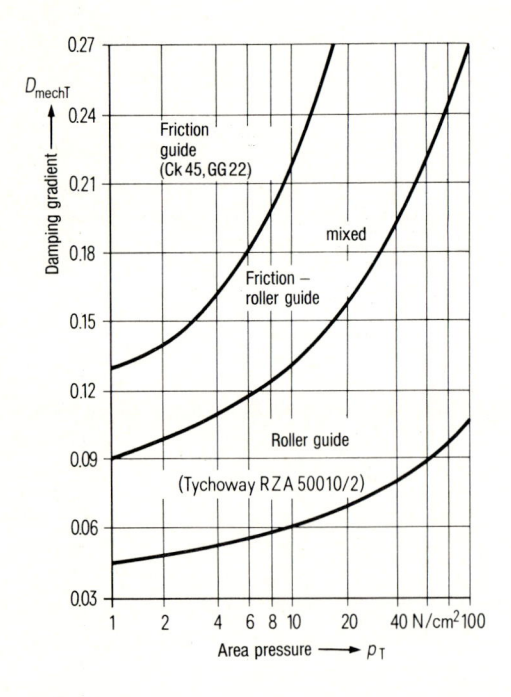

Fig. 4.40
Dependency of the table guide damping gradient D_{mech} on the area pressure p_T

ment of the damping gradient. For roller guides, these measures lead to an increase of the damping gradient only if the lubricant used for the roller bodies is a lubricant with polar additions. Such additions form a molecular film on the surface of the metal, which does not undergo any chemical bonding. They lower the friction value and increase the frictional damping. When lubricants without such additions are used, practically no dependency between damping and total stiffness can be noted.

Values achievable for the damping gradient of the table at:

friction guide $D_{\text{mech T}} = 0.2$–0.4
roller guide $D_{\text{mech T}} = 0.04$–0.15

The only factors that can influence the damping behavior of hydrostatic guides are the fluid friction and increases in the total stiffness. Since the possibilities of influencing this damping behavior through the viscosity and composition of the lubricant are limited, more weight is to be given to ensuring an especially high stiffness value.

Aerostatic guides are at the moment not widely used on machine tools. For larger machines, one can take advantage of the possibility of using an additional friction guide for damping at large machining forces. Smaller tables, as for instance those on measuring machines, are designed for high stiffness and high mechanical nominal angular frequencies.

4.4 Gears

4.4.1 Overview of Applications

Gears are installed on feed axes for the following reasons:

▷ To adapt the servo motor speed to the lower speeds of the ball screw, rack pinion, or worm drive.
▷ To reduce the external moment of inertia reflected onto the motor shaft.
▷ To increase the torque on the drive shaft.
▷ To integrate the physical position of the motor within the design of the machine.

Whether or not a drive should be equipped with a gear depends on how important one of the points mentioned above is to the design of that particular feed unit. For the common numerically controlled feed drive, the generally pertinent issues are the speed integration and the increase in the torque on the shaft. For acceleration drives, for long feed screws, and for large masses to be moved, the most significant task of the gear is to adapt the external moment of inertia (see 2.3.5.2 with equation (2.45)).

Figure 4.41 shows the mounting of the feed motor for the example of the feed screw unit.

Construction form 1 is the version with the lowest demands, but is only offers two degrees of freedom, "screw lead" and "motor speed", for the integration to the desired feed rate and the required feed force. It is used mainly for torque ranges of 15–40 Nm, at speeds of up to approx. 1,200 min^{-1}, with permanent magnet-excited DC servo motor. This version offers the largest torsion stiffness, thus the lowest friction reversing error. The backlash is zero.

Construction form 2 is frequently used because of its advantageous design possibilities. It helps achieve small external machine dimensions, reduces tilt torques and skewings of the feed unit. Timing belts are hereby preferably used, since in these cases the distance between axes can be selected larger. Good access to the motor brushes and the tacho generator is to be taken into consideration because it determines the ease with which the unit can be serviced.

Construction form 1:
Direct coupling

Construction form 2:
Predominantly timing belt drive

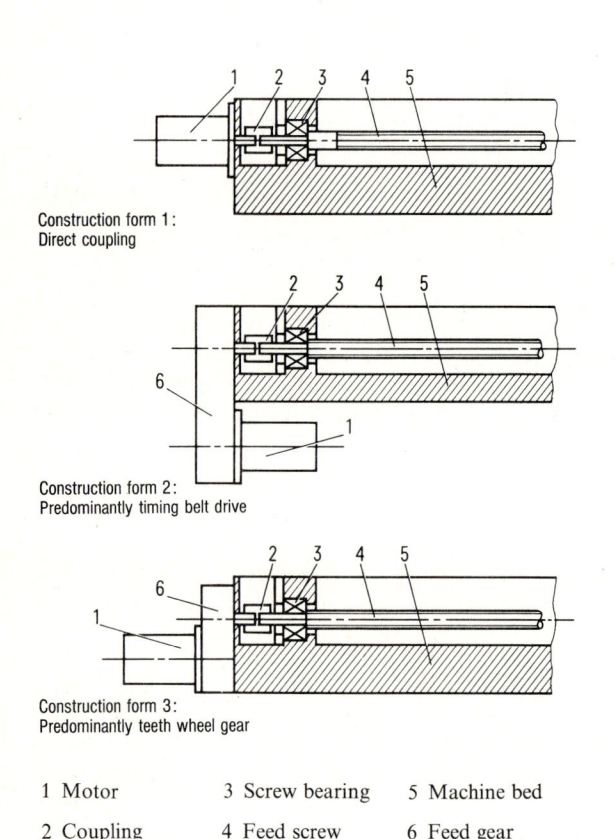

Construction form 3:
Predominantly teeth wheel gear

1 Motor　　　　　3 Screw bearing　　　5 Machine bed
2 Coupling　　　 4 Feed screw　　　　 6 Feed gear

Fig. 4.41　Mounting the feed motor on a feed table with feed screw

Construction form 3 is a version used mostly with electrically-excited DC servo motors. The gear is a one or two stage tooth wheel gear. The feed screw speed and the torque can be integrated by the selection of the gear ratio, but measures for the reduction of reversing error are necessary. Timing belt drives and drives with permanent magnet-excited servo motors can also be designed in construction form 3.

Deciding to install a gear or not, is a matter of evaluating the particulars of individual application cases. A knowledge of the basic advantages and disadvantages of gears is necessary for this purpose.

Advantages of a Gear

▷ The axes of the feed motor and feed screw do not have to be aligned, i.e. several possibilities exist for the mounting of the motor.

▷ The external moment of inertia J_{ext} is reduced onto the motor shaft according to the ratio $1/i^2$, the torque requirement on the motor shaft is reduced by $1/i$, which means that smaller motors can be used (see the equations of table 2.7 on page 128).

▷ The gear ratio i represents another parameter, besides the feed screw lead h_{Sp}, for the integration of the motor speed at required rapid traverse feed rates. This means that the manufacturers of machine tools can agree on a standardized feed screw lead (e.g. $h_{Sp} = 10$ mm), independently of motor speeds.

▷ With respect to the integration of the moment of inertia and the related dynamic of the feed drive, the gear ratio can be optimized (see section 2.2.2.6). The gear ratio determined with equation (2.45) can be utilized only within a limited area, which is determined on the one hand by the possible motor speeds, and on the other hand by the feed screw speed required.

Disadvantages of a Gear

▷ The gear is an additional construction element. If it is produced by the machine tool manufacturer, it represents for him additional design and production demands.

▷ A gear can, under circumstances, introduce additional non-linearities (backlash) into the position control loop. The resulting problems are described in section 4.2.4.3. Preventing these types of non-linearities is only partially possible, and requires design as well as time expenditure, which raises the manufacturing costs.

▷ Although all moments of inertia on the drive side of the gear can be reduced by a factor of $1/i^2$ through the gear ratio, the gear moment of inertia contributes itself to the total moment of inertia of the drive.

▷ Because the additional moment of inertia and the non-linearities introduced influence the control parameters, a more thorough adjustment of the speed regulator becomes necessary.

▷ The wear in the gear can lead to a gradual occurrence of reversing error, and thus readjustments may become necessary.

In addition, for tooth gear wheels one can list:

▷ A gear can generate higher noise levels during run and at stand still, especially when the current supply is through a line-synchronized current converter. During run, this is due to

> tooth form errors and the meshing process of the tooth flanks resulting in an alternating reversing error between the tooth wheels.

During stillstand, it is due to

> the fact that the motor oscillates within the gear play. Depending on the pulse number of the current converter, humming noises are generated with 2, 3, or 6 times the line frequency.

▷ A tooth wheel gear has higher maintenance costs (lubrication).

Although to a much lesser extent, the last two points also apply to timing belt gears. For these however, the emphasis is on:

> The timing belt gear is manufactured at lower costs than the tooth wheel gear, because its components necessitate less manufacturing accuracy.

> It is considerably less noisy; excitations from the current converter frequencies are damped heavily. Small timing belt pulleys or extremely wide belts produce higher run noises only at high speeds.

> Play can easily be eliminated with tension idlers, but by comparison to the tooth wheel gear, the motor does not have to have the capability to be shifted in radial direction.

A comparison shows, that for a tooth wheel gear the disadvantages outweigh the advantages. The noise levels generated and the expenditure of effort necessary to rid the tooth wheel gear of play, are especially bothersome for practical use. If, because of design reasons, it proves impossible to sacrifice a gear stage, then a timing belt is to be preferred over a tooth wheel gear. Sections 4.4.2 and 4.4.4 contain examples with references concerning ways to improve the nominal angular frequency through gear ratios.

4.4.2 Requirements

In order for a feed gear to be accomplished either through a tooth wheel or through a timing belt gear, it needs to meet the following requirements:

▷ small moment of inertia of the gear wheels reflected to the motor shaft
▷ high stiffness
▷ to be free of play
▷ low noise levels

Gears for special application cases are for instance those already presented in section 4.3.1.2 for the hydrostatic worm gear drive, which thus avoided the use of multi-stage gears. The friction losses are low, the stiffness very high. The so-called harmonic drive also has very high gear ratios ($i = 80...300$), but its stiffness is generally insufficient. These special versions will not be elaborated on any further here; comparisons of different designs can be found in [4.22] and [4.31].

4.4.2.1 Moment of Inertia

For the dynamic behavior of the total feed drive, it is important that the moment of inertia of the gear be low. The wheels should be made out of materials with high stiffness and low specific gravity. The moment of inertia can be further reduced through forming. The diameter of the wheels should be selected as small as possible, because they contribute to the moment of inertia with their fourth power. This applies especially to wheels mounted on the gear drive shaft, since their moments of inertia are not reduced.

This narrows the value possible for the ratio of a gear stage. The applicable ratio for a timing belt gear lies at approximately 3, since the angle of the arc on the small pulley must not be too small, and the moment of inertia of the gear wheels must not be too large. For tooth wheel gears one can go to about $i = 4$ per gear stage. This means, that at high motor speeds, like the ones of electrically excited DC servo motors, mostly tooth wheel gears must be used.

If, because of design reasons, construction form 2 (from fig. 4.41) must be used instead of version 1, then a gear ratio of $1 \leq i \leq 1.3$ is preferable over the direct ratio of $i = 1$. In this case however, the maximal motor speed must be correspondingly higher than would be required for a direct ratio. This can be deduced from the comparison of two gear stages with the values:

Gear ratio $i = 1$: d_{Gt} diameter of both gear wheels 1 and 2
Gear ratio $i > 1$: d_{Gt1} diameter of gear wheel 1 on the motor shaft
d_{Gt2} diameter of gear wheel 2 on the drive shaft

(For gear wheels denominations, see fig. 2.13 or table 2.7 on page 128.)

The moment of inertia of the gear reflected onto the motor shaft is:

▷ for gear ratio $i=1$

$$J_{Getr\,(i=1)} \sim 2 \cdot d_{Gt}^4$$

▷ for gear ratio $i>1$

$$J_{Getr\,(i>1)} \sim \frac{d_{Gt\,2}^4}{i^2} + d_{Gt\,1}^4$$

If we assume that $d_{Gt\,1} + d_{Gt\,2} = 2 d_{Gt}$, and then substitute for $d_{Gt\,2}/d_{Gt\,1} = i$, we obtain for the relationship of the moments of inertia:

$$\frac{J_{Getr\,(i=1)}}{J_{Getr\,(i>1)}} = \frac{0.125(1+i)^4}{1+i^2}$$

A good approximation for the range $1 \leq i \leq 1.3$ is:

$$\frac{J_{Getr\,(i=1)}}{J_{Getr\,(i>1)}} \approx i \qquad (4.40)$$

This means that the moment of inertia of a gear with a ratio in the range of $1 \leq i \leq 1.3$ is smaller by a factor of $1/i$ than the moment of inertia of a gear with the ratio $i=1$.

For the construction form 3 with the tooth wheel gear, a smaller total moment of inertia can be achieved by the transfer to a multi-stage gear. A study in [4.23] has demonstrated that a two-stage gear at constant gear ratio, results in a moment of inertia which is smaller by almost two thirds. The possible reduction of the wheel width in the first stage of the two-stage gear was thereby not considered. Fig. 4.42 shows a gear with $i=4$ in a single-stage version, in contrast to a two-stage arrangement.

For the two-stage version, two equal ratios with $i=2$ are presumed. We obtain for the gear moment of inertia:

▷ single-stage: $J_{Getr} \sim 17 \cdot d_{Gt\,1}^4$

▷ two-stage: $J_{Getr} \sim 6.25 \cdot d_{Gt\,1}^4$

A two stage gear however, is more demanding, because efforts will have to be made to keep the backlash (play) at low levels. For this reason, single stage gears are generally preferable.

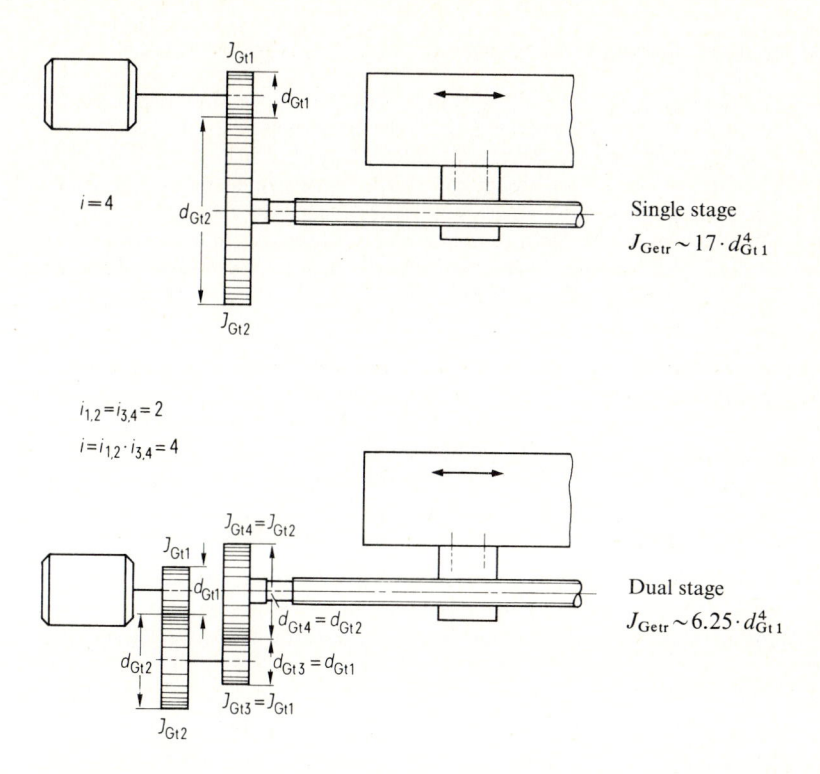

Fig. 4.42
Comparison of the moments of inertia of the single- and two-stages gears

Single stage
$J_{Getr} \sim 17 \cdot d_{Gt1}^4$

Dual stage
$J_{Getr} \sim 6.25 \cdot d_{Gt1}^4$

4.4.2.2 Torsion and Bending

In order to achieve a high total stiffness for the mechanical transmission elements, the torsional spring constant in the gear must have a value as large as possible. References about the pertinent methods of calculation are given in sections 4.3.1.2.1 and 4.3.2.2. Timing belt gears, when properly dimensioned and preloaded, have very high stiffness values. Design characteristics can be found in the manufacturers' catalogs, and in section 4.4.3.

The gear shafts of tooth wheel gears have to be as short, and as bending and torsion-stiff as possible. Shafts lying within the gear must be supported on both sides with appropriately sized roller bearings and with sufficient stiffness. Straight tooth wheels are to be given preference over helical tooth wheels, if the straight tooth version offers sufficient stiffness. According to fig. 4.32, helical tooth wheels have higher stiffness, but they require an additional axial bearing in order to accept forces in the direction of the axis.

4.4.2.3 Reversing Errors

Lack of reversing error in the transmission path between the motor and the ball screw can be best accomplished with the construction form 1 of fig. 4.41. Construction forms 2 or 3 with timing belt drives are also free of reversing error to a large extent. Rack and pinion systems or worm gear drives, however, in most cases require multi-stage tooth wheel gears. In these cases, it must be ensured by careful design and manufacturing that the total reversing error remain small. The sum of the reversing errors acts as a hysteresis in the signal flow of the position control; the effects are described in section 4.2.4.3.

On tooth wheel gears, for instance, low levels of play are achieved by using two wheels with the same number of teeth in each stage, and by twisting them against each other with a spring force, or by mounting them play-free against each other in preloaded stage (see fig. 4.43). If the gear drive pinion is situated directly on the motor shaft, the motor should then possess the capability of being shifted radially, so that the drive pinion can be adjusted onto the following gear wheel play-free. This can be accomplished, e.g. through an excentrically designed spacer ring.

The tooth wheels in feed drives must have increased accuracy (at least of quality 5 as per DIN 3967), and must be sufficiently run-in before the start-up. Under circumstances, the wheels may require several readjustments during the running-in.

The preload of the tooth wheels does not have to be designed for the maximal drive torque, because this torque does not come into play during contour controlled driving and during positioning with accordingly low speeds. When calculating the stiffness of preloaded gears however, the preload forces must be taken into account.

Straight teeth   Helical gear
Key shaft

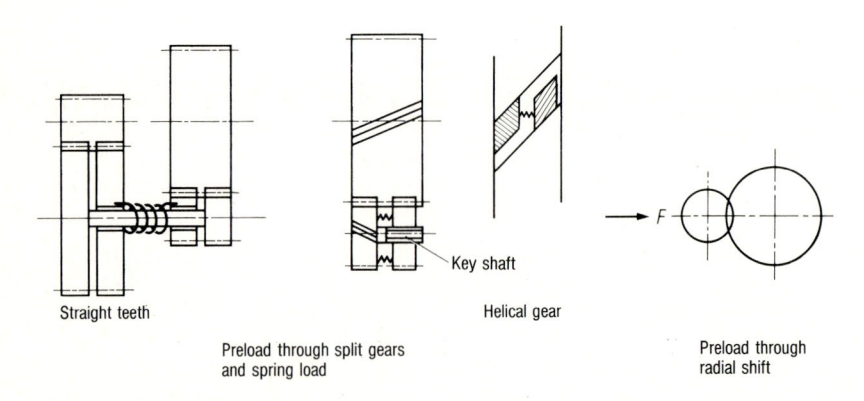

Preload through split gears and spring load

Preload through radial shift

Fig. 4.43 Possible ways of preloading the gears

Play-free connections between shafts and wheels can be accomplished by means of frictional connections, e.g. shrink seal, oil pressure seal, pressure bushings and conical seats. The force is transmitted between wheel and shaft exclusively through surface pressure. Any key ways present should only be used to fixate the wheels.

Although tooth wheel gears can basically function with continuous oil lubrication, it is more advantageous to provide them with circulating oil lubrication because the oil flow can also remove the heat generated through friction.

4.4.3 Toothed Belt Gear

Toothed belt gears provide a particularly economical way to meet the requirements of a feed drive. In regards to stiffness and positive force transmission, they are a combination of flat belt and chain. They display the wear resistance of high wear-resistant steel. The glass fiber tensile bodies most commonly used have high tensile strengths, very good bending capabilities, and low expansion factors.

The application of toothed belt gears requires that certain limitations be observed:

▷ The pulleys must consist of materials of low specific gravity (plastics, aluminium alloys), in order to achieve low moments of inertia. The smallest diameter allowed for pulleys is determined by the toothed belt manufacturer, and is based among other things on the number of active teeth and the angle of belt wrap. This smallest possible diameter should be aimed at for the driving pulley (see the calculation example on page 256).

▷ Toothed belts with circle segment tooth form (e.g. HTD belt of the Uniroyal Co.) are to be preferred.

▷ In order to avoid additional moments of inertia due to the presence of clutches, the pulley on the drive side should be mounted directly on the motor shaft, as is the case with conventional gears.

▷ In order for the toothed belt to function properly, it is necessary to do a parallel adjustment of the shaft and the pulleys. Not admissible deviations from the parallel arrangement cause rim tensions in the belt, whereby the belt runs toward the side with the higher tension and suffers heavy wear. One pulley at least must be equipped with double-sided rim disks.

▷ To compensate for the manufacturing tolerances of the toothed belt length and to prevent play, the toothed belt must be preloaded. To accomplish this, either the motor of the drive can be mounted to allow radial shift, or one can install tension rollers.

▷ For longer free belt lengths (larger than about 10 times the belt width), tension rollers are a requirement for the damping of the belt oscillations. Such oscillations can be excited by the basic oscillating frequency of the current converter. If tension rollers are not used, the gain in the speed control loop has to be reduced.

Fig. 4.44 Toothed belt gear

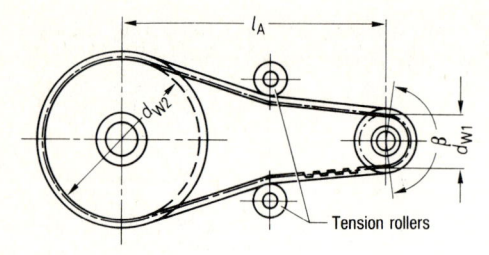

Tension rollers

α_w active diameter
β wrap angle on the small pulley

Tension rollers can be mounted either on the inside as teeth rollers, or even better, on the outside as smooth cylindrical rollers, as shown in fig. 4.44. Since the belt is put under tension by the pressure of an outside roller, which increases the wrap-around angle, this arrangement is preferable. In this case, if one whishes to prevent running noises, toothed belts with polished backs must be used. The tension rollers are not to have elastic support. An unecessarily high preload reduces the lifetime of a toothed belt, increases the radial load on the bearings, and favors running noises. An insufficient preload can cause, under heavy loads, insufficient meshing with the teeth of the pulley, and also that teeth be jumped over.

When using HTD toothed belts on a numerically controlled feed drive, the belt preload can be set to about 1.1 times the value of the circumferential forcer at the motor rated torque:

$$F_{ZV} \approx 2.2 \cdot \frac{M_{0M}}{d_{W1}} \qquad (4.41)$$

M_{0M} motor rated torque
d_{W1} active diameter of the pulley on the motor shaft

This force loads the drive side bearing of the DC servo motor. It must be double checked whether this bearing has the necessary design for the length of the motor shaft required for the mounting of the pulley.

Example:

Calculate a toothed belt gear, according to the data provided by the Flender Co. of Duesseldorf. The data contains the points to be considered for the numerically controlled feed drive.

Find a drive for the DC servo motor 1HU3 076–0AF01, with $M_{0M} = 10$ Nm, $n_0 = 3,000$ min^{-1}, i approximately 2.5, axes distance approximately 300 mm.

One calculates with a maximal transmission power according to equation (3.7), with a motor torque which corresponds to the current limit, and with a speed at which the speed-dependent current limit is activated. In the example, M_M becomes 40 Nm, and $n_M = 1,500 \text{ min}^{-1}$.

$$P = 0.1047 \cdot 10^{-3} \cdot 40 \cdot 1,500 \text{ kW} = 6.27 \text{ kW}$$

The pitch to be selected is 8 mm, as determined from a diagram. One can derive the ratio $i = 2.57$ from the selection tables (the selection should be for the smallest diameter possible).

Driving pulley	28 teeth
	active diameter $d_{W1} = 71.3$ mm
Driven pulley	72 teeth
	active diameter $d_{W2} = 183.35$ mm
Axis distance l_A	274.3 mm at 960 mm toothed belt active length, 120 teeth
or	315 mm for 1,040 mm toothed belt active length, 130 teeth

(The smaller axis distance is chosen in order to limit the free belt length).

According to the approximation formula given, the wrap-around angle on the pulley will be:

$$\beta \approx 180° - 60° \frac{d_{W2} - d_{W1}}{l_A} = 180° - 60° \frac{183.35 - 71.3}{274.3} = 155.5°$$

The meshing number of teeth on the driving pulley thus obtained is

$$Z \approx 28 \cdot \frac{155.5°}{360°} \approx 12$$

Experience shows that this number should reach the value of 12; the wrap-around angle should thus suffice.

From the calculation data, we can now determine the correction factors:

Length factor	$b_1 = 1$
Load factor	$b_2 = 1$ (single-shift operation)
Gear ratio factor	$b_3 = 0.3$

The load factor for dual-shift operation of the machine tool would be $b_2 = 1.2$. It does not need to be set any higher, because the calculation of the transmission power already used the maximal motor torque.

With these data, we can calculate the theoretical power to be:

$$P_B = P(b_2 + b_3) \cdot b_1 = 6.27(1 + 0.3) \cdot 1 \text{ kW} = 8.15 \text{ kW}$$

The necessary belt width can be determined to 50 mm from the power tables. For the driving pulley with 28 teeth, the admissible power with $n_M = 1,500 \text{ min}^{-1}$ is approximately 8.4 kW. It is sufficient.

According to equation (4.41), the belt preload to be set is approximately

$$F_{zv} \approx \frac{2.2 \cdot 10 \text{ Nm}}{71.3 \cdot 10^{-3} \text{ m}} = 307 \text{ N}$$

From the motor selection tables we can derive, that for a force attack point of 28 mm on the shaft end at 3,000 min^{-1}, the allowed shear force is about 430 N. The preload force is thus admissible.

For steel material, the moments of inertia according to equation (4.10.1) are:

$$\frac{J}{\text{kgm}^2} = 0.77 \cdot 10^{-12} \left(\frac{d}{\text{mm}}\right)^4 \cdot \frac{l}{\text{mm}}$$

Driving pulley (pulley width selected with 56 mm):

$$J_{Gt1} = 0.77 \cdot 10^{-12} \cdot 70^4 \cdot 56 \text{ kgm}^2 = 0.00104 \text{ kgm}^2$$

Driven pulley

$$J_{Gt2} = 0.77 \cdot 10^{-12} \cdot 182^4 \cdot 56 \text{ kgm}^2 = 0.0476 \text{ kgm}^2$$

and the moment of inertia of the toothed belt gear, reflected on the motor shaft:

$$J_{Getr} = J_{Gt1} + \frac{J_{Gt2}}{i^2} = \left(0.00104 + \frac{0.0476}{2.57^2}\right) \text{ kg m}^2 = 0.00824 \text{ kgm}^2$$

Since the motor has its own moment of inertia $J_M = 0.0065$ kgm^2, we must recognize that this moment of inertia is much too large. A lower moment of inertia must be achieved, by selecting materials accordingly, and by forming of the pulleys.

If aluminium alloys are used, the moment of inertia will be reduced to approximately 1/3, and if plastics are used, to about 1/6. Besides duroplasts with textile fibers as fillers, epoxy pressed wood can also be used for the pulleys (according to DIN 7707). Such materials are produced, for instance, by the Roechling Co., Haaren/Ems, who manufacture the material Lignostone, fairing very well as pulley material on toothed belt gears used on machine tools. The tension rollers mentioned earlier, which were not necessary in this example, must also be considered in connection with moments of inertia, since they too, undergo acceleration and deceleration. For the mounting of a plastic wheel on the motor shaft, it is appropriate to use a steel bushing.

The toothed belt gear thus equipped, works at rapid traverse with 3,000 min^{-1} of the driving pulley, in the range where according to the selection tables of the Flender Company reductions of lifetime can occur. Rapid traverse movements however, are limited timewise, and are driven with lower torques. Increasing the number of teeth and thus the diameter of the pulley, would intolerably increase the moment of inertia, and is not recommendable. A wider toothed belt would bring no advantage worth mentioning.

The toothed belt manufacturers give for loads with rated power, an expansion figure of approximately 0.03%. The bending of the belt teeth must additionally be considered because it determines the stiffness to a significant extent. As mentioned in chapter 6, the stiffness on a similar gear was measured at 330 N/μm. This shows that the stiffness of timing belt gears is very high, and fully sufficient for numerically controlled feed drives.

4.4.4 Advantage Limits of a Gear Ratio

4.4.4.1 Motors of Equal Frame Size

The starting point for considerations about the suitability of a gear ratio, is the achievable nominal angular frequency. If only small command value changes are considered, the nominal angular frequency of the drive can be obtained with equation (2.21) from

$$\omega_{0\,max\,A} \approx \frac{1}{T_{el\,A}} \left(1 + \frac{1}{2\frac{T_{mech\,A}}{T_{el\,A}}}\right)$$

At large command value steps, the limitations set by the current converter come into effect. $\omega_{0\,A}$ can then be determined according to equation (2.22). (See section 2.2.2.5, on page 86.)

The mechanical time constant of the drive can be derived from equation (2.13):

$$T_{mech\,A} = \frac{J_{Ges} \cdot \omega_{max\,M}}{M_{St\,A}}$$

It is proportional to the total moment of inertia.

In this equation, the total moment of inertia J_{ges} is substituted with the sum of the individual moments of inertia. Using equations (2.3) and (2.51), we get:

$$J_{Ges} = J_M + J_{Getr} + \frac{J_2}{i^2} \tag{4.42}$$

J_M motor moment of inertia
J_{Getr} total moment of inertia of the gear reflected on the motor shaft
J_2 sum of the moments of inertia reflected on shaft 2, without gear wheel 2
i gear ratio

Further, we define a mechanical reference time constant to be:

$$T_{0\,mech} = \frac{\omega_{max\,M}}{M_{St\,A}} \cdot J_M \tag{4.43}$$

and obtain as adjusted dimensional equation:

$$\frac{T_{0\,mech}}{ms} = \frac{2\pi}{60} \cdot 10^3 \cdot \frac{\frac{J_M}{kgm^2} \cdot \frac{n_{max\,M}}{min^{-1}}}{\frac{M_{St\,A}}{Nm}} \tag{4.43.1}$$

With this set of data, the mechanical time constant of the drive becomes:

$$T_{\text{mech A}} = T_{0\,\text{mech}}\left(1 + \frac{J_{\text{Getr}}}{J_M} + \frac{J_2}{J_M \cdot i^2}\right) \qquad (4.44)$$

For equal feed force and rapid traverse feed rates, either a direct drive or a drive with gears can be selected for a feed unit. If motors of the same construction series are used in both cases, the mechanical reference time constant defined in (4.43) and the electrical time constant of the two drives will not differ significantly. If we compare the nominal angular frequencies $\omega_{0\,A(oG)}$ (drive without gear) and $\omega_{0\,A(mG)}$ (drive with gear), under the assumption that $T_{0\,\text{mech(oG)}} \approx T_{0\,\text{mech(mG)}}$ and that $T_{\text{el A(oG)}} \approx T_{\text{el A(mG)}}$, we obtain:

$$\frac{\omega_{0\,A(oG)}}{\omega_{0\,A(mG)}} = \frac{1 + \dfrac{T_{\text{el A}}}{2\,T_{\text{mech A(oG)}}}}{1 + \dfrac{T_{\text{el A}}}{2\,T_{\text{mech A(mG)}}}}$$

If we expand the equation with $2 \cdot T_{0\,\text{mech}}/T_{\text{el A}}$, and add in the mechanical time constant $T_{\text{mech A(oG)}}$ and $T_{\text{mech A(mG)}}$ derived from equation (4.44), we get the relationship:

$$\frac{\omega_{0\,A(oG)}}{\omega_{0\,A(mG)}} = \frac{2\dfrac{T_{0\,\text{mech}}}{T_{\text{el A}}} + \dfrac{1}{\left(1+\dfrac{J_2}{J_M}\right)}}{2\dfrac{T_{0\,\text{mech}}}{T_{\text{el A}}} + \dfrac{1}{1+\dfrac{J_{\text{Getr}}}{J_M}+\dfrac{J_2}{J_M \cdot i^2}}}$$

It follows, that $\omega_{0\,A(mG)}$ will be larger than $\omega_{0\,A(oG)}$, if

$$\frac{1}{1+\dfrac{J_2}{J_M}} < \frac{1}{1+\dfrac{J_{\text{Getr}}}{J_M}+\dfrac{J_2}{J_M \cdot i^2}}$$

If we invert this, we obtain the relationship illustrated in fig. 4.45:

$$\frac{J_{\text{Getr}}}{J_2} < 1 - \frac{1}{i^2}. \qquad (4.45)$$

From the diagram in fig. 4.45, we can read out whether a drive with gear reduction has a higher nominal angular frequency than a direct drive. The

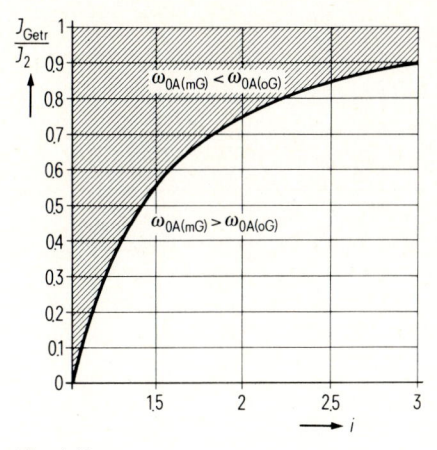

Fig. 4.45
Advantage limits of gear reduction, under the assumption that the proposed drives have equal mechanical reference and electrical time constants

presumption hereby is that both have the same mechanical reference time constants and electrical time constants. One should be aware that when the external moments of inertia are large, or the gear moments of inertia are small, a gear has a positive influence on the nominal angular frequency achievable.

4.4.4.2 Motors of Different Frame Size

Table 2.6 on page 122 presents different drives, listed for a desired feed force of 12.5 kN and a rapid traverse feed rate of 12 m/min. The previously made assumption about the mechanical reference, with electrical time constants being equal, cannot hold any longer, because the table also lists motors of different frame sizes. The nominal angular frequencies of the drives must, therefore, be derived with equations (2.21) or (2.22), and compared.

Table 4.6 shows such a comparison of two different sums of table- and work piece masses $m_T + m_W$. The presumption is that all the drives listed have 6-pulse circulating current-free current converters. The values for the gear moments of inertia are practically achievable. The ball screws are entered with 40 mm diameter and 2 m length. The value derived for the electrical time constant takes into consideration the current-dependent saturation of the inductance. Both values, $\omega_{0\,\text{max}\,A}$ and $\omega_{0\,\text{grenz}\,A}$, are given in part, for the drive nominal angular frequency.

For small command value changes, the value $\omega_{0\,\text{max}\,A}$ is actually reached. At larger values, the nominal angular frequency is limited by the control range and the current limit. This $\omega_{0\,\text{Grenz}\,A}$ value can be modified by the selection

261

Table 4.6
Comparison of drive systems according to table 2.6, for two different table- and work piece masses
(Values of $\omega_{0\,\mathrm{grenz\,A}}$ designated with * are limited through the current converter)

Drive Value, formula		① 1HU3104-0AD01 with $m_T + m_W$		② 1HU3078-00AC01 with $m_T + m_W$		③ 1HU3076-0AF01 with $m_T + m_W$		④ 1HU3078-0AC01 with $m_T + m_W$		⑤ 1HU3108-0AD01 with $m_T + m_W$		⑥ 1GS3107-5SV41 with $m_T + m_W$	
		300 kg	3,000 kg	300 kg	3,000 kg	300 kg	3,000 kg	300 kg	3,000 kg	300 kg	3,000 kg	300 kg	3,000 kg
Rated torque M_{0M}	Nm	25		14		10		14		38		6.8	
Idle speed $n_{\max M}$	min^{-1}	1,350		2,200		3,700		2,200		1,450		7,000	
Voltage constant K_E	V/1,000 min^{-1}	138		88		53		88		134		30.8	
Torque constant K_T	Nm/A	1.30		0.837		0.501		0.837		1.27		0.294	
Armature circuit resistance R_A	Ω	0.6		0.6		0.4		0.6		0.3		0.65	
Short circuit torque $M_{St.A}$, according to equation (2.12.1)	Nm	404		270		246		270		823		97.5	
Motor moment of inertia J_M	kgm^2	0.028		0.0085		0.0065		0.0085		0.045		0.002	
Mech. ref. time constant $T_{0\,\mathrm{mech}}$, acc. to (4.43.1)	ms	9.8		7.25		10.2		7.25		8.3		15	

(Table 4.6 continued)

Quotient Q, acc. to section 2.2.2.5		3.1		2.6		2.03		2.6		2.3		3.49	
Moment of inertia J_2 reflected on shaft 2, from equation (2.49), resp. (2.50)	kgm²	0.0047	0.0115	0.0047	0.0115	0.0047	0.0115	0.0042	0.00668	0.00565	0.011	0.0047	0.0115
Gear ratio i		–		1.66		2.5		–		–		5	
Gear moment of inertia J_{Getr}	kgm²	–		0.00116		0.002		–		–		0.00077	
Total moment of inertia J_{Ges} acc. to equation (4.42)	kgm²	0.0327	0.0395	0.0114	0.0138	0.00927	0.0104	0.0127	0.0152	0.0507	0.056	0.00296	0.00323
Mechanical time constant T_{mechA} acc. to equation (2.13.1)	ms	11.4	13.8	9.7	11.8	14.6	16.3	10.8	13	9.35	10.3	22	24.3
Electrical time constant T_{elA} acc. to equation (2.10.1)	ms	10		5		3.5		5		17		0.8	
Ratio T_{mechA}/T_{elA}		1.14	1.38	1.94	2.36	4.17	4.66	2.16	2.6	0.55	0.61	27.5	30.4
Drive nominal angular frequency from fig. 2.9 $\omega_{0\,max\,A}$ (equation (2.21)) or characterized with * $\omega_{0\,grenz\,A}$ (equation (2.22))	s⁻¹	144	137	250	244	314	308	244	232	113	107	158*	143*
					220*	140*	125*	238*	200*				

of current limitation and maximum feed rate. As shown in position 3, on a drive this results in a significant improvement of the nominal angular frequency. Hereby, the large moment of inertia of gear wheel 2 has a particularly large effect, due to the fact that it can only be reduced by different material selection (see the example in 4.4.3 on page 256). When comparing the drive in position 1 to the drives in positions 2 and 4, it can be observed that the higher nominal angular frequency of 2 and 4 is due to the low electrical time constants.

In combination with a small motor, a gear stage can have favorable effects on the achievable nominal angular frequency of the drive. It is very sensible to do such calculations in the design stage of the machine tool, because a higher nominal angular frequency of a drive will result in a better command value behavior in the position control loop.

4.4.4.3 Disturbance Transient Response

The disturbance transient response was derived with equation (1.51) where the gear ratio i, together with the drive's nominal angular frequency, were, in the denominator, the deciding variables for the determination of the magnitude of the deviation at jumping load changes. In the practical operation of a machine tool however, these have no meaning, so that checks through the nominal angular frequency of the drive are, by themselves, sufficient for the supervision of the command transient response behavior. Optimization according to the disturbance transient response is only necessary in special application cases.

4.4.4.4 Comparison of Two Motors

In order to explain the decision process concerning optimizations, let us compare the direct drive in position 1 of table 4.6, to the drive with gear of position 3. The characteristic values of both DC servo motors, 1HU3 104-0AD01 in position 1 and 1HU3 076-0AF01 in position 3, are given in table 4.6. If we set the nominal angular frequency $\omega_{0\,max\,A}$ of drive 1, calculated with (2.21), in a ratio with the nominal angular frequency $\omega_{0\,grenz\,A}$ of drive 3, calculated with (2.22), we obtain:

$$\frac{\omega_{0\,A1}}{\omega_{0\,A3}} = \frac{T_{elA3}}{T_{elA1}} \frac{1+\dfrac{T_{elA1}}{2T_{mechA1}}}{\dfrac{Q_3 \cdot T_{elA3}}{T_{mechA3}}}$$

If we substitute the mechanical reference time constant defined in equation (4.43) for the mechanical time constant, the ratio of the nominal angular frequencies will become:

$$\frac{\omega_{0A1}}{\omega_{0A3}} = \frac{T_{elA3}}{T_{elA1}} \cdot \frac{1 + \dfrac{T_{elA1}}{2T_{0mech1}\left(1+\dfrac{J_2}{J_{M1}}\right)}}{\dfrac{Q_3 \cdot T_{elA3}}{T_{0mech3}\left(1+\dfrac{J_{Getr}}{J_{M3}}+\dfrac{J_2}{J_{M3}\cdot i^2}\right)}}$$

or

$$\frac{\omega_{0A1}}{\omega_{0A3}} = \frac{T_{0mech3}}{Q_3}\left(1+\frac{J_{Getr}}{J_{M3}}+\frac{J_2}{J_{M3}\cdot i^2}\right)\cdot\left(\frac{1}{T_{elA1}}+\frac{1}{2T_{0mech1}\left(1+\dfrac{J_2}{J_{M1}}\right)}\right)$$

If we solve for the moment of inertia of the gear J_{Getr}, we get:

$$J_{Getr} = \frac{2Q_3 \dfrac{\omega_{0A1}}{\omega_{0A3}} \cdot \dfrac{T_{0mech1}}{T_{0mech3}} \cdot T_{elA1}\left(J_{M3}+\dfrac{J_{M3}}{J_{M1}}\cdot J_2\right)}{2T_{0mech1}\left(1+\dfrac{J_2}{J_{M1}}\right)+T_{elA1}} - J_{M3} - \frac{1}{i^2}J_2$$

In order to be able to compare the two motors, we set $\omega_{0A1}/\omega_{0A3}=1$, and use the values from table 4.6 for Q_3, i, T_{elA1}, T_{0mech1}, T_{0mech3}, J_{M1}, and J_{M3}:

$$J_{Getr} = \frac{0.254\,\text{kgm}^2 + 9.06\,J_2}{29.6 + 700\,J_2 \cdot \dfrac{1}{\text{kgm}^2}} - 0.0065\,\text{kgm}^2 - 0.16\,J_2$$

The graph of this function is shown in fig. 4.46.

The strong influence of the gear moment of inertia J_{Getr} is quite apparent. Even if J_2 were around 0.001 kg m², the nominal angular frequency will be $\omega_{0A3} \geqq \omega_{0A1}$, only if J_{Getr} < approximately 0.002 kg m². For J_2 values of more than about 0.02 kg m², the direct drive will always have the higher nominal angular frequency.

This comparison shows that the moment of inertia of a gear must not exceed a certain limit value, if the nominal angular frequency to be achieved is to be at least equal to that of direct drive, with given motors and moments of inertia J_2. The graph of this type of function can be derived in dependence on the time constants T_{elA} and T_{0mech} of the drive. No flat statement can be derived from this example.

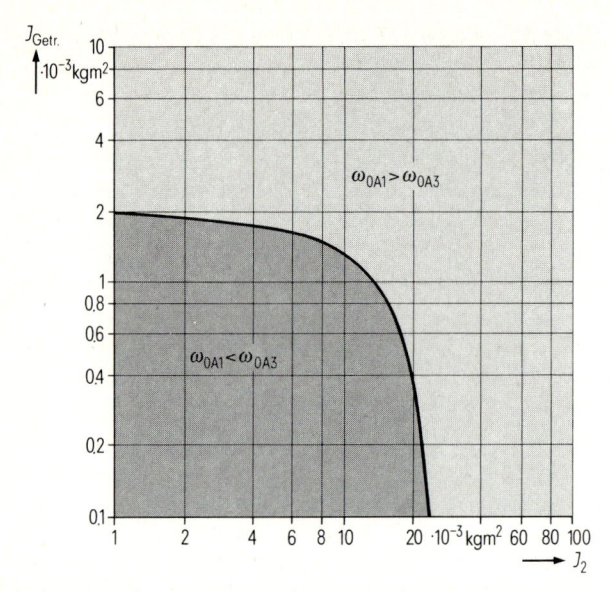

Fig. 4.46
Advantage limits for the drive systems in positions 1 and 3 of table 4.6, in dependence of the moment of inertia J_2 and the gear moment of inertia J_{Getr}

4.5 Couplings

Couplings are used to connect two shaft ends; in the process, shaft offsets, angular errors, and axial distances must generally be compensated. With no exception feed drives are equipped with torsion stiff couplings, which therefore must meet high requirements in respect to:

▷ torsion stiffness
▷ freedom of play
▷ low moment of inertia.

A distinction is made between *positive couplings*, for instance:

▷ versions according to the sliding pad principle
▷ cross-link couplings

and *friction couplings*, for instance:
▷ split bushing
▷ corrugated pipe couplings
▷ spring disk coupling.

The requirements mentioned above can be best met by friction couplings. If in exceptional cases positive couplings are used, particular attention has to be given to the moment of inertia and the play tendency.

When connecting shaft ends to couplings, the friction coupling with conical or cylindrical clamp elements is also preferable to positive coupling. A twist-lock achieved with key ways remains play-free, only if no shock loads occur. At large velocity changes, as occur on contour controlled machine tools, these key ways can easily undergo plastic deformation. The reversing error thus generated disturbs the behavior of the position control, and can be removed only by replacing the entire connection.

Figure 4.47 shows two examples of friction couplings. Reliable, commercially available, are for instance the bellow couplings produced by the Jakob Co. of Kleinwallstadt, and the spring disk couplings manufactured by the Flender Co., Bocholt (construction form Arpex).

Couplings are dimensioned according to the torque to be transmitted, the stiffness, and the shaft diameter. The formulas for the dimensioning of couplings can be found in the catalogs of the manufacturers. The safety factor that must be considered depends on load type and clutch design. The values can be found in the data sheets for the clutch. The characteristical values for feed drive safety factors are 1.3–1.6. Further references are presented in [4.32].

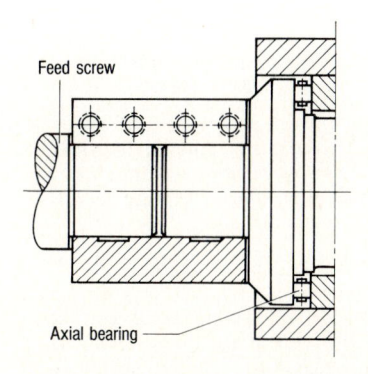

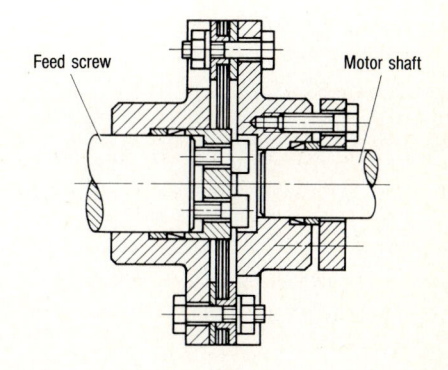

Fig. 4.47 Examples of friction couplings

Disengageable safety clutches have lately also become comercially available. They limit the torque at overloads or when the feed slide gets blocked, by breaking the connection between motor and machine. The mechanical interlock allows reengagement in the same position, so that the fixed reference points set for the numerical control will not be lost [4.33].

4.6 Bibliography

[4.1]   Kopperschläger, F.D.: Über die Auslegung mechanischer Übertragungselemente an numerisch gesteuerten Werkzeugmaschinen. Aachen, Dissertation an der Technischen Hochschule, 1969

[4.2]   Augsten, G.; Schmid, D.: Einfluß von Spiel und Reibung auf Konturfehler bahngesteuerter Werkzeugmaschinen. Steuerungstechnik 2 (1969), Heft 3, Seite 103 bis 108

[4.3]   Stute, G. (Herausgeber): Die Lageregelung an Werkzeugmaschinen. Begleittext zum gleichnamigen Seminar. Stuttgart, 3. Auflage, Selbstverlag des Instituts für Steuerungstechnik der Werkzeugmaschinen und Fertigungseinrichtungen der Universität Stuttgart, 1975

[4.4]   Bühler, W.: Antriebssysteme und ihre Kenngrößen. Werkzeugmaschine international, 1971, Nr. 2, Seite 21 bis 25

[4.5]   Opitz, H.: Auslegung von Vorschubantrieben für NC-Maschinen. Bericht über die VDW-Konstrukteur-Arbeitstagung am 7. und 8. Februar 1969. Selbstverlag des Laboratoriums für Werkzeugmaschinen und Betriebslehre der Rheinisch Westfälischen Technischen Hochschule Aachen 1969

[4.6] Schmid, D.: Numerische Bahnsteuerung; Beitrag zur Informationsverarbeitung und Lageregelung. Berlin-Heidelberg-New York, Springer-Verlag, 1972

[4.7] Weck, M.: Werkzeugmaschinen Band 2, Konstruktion und Berechnung. Düsseldorf, VDI-Verlag, 1979

[4.8] Köhler, G.: Statische, dynamische und thermische Steifigkeit von Fräsmaschinen. Industrieanzeiger 93 (1971), Heft 51

[4.9] Opitz, H.: Vorschubantriebe, Lagerungen, Führungen. Bericht über die VDW-Konstrukteur-Arbeitstagung am 27., 28. und 29. Oktober 1966. Selbstverlag des Laboratoriums für Werkzeugmaschinen und Betriebslehre der Rheinisch Westfälischen Technischen Hochschule Aachen 1966

[4.10] Stöferle, D.; Dräger, H.-J.: Untersuchungen des Dämpfungsverhaltens von Werkzeugmaschinenschlitten auf Gleit- und Wälzführungen. Frankfurt, VDW-Forschungsbericht Nr. 0405, Mai 1970, Heft 1, Heft 2

[4.11] Jorden, W.: Untersuchungen an einem Lageregelkreis für Werkzeugmaschinen unter besonderer Berücksichtigung des Spiels. Fortschritt-Berichte VDI-Zeitschrift, Reihe 2, Nr. 20, Dezember 1969

[4.12] Domrös, D.: Über das Verschleiß- und Reibungsverhalten von Werkzeugmaschinen-Gleitführungen. Aachen, Dissertation an der Technischen Hochschule, 1966

[4.13] Dutcher, J.L.: Maschinengestaltung und Regelantrieb für numerische Steuerungen. Firmenschrift der General Electric (GET 3210.2)

[4.14] Klotter, K.: Technische Schwingungslehre, 1. Band, Seite 73 bis 77. Berlin-Heidelberg-New York, Springer-Verlag, 1951

[4.15] Hütte I: Theoretische Grundlagen, 28. Auflage, 4. Abschnitt, Schwingungen. Berlin, Verlag von Wilhelm Ernst und Sohn, 1955

[4.16] Wolters, P.: Rechnerunterstützte Dimensionierung von Vorschubantrieben für numerisch gesteuerte Werkzeugmaschinen. Aachen, Dissertation an der Technischen Hochschule, 1976

[4.17] Arafa, H.: (siehe Literaturhinweis [2.12])

[4.18] Stuck, K.: Untersuchungen über die Federeigenschaften der vorgespannten Teile von Schraubenverbindungen. Aachen, Dissertation an der Technischen Hochschule, 1968

[4.19] Weyand, M.: Hauptspindellagerungen von Werkzeugmaschinen. Reibungs- und Temperaturverhalten der Wälzlager. Aachen, Dissertation an der Technischen Hochschule, 1969

[4.20] Wiche, E.: Die radiale Federung von Wälzlagern bei beliebiger Lagerluft. Konstruktion, 1967, Heft 5, Seite 184 bis 192. (Druckschrift WTS. 93 der Kugellagerfabriken GmbH, Schweinfurt)

[4.21] Bosch, M.: Über das dynamische Verhalten von Stirnrad-Getrieben unter besonderer Berücksichtigung der Verzahnungsgenauigkeit. Aachen, Dissertation an der Technischen Hochschule, 1965

[4.22] Weck, M.: Werkzeugmaschinen Band 3, Automatisierung und Steuerungstechnik. Düsseldorf, VDI-Verlag, 1978

[4.23] Herold, H.-H.; Maßberg, W.; Stute, G.: Die numerische Steuerung in der Fertigungstechnik. VDI-Verlag GmbH, Düsseldorf, 1971

[4.24] Autorenkollektiv: Die Auslegung von Vorschubantrieben für NC-Maschinen. Industrie-Anzeiger 90 (1968), Heft Nr. 67

[4.25] Tönshoff, H.K.: Auswahl und Berechnung von Führungssystemen für Werkzeugmaschinen. Die Maschine, 1981, Heft 3, Seite 15 bis 18

[4.26] Opitz, H.; Bongartz, B.: Untersuchungen des Reibungs- und Verschleißverhaltens von Werkzeugmaschinen-Gleitführungen. Frankfurt, VDW-Forschungsbericht, April 1969

[4.27] (ohne Verfasserangabe), Wälzführungen, die an Werkzeugmaschinen eingesetzt sind. Die Maschine, 1981, Heft 1, Seite 15 bis 18, Heft 2, Seite 15 bis 19

[4.28] Wiemer, A.: Luftlagerung. Berlin, VEB Verlag Technik, 1968

[4.29] Blondeel, E.; Snoeys, R.; Devrieze, L.: Aerostatic Bearings with infinite Stiffness. Annals of the C.I.R.P., Vol. 25/1/1976

[4.30] Bongartz, B.: Die Tragkraftkomponente der Gleitführung und ihr Einfluß auf das Reibungsverhalten im Bereich der Mischreibung. Aachen, Dissertation an der Technischen Hochschule, 1970

[4.31] Harmonic Drive, Firmenschrift der Firma Harmonic Drive Systems GmbH, Langen

[4.32] Jakob, L.: Drehstarre, flexible Kupplung. Konstruktion, Elemente, Methoden, 1979, Heft 5

[4.33] Jakob, L.: Sicherheitskupplungen in Achs-Antrieben von NC-Maschinen, Konstruktion, Elemente, Methoden, 1979, Heft 7

Simon, W.: Die numerische Steuerung von Werkzeugmaschinen, 2. Aufl. München, Carl Hanser Verlag, 1971

Palmgren, A.: Grundlagen der Wälzlagertechnik. Franksche Verlagshandlung, Stuttgart, 1964

Danek, P.; Polacek, M.; Spacek, L.; Tlusty, I.: Selbsterregte Schwingungen in Werkzeugmaschinen. Berlin, VEB Verlag Technik, 1962

Stute, G.; Schmid, D.: Typische Konturfehler bei der Werkstückbearbeitung mit numerischen Bahnsteuerungen. Annals of the C.I.R.P., Vol. XVIII, 1970, S. 531 bis 540

Niemann, G.; Ehrlenspiegel, K.: Anlaufreibung und Stick-Slip bei Gleitpaarungen. VDE-Zeitschrift, 105, 1963, Heft 5

DIN 1319, 1974: Grundbegriffe der Meßtechnik

DIN 5479, Mai 1978: Übersetzung bei physikalischen Größen, Begriffe, Formelzeichen

DIN 69051: Kugelgewindetriebe:
    Teil 1, Entwurf Juni 1978: Begriffe und Benennungen
    Teil 2, Entwurf August 1978: Nenndurchmesser und Nennsteigungen
    Teil 3, Entwurf Juni 1978: Abnahmebedingungen
    Teil 4, Entwurf Juni 1978: Berechnung der Tragfähigkeit
    Teil 5, Entwurf November 1978, Anschlußmaße

VDI/VDE-Richtlinie 2185, Mai 1963: Formulierung der Anforderungen an geregelte Antriebe

5 Command Value Modification

5.1 Purpose of the Command Value Modification

On numerically controlled machine tools, the command value modification of position command values serves to reduce the dynamic contour deviations which result from distortions in the signal transfer behavior of the position control loop. In this process, the position command values which generate the smallest position increments, are adjusted in their course over time to the transfer behavior of the position control loop. With the aid of command value modifiers, less dynamic feed drives can deliver the same accuracy as highly dynamic feed drives would without command value modification.

Besides the improvement in the accuracy of contouring, the command value modification also causes a reduction in the stresses on the drive system, including the mechanical transmission elements. In this manner, the position loop gain K_v can be kept constant over the total velocity range. No supplementary devices are necessary for its reduction at higher velocities.

The improvement in contour accuracy and the reduction in the stress on the drive, achieved with the command value modification, result however, in an increase of the movement duration. The magnitude of this time interval increase depends on the size of the command value modification. Since the time behavior of the position command value is modified by means of a delay, we can also refer to a *position command value smoothing*.

5.2 Smoothing the Command Values

Figure 5.1 shows the command contours and the command values of both X and Y directions of motion, for linear, parabolic, and circular movements. Break points are clearly recognizable at the transitions from one motion segment to the next.

During start and stop processes, and during discontinuous contour direction changes, the smoothing of the command values causes the break points in the corresponding distance over time functions to be replaced by continuous transitions. Figure 5.2 presents the unsmoothed position command value, in contrast with the smoothed position command value of an axis, for a linear positioning process. As one can see, the break points at the start and at the end of the positioning process are replaced by smooth transitions. For acceleration processes, this position command value smoothing means a temporary replacement of the linear time dependency of the command value $x_s(t)$ with a higher order time dependence. This results in tangential transitions at the beginning and at the end of the distance vs. time functions [5.1].

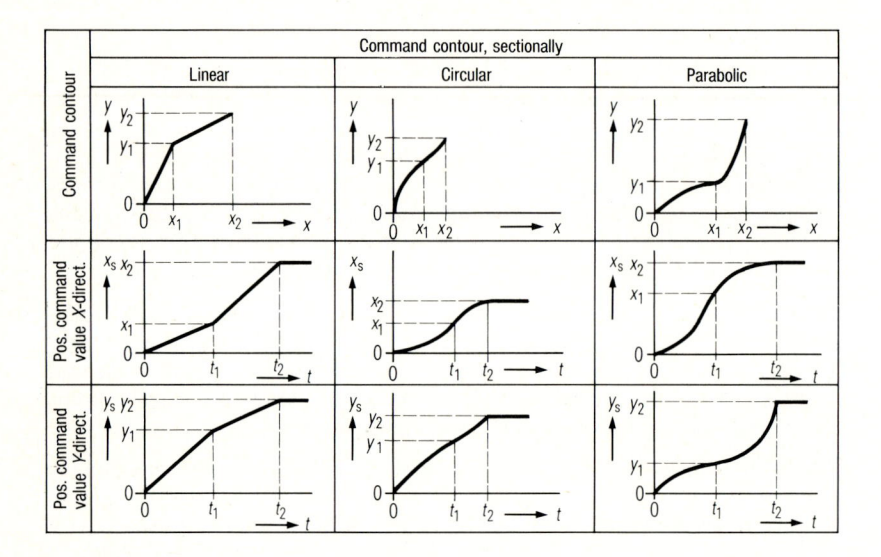

Fig. 5.1
Position command values of the X and Y axes when generating contours out of the smallest increments, without command value modification

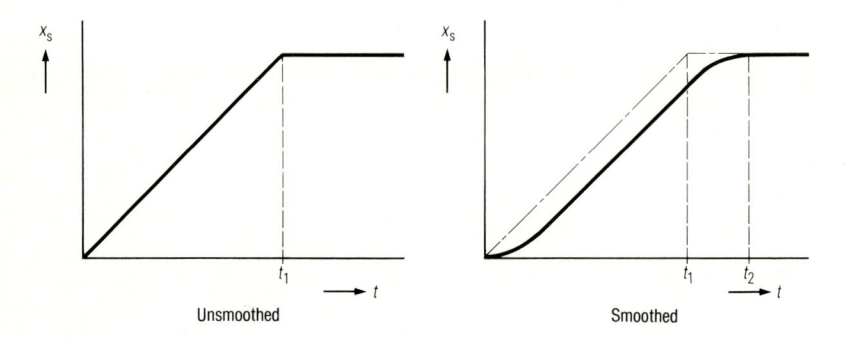

Fig. 5.2
Position command values x_s of a linear positioning process

In case of the unsmoothed position command value, the first derivative of the command value – the command value of the contour velocity v_f – is sectionally constant. It is referred to as a *velocity-controlled position command value generation*. If we execute a command value modification by smoothing the position command value, the second derivative of the position command value – the acceleration command value a_f – during a velocity change, will additionally be limited to the constant limit value $\pm a_{f0}$. This is referred to as an *acceleration-controlled position command value generation*. Figure 5.3 shows the course over time of the position command value and of the command values generated in the two processes.

For the acceleration-controlled position command value generation, the course of the position command value $x_s(t)$ has an "S" shape, instead of the ramp form of the graph obtained for the velocity-controlled command value generation (linear positioning). Between t_1 and t_2, the command velocity is constant. During the acceleration phase up to t_1, respectively the deceleration phase between t_2 and t_3, the position command value function $x_s(t)$ displays a quadratic relationship.

For a numerically controlled machine tool, this position command value smoothing, with limitation of the command values for acceleration and velocity, is generally sufficient for the achievement of a good dynamic behavior. At contour direction changes, in order to avoid contour deviations caused by command value smoothing, it is necessary to divide the entire travel process

Velocity-controlled position command value generation

Acceleration-controlled position command value generation

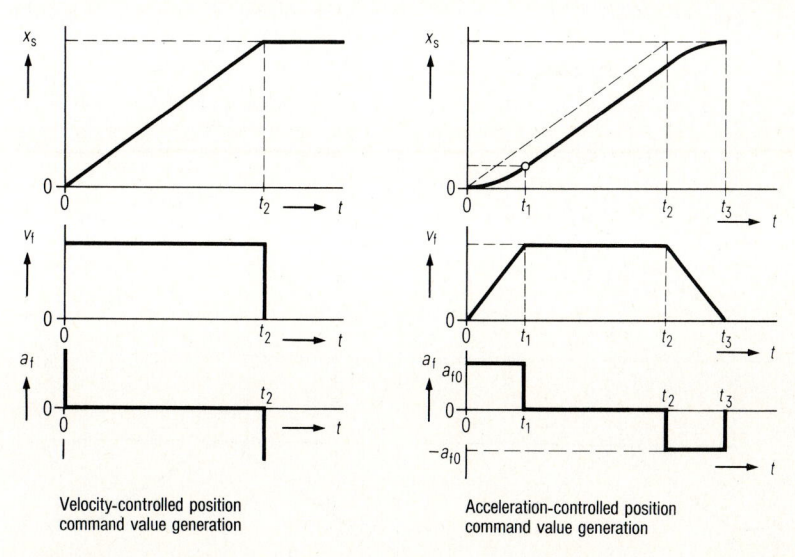

Fig. 5.3
Graph of the position command value with limited velocity (left), and limited velocity and acceleration (right)

into consecutive positioning subprocesses within the given contour points. For this, the components of the command velocity must be, temporarily, simultaneously set back to zero in all the axes. As an example, fig. 5.4 shows the courses of the position command value and of the command velocity v_f, with the limitation of the command acceleration, for the generation of a 90° corner out of two straight lines.

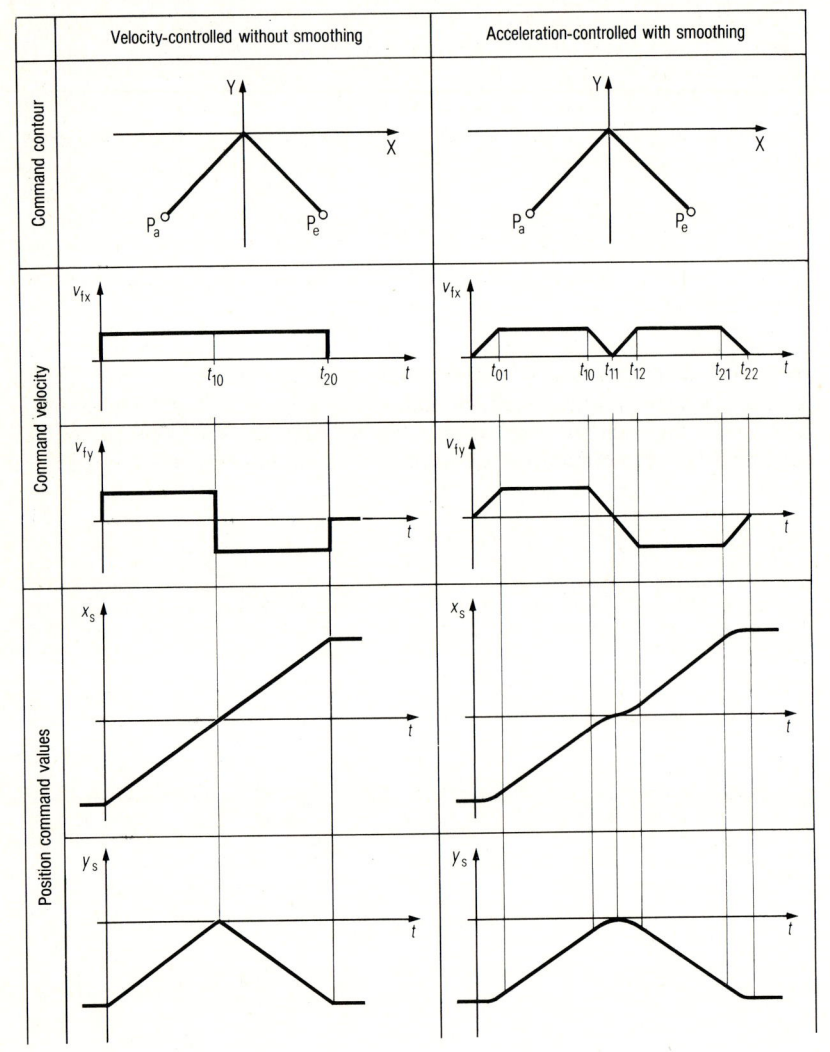

Fig. 5.4
Position command values and command values of velocity, left without, and right with smoothing – at the traveling of a 90° corner

5.3 Effects of Command Value Smoothing

On numerical controls, the most frequently programmed smallest path increments are straight lines and arcs (circular). However, since the computing algorithms used at linear interpolations for command value determination are the same as those generally used for determination of command values at circular interpolations, it makes sense to analyze the effects of position command value smoothing on segmented linear movements, especially at travels around corners. In this manner it is possible to establish a relationship between the limit value of the command acceleration a_{f0}, and the contour deviations and load characteristic values. The latter are determined from the presence of distortions in the signal transfer behavior of the position control system (see 1.2.3.4 and 4.2.4.3).

5.3.1 Stress and Contour Deviation Characteristics

At traveling around corners, overshoot and corner deviations can occur, as shown in fig. 5.5. The contour deviation area is contained between the desired command contour and the actual contour obtained.

When traveling around a 90° corner, the characteristic values for the evaluation of the position control are the corner deviation e_E and the overshoot width $\ddot{u}_a$. For linear position control, the latter is identical to those observed in positioning processes, where overshoots are generally not tolerated (see 1.3.2).

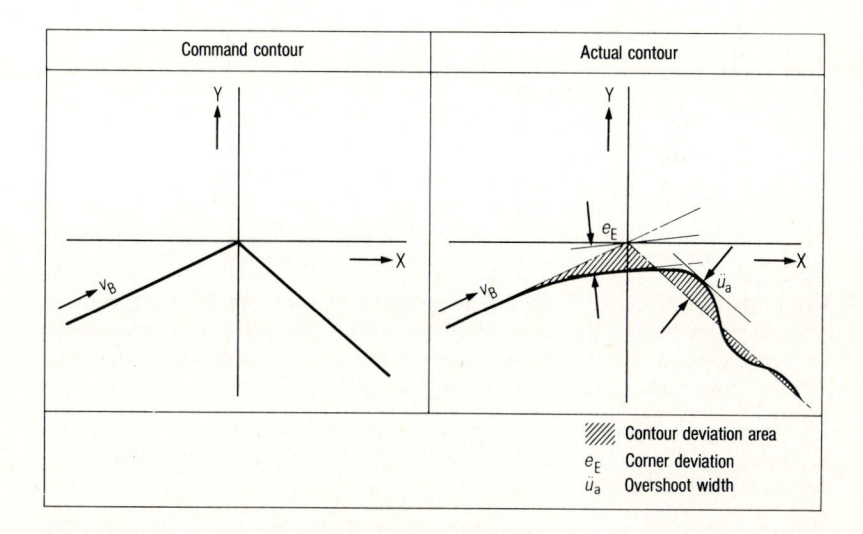

Fig. 5.5 Contour deviations at travel around a corner

The maximal value of the actual acceleration $a_{i\,max}$ is used as stress characteristic value in positioning processes. This value indicates the measure of dynamic load occurring in the drive system and transmission elements of the position control system. When the tolerances and the load limits are given, the contour deviation e_E, $\ddot{u}_a$, and the stress characteristic value $a_{i\,max}$ offer the possibility of determining a suitable value for the limit of the acceleration command value a_{f0}.

5.3.2 Contour Movements

Figure 5.6 shows the course of actual contours at traveling around a corner, with and without position command value smoothing. The proposed position control loops thereby possess linear behavior, but different control system structures.

The course of the contour is shown in a coordinate system based on the value s_H, where s_H is a distance variable and is designated as auxiliary distance:

$$s_H = \frac{v_B}{\omega_{0A}} \tag{5.1}$$

v_B contour velocity
ω_{0A} nominal angular frequency of the drive

For a position control loop of basic structure, the control system has the behavior of a 2nd order delay element (PT_2-element) with integral element switched in series. In a control with dead time, this serial connection is complemented with a dead time element (see fig. 1.19). For the position control with "2nd oscillator inside", the dead time element is replaced by another oscillation element (position control loop structure for direct position measurement, see fig. 4.13). In position control with "2nd oscillator outside", the transfer behavior of the position control is determined through the serial connection of a control loop with basic structure and an external oscillating element (position control loop structure for indirect position measurement, see fig. 4.9).

The dotted curves shown in fig. 5.6 characterize the course of the actual contour for velocity controlled position command values. The solid lines characterize the actual contour course of movements executed with acceleration controlled position command values. Both are programmed without intermediate stops.

A comparison of the two curves reveals clearly the reduction of the contour deviations of the acceleration controlled, vis-a-vis the velocity controlled position command value generation.

The corner deviations can be reduced further, if a short dwell time t_V is introduced between the single positioning processes. For this, the velocity command

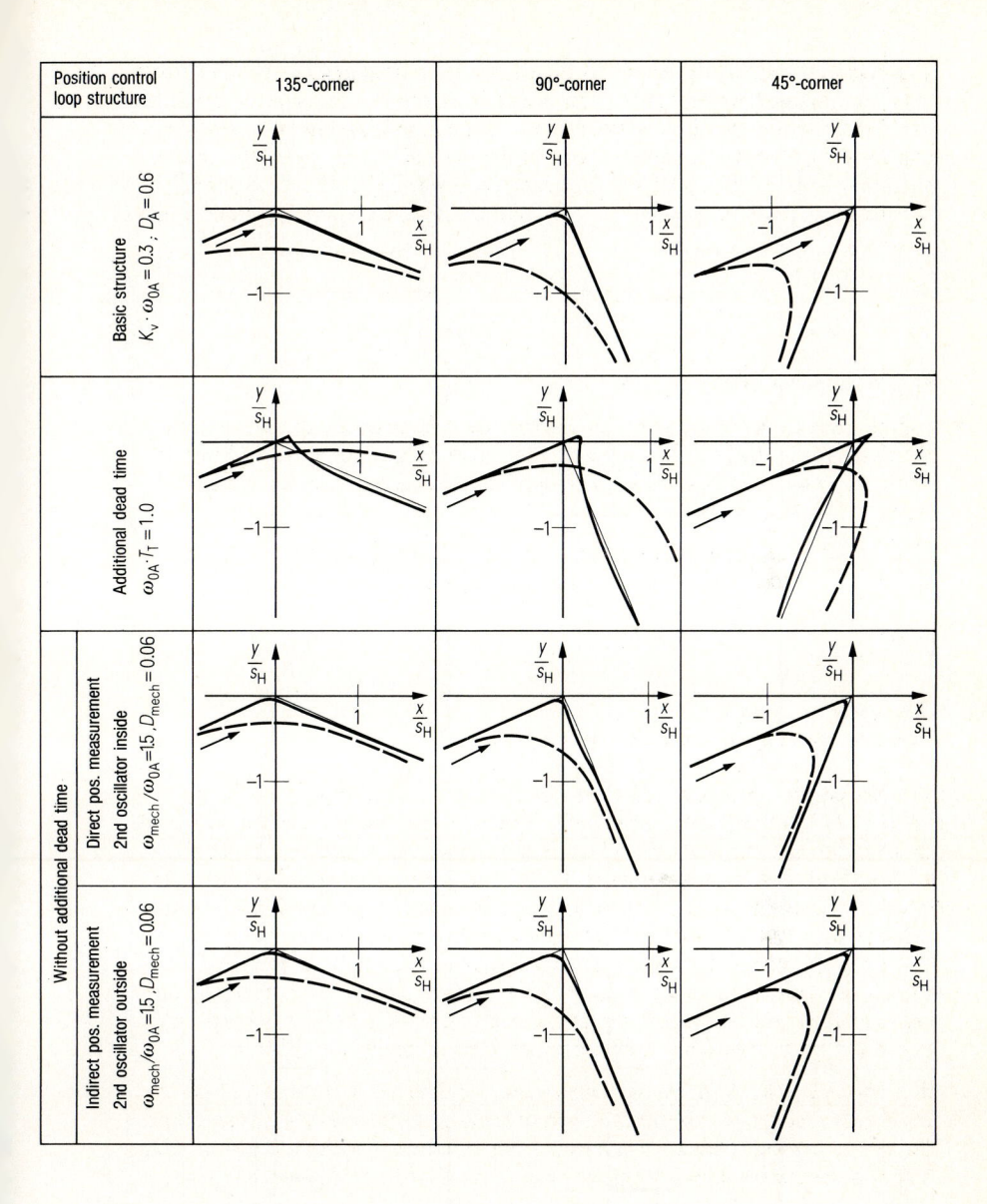

Fig. 5.6
Actual contour curve at traveling around corners with linear position control loops, and acceleration (——), resp. velocity (----) controlled position command value generation

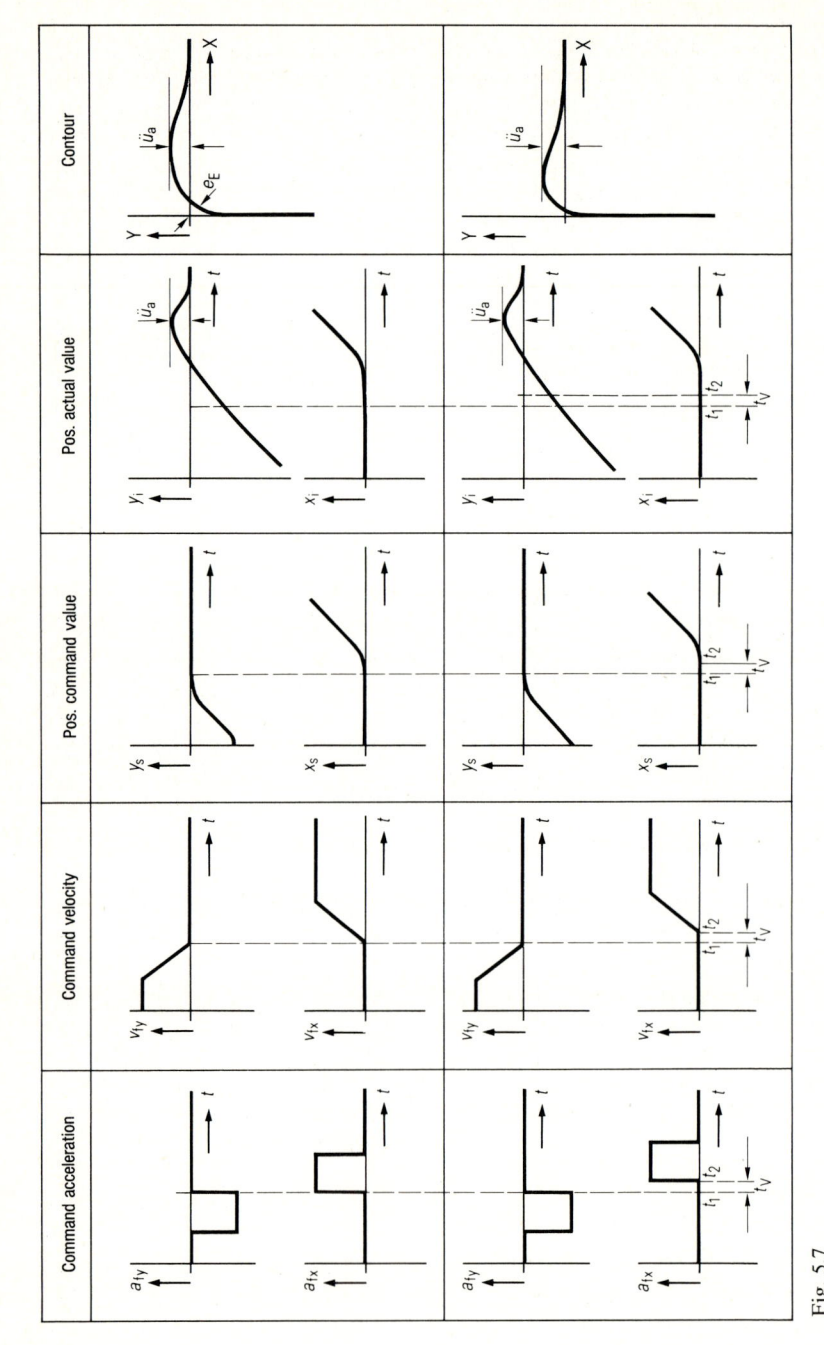

Fig. 5.7
Signal progress for the acceleration controlled position command value generation with (below) and without (above) dwell time, at the traveling of a 90° corner

value must be held at zero in the interval between the deceleration and acceleration phases of two consecutive positioning processes. The duration of the dwell time is selected according to the tolerance band with which the corner points desired are to be reached. Figure 5.7 shows the signal course of the command and actual values, when generating an axis-parallel 90° corner with a tolerance dependent dwell time inserted between the two movement segments. The structure of the position control loop is presumed to be basic, according to fig. 5.6. It should be noted, that the corner deviation is considerably reduced while the overshoot width remains unchanged.

For position control loops with reversing errors, position command value smoothing also causes reductions of contour deviations, and especially for corner deviations. Reductions in the overshoot width however, are limited by the presence of the reversing error (also see 4.2.4.3).

5.3.3 Diagrams for Contour Deviation and Stress Characteristics

The dependence of the characteristic values e_E and $ü_a$ on the limit value of the command value of the acceleration a_{f0} is not linear, and can be determined only from diagrams. The corresponding graph curves are influenced by the parameters of the position control loop. For position control loops with dead times, whose block diagrams are shown in fig. 1.19, these parameters are:

ω_{0A} the nominal angular frequency of the drive
D_A the damping gradient of the drive
K_v the position loop gain
T_T the dead time
(At position control loops with basic structure there is no dead time.)

In order to reduce the parameters, the characteristic values e_E, $ü_a$, and $a_{i\,max}$ in the diagrams are normalized. They are shown over the normalized and referenced limit value of the acceleration a_{f0} command value. The normalizing parameter for $ü_a$ and e_E is the auxiliary distance $s_H = v_B/\omega_{0A}$, according to equation (5.1). The normalizing parameter for the acceleration values $a_{i\,max}$ and a_{f0} is the auxiliary acceleration a_H, which is defined as follows:

$$a_H = \Delta v_B \cdot \omega_{0A} \qquad (5.2)$$

Δv_B maximum programable velocity change of the numerical control (without directional change)
ω_{0A} drive nominal angular frequency

The normalized limit value of the acceleration command value a_{f0}/a_H is referenced to the value K_v/ω_{0A}.

The parameters remaining are the referenced position loop gain K_v/ω_{0A}, the referenced dead time $\omega_{0A} \cdot T_T$, and the damping gradient D_A.

For the position control loop with basic structure and dead time shown in fig. 1.19, the normalized corner deviation e_E/s_H, the normalized overshoot width $\ddot{u}_a/s_H$ for the traveling of a 90° corner, and the normalized maximal actual acceleration $a_{i\,max}/a_H$ for start or stop processes, are shown in figs. 5.8 through 5.13. The diagrams of the normalized corner deviations refer to a continuous transition from deceleration to acceleration phases, and disregard the possibility of reducing the deviations through tolerance-dependent dwell times.

The dotted lines in the diagrams show the limit values of the normalized characteristic values e_E/s_H, $\ddot{u}_a/s_H$ or $a_{i\,max}/a_H$, which are approached by the curves of $a_{f0}/a_H \to \infty$. The tables inserted in the diagrams show the corresponding numerical values. These limit values for the characteristical values, are reached with velocity controlled command value generation.

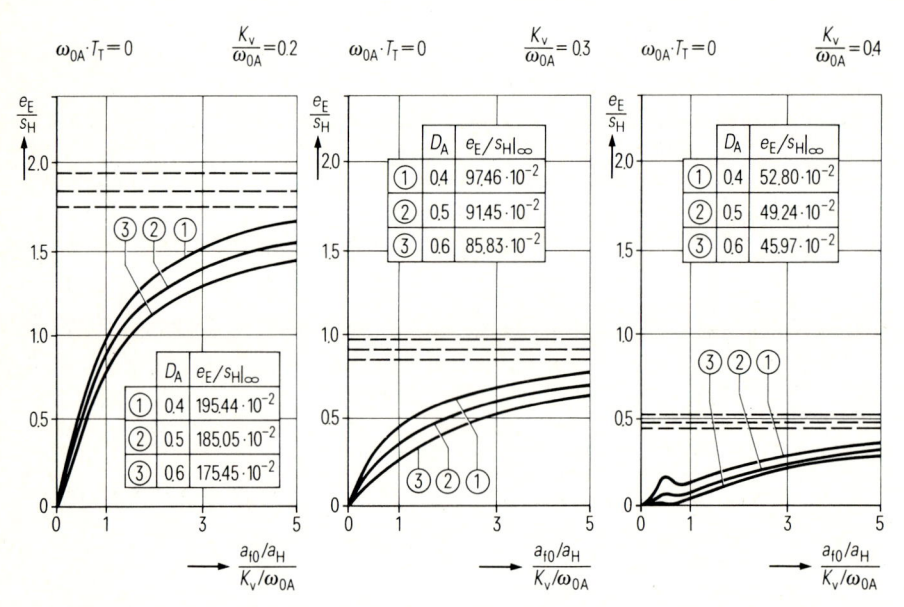

D_A	$e_E/s_H|_\infty$
① 0.4	$195.44 \cdot 10^{-2}$
② 0.5	$185.05 \cdot 10^{-2}$
③ 0.6	$175.45 \cdot 10^{-2}$

D_A	$e_E/s_H|_\infty$
① 0.4	$97.46 \cdot 10^{-2}$
② 0.5	$91.45 \cdot 10^{-2}$
③ 0.6	$85.83 \cdot 10^{-2}$

D_A	$e_E/s_H|_\infty$
① 0.4	$52.80 \cdot 10^{-2}$
② 0.5	$49.24 \cdot 10^{-2}$
③ 0.6	$45.97 \cdot 10^{-2}$

Fig. 5.8
Acceleration-controlled position command value generation. Corner deviation at the traveling of a 90° corner, without dead time

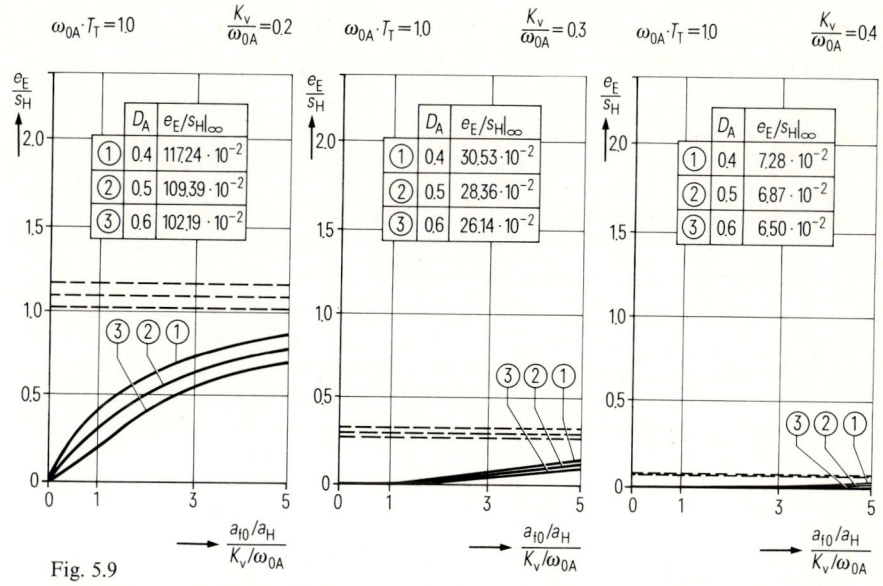

Fig. 5.9
Acceleration-controlled position command value generation. Corner deviation at traveling of a 90° corner, with dead time

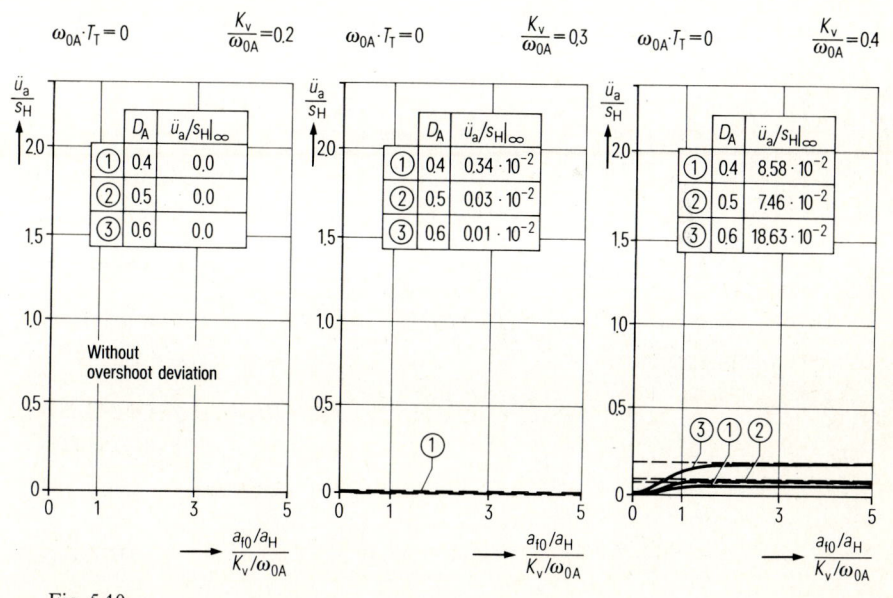

Fig. 5.10
Acceleration controlled position command value generation. Overshoot width during stopping process, without dead time

Fig. 5.11
Acceleration controlled position command value generation. Overshoot width during stopping process, with dead time

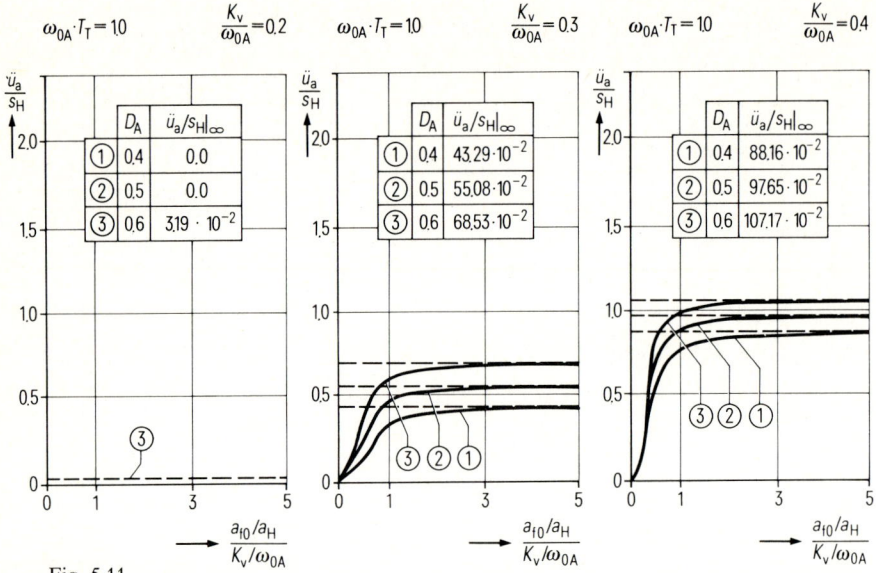

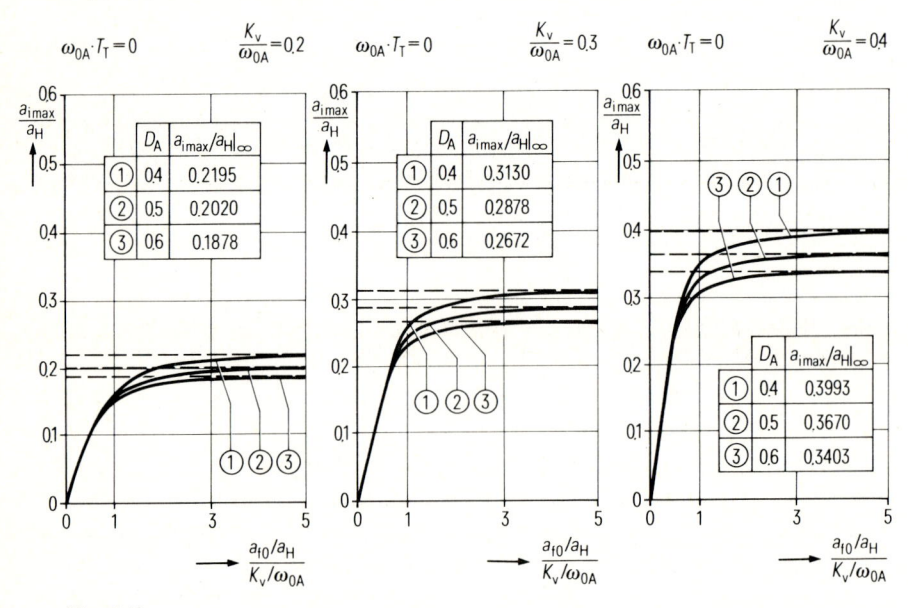

Fig. 5.12
Acceleration controlled position command value generation. Maximal actual acceleration (deceleration) during starting (stopping) processes, without dead time

Fig. 5.13
Acceleration controlled position command value generation. Maximal actual acceleration (deceleration) during starting (stopping) processes, with dead time

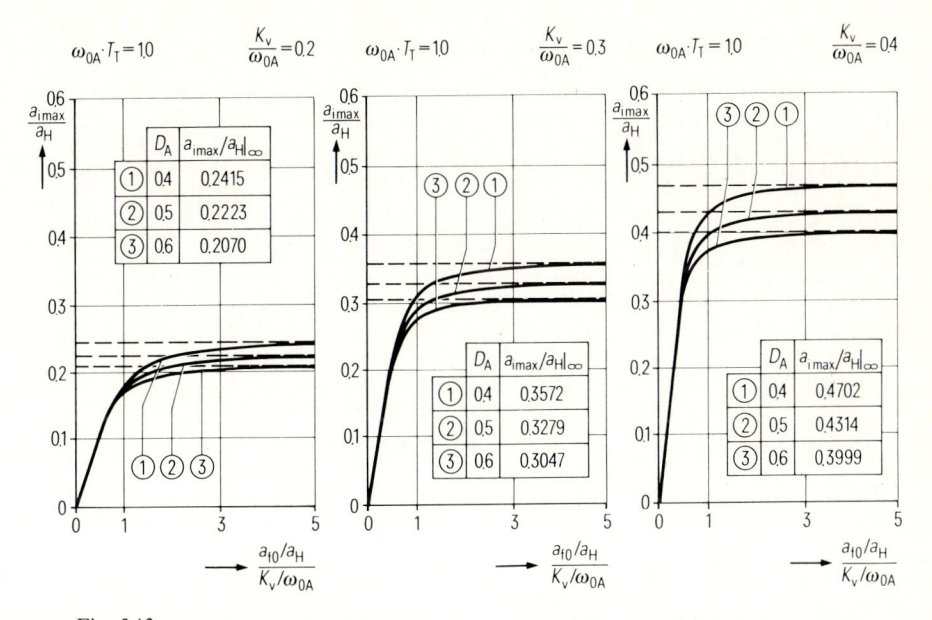

From the diagrams (figs. 5.12 and 5.13) for the normalized maximal acceleration $a_{i\,max}/a_H$, we can derive that when $K_v/\omega_{0A} = 0.2$, for values of:

$$a_{f0} \leq \tfrac{1}{2} \cdot K_v \cdot \Delta v_B \tag{5.3}$$

the maximal value of the actual acceleration $a_{i\,max}$ and the limit value of the command value for the acceleration a_{f0}, coincide approximately, so that:

$$a_{i\,max} \approx a_{f0} \tag{5.4}$$

The maximal actual acceleration is only approximately half of that for pure velocity control (see the example in section 4.3.1.2.7).

If we compare the diagrams with and without dead time effects, we observe that the corresponding graph curves are similar. The corner deviations decrease with increasing dead times, and the overshoot widths increase with the maximal values of the acceleration stress.

Further diagrams, also referring to velocity distortions, can be found in [5.2].

5.4 Determining the Command Value for Acceleration

The control parameters known and the given limits for the overshoot width $ü_a$, the corner deviation e_E, or the maximal permitted actual acceleration $a_{i\,max}$, all presented in the diagrams in 5.3.3, allow the approximate determination of the limit value for the command value of the acceleration $a_{f\,0}$.

The maximum admissible actual acceleration is determined by the current limit set, to protect the motor and the machine. This drive-delivered possible limit acceleration $a_{i\,Grenz}$ becomes for the feed screw drive, according to equations (2.43), (2.41), and (2.6):

$$a_{i\,Grenz} = \frac{(M_{Grenz\,M} - M_L) \cdot h_{Sp}}{2\pi \cdot i \cdot J_{Ges}} \tag{5.5}$$

and for rack and pinion drives, according to equation (2.44):

$$a_{i\,Grenz} = \frac{(M_{Grenz\,M} - M_L) \cdot r_{Ri}}{i \cdot J_{Ges}} \tag{5.6}$$

$M_{Grenz\,M}$ motor torque with current limitation
M_L load torque ($M_L = \Sigma M_R + M_V$)
h_{Sp} screw lead
r_{Ri} pinion radius
i gear ratio
J_{Ges} total moment of inertia reflected on the motor shaft

The limit acceleration can be calculated from the data for the feed drive. The maximal actual acceleration of the feed table cannot exceed this limit value. It has to hold that:

$$a_{i\,max} \leq a_{i\,Grenz} \tag{5.7}$$

The following procedure is suggested for the determination of the permissible limit value for the acceleration command value $a_{f\,0}$:

▷ Determine the position control loop parameters and the drive's characteristic data (see table 5.1).
▷ Establish the acceptable values for the corner deviation e_E, the overshoot width $ü_a$, as well as the value for the maximal actual acceleration $a_{i\,max}$; normalize the data necessary for use with the diagrams (see table 5.2)
▷ Determine from the diagrams the possible referenced limit values of the command value for the acceleration (fig. 5.8–5.13), and then derive the smallest possible value and the absolute limit for the command value of the acceleration $a_{f\,0}$ (see table 5.3).

(The values for the drives in positions 1 and 3 from table 4.6 are used as an example)

Example:

Determine the limit value of the command acceleration that could be set in the numerical control of drives 1 and 3 from table 4.6:

▷ Drive ①: Direct drive with DC servo motor 1HU3 104-0AD01

▷ Drive ③:
   Drive with gear with $i=2.5$, and DC servo motor 1HU3 076-0AF01

Table 5.1
Summary of the position control loop variables, for the determination of the limit value for the command value of acceleration a_{f0}

Symbol	Designation	Unit	Equations	Drive ①	Drive ③
ω_{0A}	Drive nominal angular frequency	s^{-1}	(2.21), (2.22)	137	125
D_A	Damping gradient of the drive	–	–	0.5	0.5
T_T	Dead time	ms	–	7.67	7.67
K_v	Position loop gain	s^{-1}	–	27.5	25
$M_{\text{Grenz M}}$	Motor torque with current limitation	Nm	–	100	40
M_L	Load torque	Nm	(2.25)	4	1.2
h_{Sp}	Feed screw lead	mm	–	10	10
i	Gear ratio	–	–	–	2.5
J_{Ges}	Total moment of inertia	kgm²	(2.3), (2.51)	0.0395	0.0104
$a_{i\,\text{Grenz}}$	Limit acceleration	m/s²	(5.5), (5.6)	3.87	2.38
v_B	Contour velocity	m/min	–	3	3
Δv_B	Maximal programable velocity change	m/min	–	12	12
s_H	Auxiliary distance	mm	(5.1)	0.365	0.4
a_H	Auxiliary acceleration	m/s²	(5.2)	27.4	25

Table 5.2
Limit values and normalized, resp. referenced values for applying diagrams of section 5.3.3 (figs. 5.8 through 5.13)

Symbol	Designation	Unit	Equations	Drive ①	Drive ③
e_E	Corner deviation	mm	–	0.1	0.1
$\ddot{u}_a$	Overshoot width	mm	–	ca. 0	ca. 0
$a_{i\,max}$	Maximal actual acceleration	m/s²	Note equation (5.7)	2	2
e_E/s_H	Normalized corner deviation	–	–	0.275	0.25
$\ddot{u}_a/s_H$	Normalized overshoot width	–	–	ca. 0	ca. 0
$a_{i\,max}/a_H$	Normalized maximal actual acceleration	–	–	0.073	0.08
K_v/ω_{0A}	Referenced position loop gain	–	–	0.2	0.2
$\omega_{0A} \cdot T_T$	Referenced dead time	–	–	ca. 1	ca. 1

Table 5.3
Summary of the resulting values to be compared for the limit value for the acceleration, and results

Evaluation criterion	Normalized, resp. referenced nominal acceleration	Example drive ①	Example drive ③
e_E/s_H	$\dfrac{a_{f0}/a_H}{K_v/\omega_{0A}}$	≤ 0.8	≤ 0.7
$\ddot{u}_a/s_H$	$\dfrac{a_{f0}/a_H}{K_v/\omega_{0A}}$	$\leq \infty$	$\leq \infty$
$a_{i\,max}/a_H$	$\dfrac{a_{f0}/a_H}{K_v/\omega_{0A}}$	≤ 0.37	≤ 0.41
Result:	Maximal admissible value	≤ 0.37	≤ 0.41
Setting value:	a_{f0} m/s²	$\leq$ approx. 2	$\leq$ approx. 2

The characteristic data are summarized in table 5.1, and the desired characteristic values in table 5.2. From the results listed in table 5.3, it can be seen that the actual acceleration given represents the limit value for the command value of the acceleration to be set. Since $a_{f0} < \frac{1}{2} K_v \cdot \omega_{0A}$ and $K_v = 0.2 \cdot \omega_{0A}$, the equality given in equation (5.4) for the occurring actual acceleration and the command acceleration can also be observed.

As long as K_v/ω_{0A} remains smaller than 0.2, no overshooting will occur even at longer dead times. This means, that it is not necessary to limit the acceleration in reference to the value of $\ddot{u}_a$.

If we were to establish the value of the admissible corner deviation at 0.01 mm, the limit value that would result for drive ① would be approximately 0.43 m/s², and for drive ③, approximately 0.35 m/s². In this case however, it would make sense to program a dwell time t_v, because that too would reduce the corner deviation. Some numerical controls also allow the programing of an "exact stop". The corner deviation is then determined by the set tolerance limit.

5.5 Conclusions

Position command value smoothing brings advantages during starting and stopping processes, and at directional changes. For one, it offers the possibility to reduce the dynamic contour deviations that occur on numerically controlled machine tools; second, it makes possible the utilization of less dynamic drives, for the same accuracy of contour that would be expected of more dynamic drives. This is not true for the mechanical transmission elements, whose stiffness cannot be reduced without increasing the reversing error, and thus decreasing the mechanical nominal angular frequency of the drive $\omega_{0\,mech}$.

5.6 Bibliography

[5.1] Stute, G.; Stof, P.: Beeinflussung der Konturgenauigkeit von numerisch bahngesteuerten Werkzeugmaschinen durch das dynamische Verhalten der Führungsgrößenerzeugung. Frankfurt, VDW-Bericht 1006, Teil III, 1976

[5.2] Stof, P.: Untersuchungen über die Reduzierung dynamischer Bahnabweichungen bei numerisch gesteuerten Werkzeugmaschinen. Berlin-Heidelberg-New York, Springer-Verlag, 1978

Stof, P.: Bahnbewegung numerisch gesteuerter Werkzeugmaschinen mit beschleunigungsgesteuerter Führungsgrößenerzeugung.
HGF Kurzberichte, Industrieanzeiger 99 (1977), Heft 37, Seite 662 bis 663 und Heft 41, Seite 733 bis 734

6 Measurements on Feed Drives

6.1 Substantiation, Aim

In order to judge an installed feed drive, and make statements regarding the quality and accuracy of the superceeding position control loop, it is necessary to conduct measurements on that feed drive and the position control loop. The purpose of these measurements is to obtain the parameters of the position control loop and the speed control loop, and of the mechanical transmission elements:

Position Control Loop

K_v Adjustable position loop gain

Speed Control Loop

ω_{0A}, f_{0A} Nominal angular frequency, resp. nominal frequency of the drive
D_A Damping gradient of the drive

Mechanical Transmission Elements

$2\varepsilon_u$ Reversing error (back lash)
$\omega_{0\,mech}, f_{0\,mech}$ Nominal angular frequency, resp. nominal frequency of the mechanical transmission elements
D_{mech} Damping gradient of the mechanical transmission elements

Methods to obtain the mentioned measurement values, and conclusions based upon these, are presented below. The selected measuring procedures take into account that simple measurement instruments might be used. These are:

▷ a variable DC source with switch and polarity switch-over (battery box)

▷ two voltmeters (at least 20 kΩ/V, moving-coil instrument)

▷ an endless potentiometer or a 10-turn potentiometer (measuring potentiometer) with mounting bracket and coupling for mounting on the DC servo motor (see below)

▷ a shunt

▷ a resistor approximately 22 kΩ/0.33 W

▷ two dial indicators (resolution 0.01 mm and 0.001 mm),

▷ a load cell,

▷ a recorder (e.g. ink jet recorder or storage oscilloscope)
▷ a moving iron amperemeter to determine the form factor (e.g. up to 30 A_{eff}).

(Concerning the measuring potentiometer: since only small angular movements have to be recorded, a friction coupling with a conical rubber head – as that on a hand held tachometer – is sufficient.)

With simple measuring procedures, some measurements can only be approximated and some cannot be determined at all (e.g. D_{mech}, or the progression of the table position over time at low feed rates). The behavior of a feed drive over time (i.e. values $f_{0\,mech}$, D_{mech}, $f_{0\,A}$, D_A) can be determined exactly only by recording of the frequency response curve (Bode diagram). To record the positioning response, positioning measurement is necessary.

The following instruments are necessary additionally:
    a sine generator with adjustable drift (approx. 1–200 Hz)
    an acceleration transducer
    an inductive position transducer
    under circumstances, a frequency response curve recorder (however, at least a two channel high speed recorder with 1 m/s paper feed).

With the simple measuring procedures presented here, the judgements and conclusions necessary for the optimal operation of the single components can be reached. These measuring procedures are also sufficiently detailed to be used during start-up.

6.2 Measuring Procedures

6.2.1 Measurements on the Mechanical Transmission Elements

Figure 4.2 shows the elements belonging to the mechanical transmission system. These include the complex of mechanical parts between the shaft of the DC servo motor and the point at which the tool touches the part, along the part contour. Depending on the position control, a part of this complex is inside or outside the position control loop. An overview of this is presented in figs. 1.18 and 4.1. The mechanical transmission elements are generally 2nd order delays, with a nominal angular frequency $\omega_{0\,mech}$ and damping gradient D_{mech} (see 4.2).

Figure 6.1 shows the arrangement of the measuring instruments on the mechanical transmission elements. The threaded screw drive is used as an example for the data to be gathered here. Similar arrangements of the measuring points will be found for other types of drives (rack and pinion systems, or worm gears), so that a separate discussion is not necessary.

The reversing error (back lash) and the total stiffness can be determined with the simple measuring instruments listed above. With these, conclusions can be drawn about the part accuracy achievable, and the nominal angular frequency of the mechanical transmission elements.

6.2.1.1 Reversing Error (Back Lash)

The total reversing error $2\varepsilon_u$ is, according to section 4.2.4.3, the sum of the frictional reversing error and the mechanical back lash. Depending on the position control, only part of the total reversing error is inside the position control loop, while the other part is outside of it (indirect position control according to fig. 4.9). The total reversing error lies entirely within the position control loop when direct position control is used, according to fig. 4.13. For

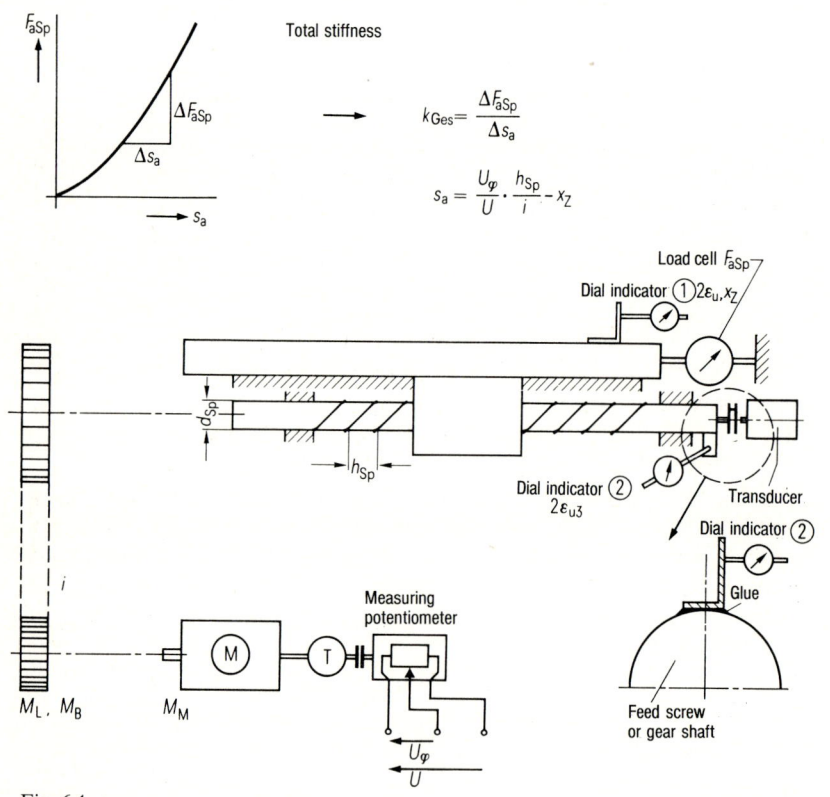

$$k_{Ges} = \frac{\Delta F_{aSp}}{\Delta s_a}$$

$$s_a = \frac{U_\varphi}{U} \cdot \frac{h_{Sp}}{i} - x_Z$$

Fig. 6.1
Measuring arrangement for measuring the reversing errors and the total stiffness of the mechanical transmission elements

indirect position control, for which the position control transducer is fixed to the motor shaft, there is no measurable reversing error within the position control loop. For the direct position control, for which the transducer is attached to the machine table, normally there is no reversing error outside the position control loop. If however, the guide ways are in poor condition or the transducer has been improperly mounted, a considerable amount of reversing error can be produced outside the position control loop.

Because of this, the following subdivisions have been established in reference to fig. 1.23:

$2\varepsilon_{u3}$ Reversing error in the mechanical transmission elements between the motor shaft and the position control transducer

$2\varepsilon_{u1}$ Reversing error in the mechanical transmission elements between the position control transducer and the feed table.

For expediency, the total reversing error $2\varepsilon_u$ of all mechanical transmission elements inside and outside the position control loop should be measured first, and the reversing error $2\varepsilon_{u3}$ of the mechanical transmission elements within the position control loop should be measured afterwards. The reversing error $2\varepsilon_{u1}$ of the mechanical transmission elements outside of the position control loop is then given by the difference:

$$2\varepsilon_{u1} = 2\varepsilon_u - 2\varepsilon_{u3} \tag{6.1}$$

The dial indicator ① in fig. 6.1 can measure the total reversing error $2\varepsilon_u$. Dial indicator ② measures the reversing error $2\varepsilon_{u3}$ within the position control loop. (For a rack and pinion system or a worm gear drive, the actuator for the dial indicator 2 is mounted to the shaft to which the position control transducer is coupled.)

Measuring procedure for the total reversing error $2\varepsilon_u$:

▷ The measuring potentiometer is fixed on the motor shaft. This will result in a voltage U_φ proportional to the angular position of the motor shaft. A drift-free voltage source U must be used. Voltage U_φ is measured.

▷ Dial indicator ① with a resolution of 0.01 mm is mounted to the machine table.

▷ The machine table is moved by approximately $+300$ μm under numerical control.

▷ The voltage on the measuring potentiometer should be noted; it is the value U_φ. The dial indicator is set to zero.

▷ The machine slide is moved again by $+300$ μm.

▷ The machine slide is moved now in the $(-)$ direction (jogging in small increments), until the previously noted voltage U_φ can be measured again.

▷ The value displayed by the dial indicator ① represents the total reversing error $2\varepsilon_u$.

▷ Measurements as above are repeated at different slide positions.

For the direct position control, the measured reversing error is active within the position control loop. A statement as to how this reversing error influences the behavior of the position control, can be made based on the description in 4.2.4.3.

Figure 6.2 shows a measuring diagram, recorded with an X/Y plotter. The movement of the machine slide is shown horinzontally, and the angular movement of the motor, proportional to the voltage U_φ, is shown vertically. At reversal point, the command steps move only the motor; the slide remains stationary until the total reversing error has been eliminated. In this example, the lathe slide has a reversing error of approximately 12 µm.

For relatively large reversing errors, $2\varepsilon_u$, can be measured without measuring potentiometer and with the drive switched off, as follows:

▷ The motor shaft is manually turned in one direction until the dial indicator 1 begins to move.
▷ The position of the motor shaft is marked.
▷ The motor shaft is turned in the opposite direction until the dial indicator also starts to move in the opposite direction.
▷ The angular position difference $\sphericalangle \beta$ is checked, and recomputed into a linear movement on the slide. (For this purpose, the ratio $\Delta U_\varphi/U$ is replaced by $\sphericalangle \beta/360°$ in equations (6.3) and (6.4).)

If the position transducer is situated directly on the motor shaft, no measuring potentiometer is needed; the reversing error $2\varepsilon_{u\,1}$ outside the position control loop can be directly measured. As in the measuring procedures for $2\varepsilon_u$, the

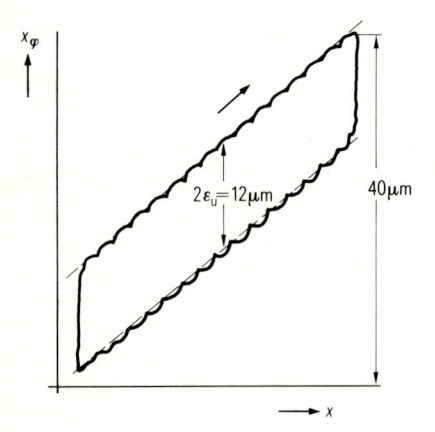

Fig. 6.2
Total reversing error measured on a feed slide (increment size of 2 µm)

motor will be moved under numerical control first in one direction, and then returned to the original position. Dial indicator 1 will show the reversing error.

For indirect position controls on a lead screw or a gear shaft, measurements of the reversing error $2\varepsilon_{u3}$ are carried out according to the following scheme.

Measuring procedures for reversing error $2\varepsilon_{u3}$ in the position control loop:

▷ The measuring potentiometer is mounted to the motor shaft.
▷ A cam made of sheet metal is glued to the feed screw or the gear shaft (see fig. 6.1). Dial indicator 2 (resolution 0.01 mm) is set against the cam. (It should be checked that the dial indicator is not damaged during the subsequent movements of the machine.)
▷ Under numerical control, the machine slide is moved by approximately $+300$ µm.
▷ Mark voltage on measuring potentiometer $U_{\varphi 1}$.
▷ Machine is moved in the $(-)$ direction (e.g. by a 2 µm increment under numerical control).
▷ As soon as dial indicator 2 starts moving in the opposite direction, the voltage $U_{\varphi 2}$ on the measuring potentiometer should be marked.
▷ The reversing error $2\varepsilon_{u3}$ is calculated from the voltage difference:

$$\Delta U_\varphi = |U_{\varphi 1} - U_{\varphi 2}| \qquad (6.2)$$

For feed screw drives the result becomes:

$$2\varepsilon_{u3} = \frac{\Delta U_\varphi}{U} \cdot \frac{h_{Sp}}{i} \qquad (6.3)$$

For rack and pinion drives:

$$2\varepsilon_{u3} = \frac{\Delta U_\varphi}{U} \cdot \frac{2\pi \cdot r_{Ri}}{i} \qquad (6.4)$$

U total voltage on the potentiometer

$U_{\varphi 1}, U_{\varphi 2}$ voltages measured on the potentiometer after the rotation of the motor shaft (for multiple turn potentiometers, ΔU_φ must be multiplied by the total number of turns)

h_{Sp} lead of feed screw
i gear ratio
r_{Ri} pinion radius

Reversing error $2\varepsilon_{u1}$, active outside of the position control loop, can be calculated from the equation (6.1). If no measuring potentiometer is available, a pointer for measuring the angular movement can be mounted to the motor shaft. The ratio $\Delta U_\varphi/U$ is replaced by the angle $\sphericalangle \beta/360°$, where $\sphericalangle \beta$ represents the angular movement of the motor shaft.

As shown in section 4.2.4.3, when the position control loop is closed, the motor moves back and forth within the reversing error (hunting). This is an undesirable effect during the above mentioned measuring procedure. Therefore, the two values of distance and voltage should be read simultaneously on the potentiometer, and care should be taken that the oscillation is not included in the calculations.

For indirect position control systems, the allowable amount of reversing error $2\varepsilon_{u3}$ inside the position control loop depends on the amount of reversing error outside the position control loop. $2\varepsilon_{u3}$ should be less than approximately 30% of the outside reversing error. For direct position control systems, the positioning error caused by overshoot or oscillations may be taken as half of the reversing error within the position control loop. Please note the reduction in the position control loop gain of up to 60% of the optimal value, as it is shown in fig. 4.14. A minimal reversing error is desired.

Reversing error outside the position control loop $2\varepsilon_{u1}$, can, for the most part, be compensated for by the numerical control. A load-dependent reversing error however, cannot be compensated for. The machining force, together with the elasticity of the mechanical transmission elements, result in an additional error between the position control transducer and the part, respectively the tool; this error is fully reflected in the part. Adding in the non-compensated amount of the reversing error $2\varepsilon_{u1}$, the position error at the part can be calculated from:

$$|x_s - x_i| \approx \sqrt{(2\varepsilon_{ux})^2 + (2\varepsilon_{uy})^2} \tag{6.5}$$

$2\varepsilon_{ux}$ resulting reversing error on the X-axis
$2\varepsilon_{uy}$ resulting reversing error on the Y-axis

For movement of drives with more than two axes, the squared reversing errors of the other axes should be added under the square root sign.

6.2.1.2 Total Stiffness, Nominal Angular Frequency

The requirements on the total stiffness of the mechanical transmission elements are determined on one side by the necessary nominal angular frequency $\omega_{0\,mech}$, and on the other side by the allowable friction reversing error $2\varepsilon_{uR}$. For direct position control, the min. value per 1,000 kg of mass moved is $k_{Ges} \geq 100\,\text{N}/\mu\text{m}$. For indirect position control, a total stiffness up to tenfold higher is necessary to keep the reversing error within limits (see section 4.2.2 and the examples for 4.2.4.2 and 4.3.2.2.5). The total stiffness is given by:

$$k_{Ges} = \frac{\Delta F_{aSp}}{\Delta s_a} \tag{6.6}$$

F_{aSp} feed force in axial direction of the feed screw
s_a axial position change because of elasticity

(The same values apply for rack and pinion systems.)

The mechanical nominal angular frequency $\omega_{0\,mech}$ can be calculated according to equation 4.1 to:

$$\omega_{0\,mech} = \sqrt{\frac{k_{Ges}}{m_T + m_W}} \tag{6.7}$$

k_{Ges} total stiffness
m_T table mass
m_W work piece mass

(The mechanical nominal frequency $f_{0\,mech}$ can be calculated from the equation $\omega = 2\pi \cdot f$.)

The measuring arrangement in fig. 6.1 is used to determine the values for ΔF_{aSp} and Δs_a, as follows:

▷ The position control loop is closed and ready for operation
▷ The measuring potentiometer is mounted to the motor shaft
▷ The load cell is mounted between the machine table and an immobile part of the machine
▷ In jogging mode (e.g. in 10 µm increments), the machine table is moved towards the load cell.
▷ After each increment the reading will be
  potentiometer voltage U_φ
  distance on dial indicator ① x_z
  force on load cell F_{aSp}

▷ Diagram $F_{aSp}=f(s_a)$ is drawn according to the measuring values. The axial shift distance s_a for feed screw drives is,

$$s_a = \frac{U_\varphi}{U} \cdot \frac{h_{Sp}}{i} - x_z \tag{6.8}$$

and for rack and pinion systems it is

$$s_a = \frac{U_\varphi}{U} \cdot \frac{2\pi \cdot r_{Ri}}{i} - x_z \tag{6.9}$$

U_φ voltage on potentiometer, caused by the angular motion of the motor shaft
For a multi-turn potentiometer, U_φ must be multiplied by the number of turns. Beginning value for U_φ is zero
U total voltage on the potentiometer
h_{sp} lead of the feed screw
i gear ratio
r_{Ri} radius of the pinion
x_z distance on dial indicator ① (corresponds to the elastic deformation of the load cell).

▷ The total stiffness k_{Ges} of the mechanical transmission elements can be determined according to equation (6.6), from the slope of the force-deformation diagram.

▷ The mechanical nominal angular frequency $\omega_{0\,mech}$ can be determined according to equation (6.7), knowing the table mass and the work piece mass $m_T + m_W$, and the total stiffness k_{Ges}.

▷ If the position control transducer is mounted directly to the motor shaft, no measuring potentiometer is necessary. The distance s_a is calculated from the value shown on the NC, less x_z.

According to table 4.1, the first mechanical nominal angular frequency $\omega_{0\,mech}$ should be 2 to 3 times higher than the nominal frequency of the drive $\omega_{0\,A}$. This corresponds to values of approximately $120\ s^{-1}$ to $540\ s^{-1}$, thus nominal frequencies $f_{0\,mech}$ of approximately 20 Hz to 90 Hz, depending on the type of feed drive used and on the required position loop gain (see 4.2.1).

6.2.1.3 Friction Characteristics

A knowledge of the friction characteristics gives indications about the possible stick-slip effects, and about the portion contributed by the friction errors to the total reversing error. The fundamental possible friction characteristics are presented in fig. 4.7.

To record the friction characteristics, the motor current is measured according to the measuring arrangement in fig. 6.4.

▷ A shunt is placed in the armature circuit to measure the motor current
▷ Different feed rates v_V are applied in both directions through a reference voltage generator

▷ After all oscillations have subsided, the motor current $I_d = I_{MR}$ is measured from the voltage drop across the shunt

At very low speeds and at stand still, for the so-called break-away torque, the current can be determined with a reference voltage setting on the drift potentiometer.

▷ For expediency, the motor current is converted into motor torque:

$$\Sigma M_R = \frac{I_{MR}}{I_{0M}} \cdot M_{0M} \tag{6.10}$$

I_{MR} motor current measured for friction and losses
I_{0M} rated motor current
M_{0M} rated motor torque

▷ The sum of the torques for friction and losses ΣM_R is shown over the feed rate in a diagram.

As determined by equation (4.8), the frictional torque should be in the range $0.2\, M_{0M} \leq \Sigma M_R \leq 0.3\, M_{0M}$. Figure 6.3 shows current values I_{MR} measured on the slide of a lathe. The asymmetry of the two movement directions, and a high static friction with down sloping characteristic in the (+) direction, are immediately obvious.

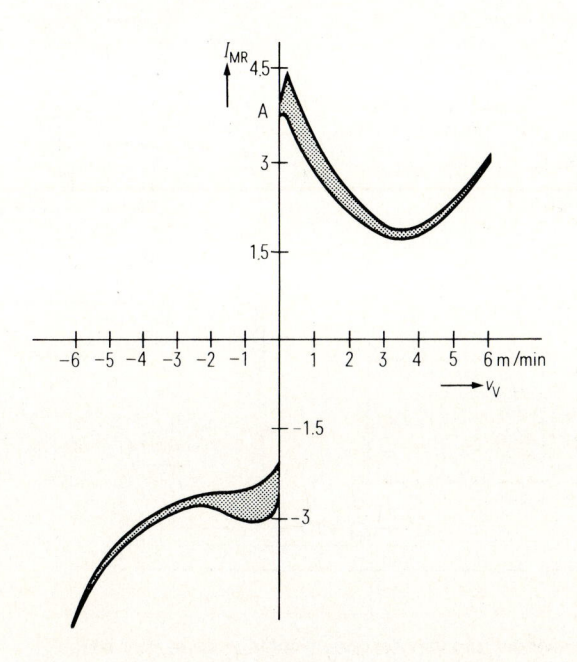

Fig. 6.3   Friction characteristics measured on a lathe slide

In this example, the rated motor current is 25 A. In the (−) direction, the amount and characteristic of the friction torque look therefore very well. In the (+) direction however, an unsatisfactory situation is to be expected. The down sloping characteristic is the reason for the stick-slip effect observed. The resulting intermittent sliding at low speeds can only be improved by lubricating the guide ways, by preloading the gibs less, or by using better suited materials for the guide ways.

6.2.2 Measurements on the Speed Control Loop

The behavior of the speed controlled DC servo motor is influenced by the mechanical transmission elements coupled to it. Measurements should thus be conducted in conjunction with the machine tool; an idling motor has generally more favorable nominal values.

Figure 6.4 shows the measuring arrangement for recording these nominal values.

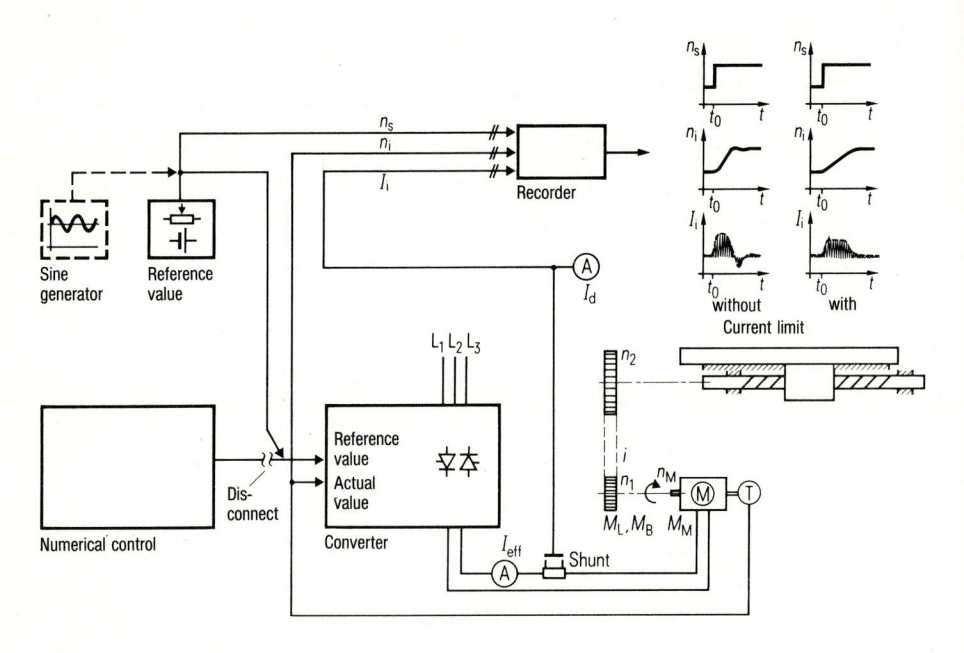

Fig. 6.4
Measuring arrangement for recording the step response, measuring the current limit, the nominal angular frequency of the feed drive, and the frictional characteristic

6.2.2.1 Step Response, Transient Response

The response time T_{An}, the delay time T_u, and the overshoot width $ü_a$ of a feed drive can be, according to fig. 1.11, derived from the transient response of the feed drive. The response time in the lower speed range, where limits are not reached, is a measure for the nominal angular frequency of the feed drive. The damping gradient can be determined from the overshoot width. The dead time within the speed control loop is determined by the pulse number of the converter. Since the delay time varies significantly according to the step response, especially because of the moment of the reference step, the calculations are made with the values given in table 2.3. Also, the delay time includes part of the current rise time.

The measuring procedure for recording the step response is:

▷ The position control loop is opened
▷ A variable DC source is connected through a switch to the reference input of the speed regulator
▷ Signals for the following will be recorded

  n_s speed reference value
  n_i actual speed value
  I_i actual current value

(The actual current value can also be derived from a suitable measuring point within the converter. This prevents problems due to non-floating values from surfacing.)

Reference value steps from zero to values of the following series are recommended:

0.1–0.2–0.5–1–2–5–10–20–50–100% of the maximum speed.

The response time T_{An} and the overshoot width $ü_a$ are shown on the oscillograms of the jump response below the current limit. The step responses to be evaluated are generally between 1% and 10% of the maximum speed.

The damping gradient can be calculated for the ideal case of a 2nd order delay, from the overshoot width. If this is the case, the damping gradient is

$$D_A = \sqrt{\frac{1}{1+\left(\frac{\pi}{\ln 100/ü_a}\right)^2}} \tag{6.11}$$

$ü_a$ overshoot width in %.

The course of this function is shown in fig. 6.5.

Since the drive does not always behave as an ideal 2nd order delay, D_A cannot be exactly derived from the measured overshoot width with equation (6.11). As shown in 6.3.1, the damping gradient can be determined from the frequency response (compare equation (6.16)). Measured overshoot widths for thyristor controllers of 20–30% correspond to D_A values of approximately 0.6–0.5.

For a damping gradient $D_A \approx 0.5$–0.6, and a behavior approaching that of a 2nd order delay, the nominal angular frequency of a feed drive can be calculated:

$$\omega_{0A} \approx \frac{2{,}5 \ldots 2{,}7}{T_{An} - T_u} \tag{6.12}$$

T_{An} response time
T_u delay time

Since the numerator strongly increases with the increasing damping gradient, it is advisable to double check this computed value with the procedure described under 6.2.2.4.

For a feed drive nominal angular frequency range of

$$\omega_{0A} = 100 \text{ s}^{-1} \text{ to } 400 \text{ s}^{-1},$$

consequent response times between

$$T_{An} = 35 \text{ ms and } 8 \text{ ms}$$

are necessary, when the delay times are presumed to be in the range

$$T_u = 10 \text{ ms to } 1 \text{ ms}.$$

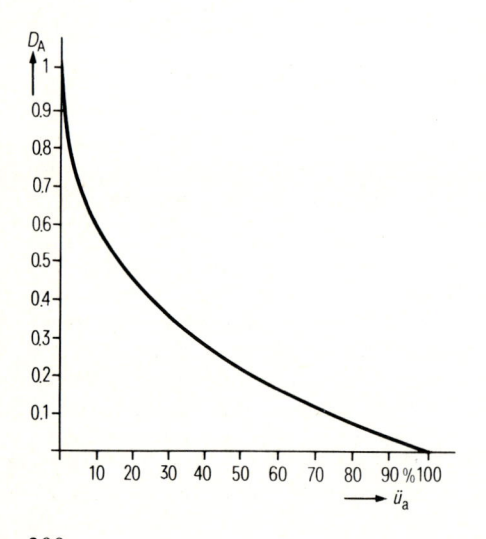

Fig. 6.5
Damping gradient of a 2nd order delay as function of overshoot width

For small reference steps in range of 0.1% to 1% of the maximum speed, attention should be paid that the response times do not increase too much, and that the reversing times for a $\pm$ reference change do not exceed a value of approximately 3 times the response time (also see 3.4.6 and fig. 3.22).

6.2.2.2 Current Limit

The converter is equipped with an adjustable current limit (see 3.4.2), in order to protect the servo motor and the mechanical transmission elements. During dynamic actions within the machining range, in order to prevent contour deviations due to interferences of the current limit, the current limit should not be active. The decisive factor here is the position loop gain set in the position loop. If the reference value is affected, reaching the current limit can be avoided by limiting the reference value acceleration a_{fo} (see 5.4 and 3.4.2).

The setting for the current limit can be checked by evaluating the step response at higher speeds (measuring circuit according to fig. 6.4). Step responses between 10–100% of the maximal speed should be evaluated. In the speed range where the current limit starts interfering, the actual current value will be limited (see step responses in the upper right corner of fig. 6.4). The ramp-up time will linearly increase with the magnitude of the reference step (see also fig. 3.22).

Depending on the size of the external inertia of the drive, the current limit is reached with a step at approximately 10–20% of the maximal speed.

6.2.2.3 Motor Heat

The selection tables for the DC servo motors show the rated torques for a form factor of the current of 1.05 and a heating by 100 K, at an ambient temperature of 40 °C (VDE 0530). For feed drives, motor heating can usually be disregarded, since the sizing is normally done according to the maximum machining force, which generally does not represent a continuous load on the feed drive.

In cyclic operations (e.g. press feeders, nibbling machines, and punch presses), heating up to the limit temperature is to be expected because of the periodic acceleration and deceleration modes. In such cases, during a test run, the commutator temperature and the surface temperature on the motor frame should be measured and plotted in a graph against time. The winding temperature can be calculated from the increase in the resistance of the armature winding, and can also be recorded. Experience indicates that the temperature increase in the winding exceeds that of the commutator by about 20 K, so that at 80 K commutator temperature, the rated temperature of the winding is reached.

The squared form factor of the current is a measure for the additional losses due to current ripple in the motor, as for instance is the case with thyristor converters (see 3.1.3). It can be calculated according to equation (3.2), by measuring the effective value I_{eff} and the DC mean value I_d of the armature circuit. For this purpose, a moving iron current instrument must be switched in series with the shunt, as shown in the measuring arrangement (fig. 6.4). This instrument shows the effective value of the current I_{eff}, while the moving coil instrument at the shunt shows the DC value I_d. With a proper load on the feed drive by the aid of a machining force, the form factor F_i for different feed rates can be determined.

Higher form factors affect the commutation. Observation of the brush arcing reveals, whether or not additional commutator load at acceleration and braking with maximal speed is still allowed. The blue pearl sparks are not critical; arcing with yellowish sparks along the commutator must be avoided. For critical load cases a long term test should be conducted, to determine the life expectancy for the brush.

6.2.2.4 Nominal Angular Frequency of the Feed Drive

The dynamic response of the feed drive can be determined exactly, only by recording the frequency response curve. However, if the feed drive acts like a 2nd order delay, i.e. there is no reaction of the mechanical transmission elements, $\omega_{0\,\text{mech}} > 2\omega_{0\,\text{A}}$, then the nominal angular frequency of the feed drive can be determined accurately enough with the following procedure. The measuring circuit is set up according to fig. 6.4. The sine generator gives the reference values:

▷ The position control loop is opened
▷ The signal on the reference value input of the speed control loop will consist of a sinusoidal signal of constant amplitude and variable frequency, and a DC signal. The DC signal supresses the effect of the converter dead time and that of the reversing error in the transmission system, and reduces wear on the guide of the feed slide. Suitable values for the signal voltages are:

    DC signal about 2–5% of the maximal reference voltage,
    superimposed sine value about 60–80% of the DC voltage

The reference voltage selected must in any case be small enough to prevent the drive from reaching the current limit.
▷ The frequency of the sinusoidal signal is changed in steps. At low paper feed of the recorder, or at slow sweep rate of the storage oscilloscope, the actual speed value n_i is recorded

▷ The nominal frequency of the drive f_{0A} can be approximated from the recording:

At f_{0A} the tacho voltage drops markedly below the value it held at lower frequencies ($f \ll f_{0A}$).

An example of such measurement is shown in fig. 6.6. The nominal frequency f_{0A} can be taken as approximately 23 Hz. The superimposed sine voltage was supressed in order to set the frequency steps. The resulting nominal frequency of the feed rate is

$$\omega_{0A} = 2\pi \cdot f_{0A} \qquad (6.13)$$

f_{0A} drive nominal frequency

In the example, the nominal frequency of the drive becomes

$$\omega_{0A} \approx 144 \text{ s}^{-1}.$$

If a high speed recorder is available, the reference and actual speed values can be recorded and evaluated at high paper feeds. The amplitude ratio n_i/n_s is calculated and normalized to the value at low frequency (e.g. at 1 Hz n_{i1}/n_{s1}) and plotted over the frequency in a double logarithmic grid.

The phase shift is also plotted over the frequency in a single logarithmic grid. This results in the Bode diagram as explained in section 1.1.3.4.

At suitable recording amplitude and stretched time scale, the values can be evaluated up to 60 Hz. If there are distortions in the actual value signal, as for instance generated by the pulsing currents of the thyristor converters, the curve of the fundamental oscillation can be approximated.

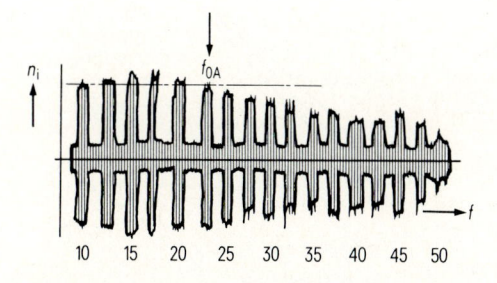

Fig. 6.6
Speed actual value, as function of a sinusoidal reference signal of variable frequency and constant amplitude

6.2.3 Measurements on the Position Control Loop

The position control loop consists of all components contributing to the signal flow, from position command to actual position values. The block diagram is shown in fig. 1.23. It becomes apparent that the properties of the electrical drive and of the mechanical elements considerably affect the control response of the position control.

6.2.3.1 Contour Deviations

Contour deviations occur because of distortions in the position control loop. One distinguishes:

▷ Linear signal distortions. These result from the control technology design of the position control.
▷ Non-linear signal distortions. They stem mainly from the construction design of the mechanical transmission elements.

Qualitative and/or quantitative statements about contour deviations can be made only concerning test contours (e.g. straight line and circle runs), and concerning only one cause. A prognosis concerning the magnitude of a contour deviation stemming from multiple causes is not possible, due to the fact that single contour errors can both add up, as well as cancel each other out.

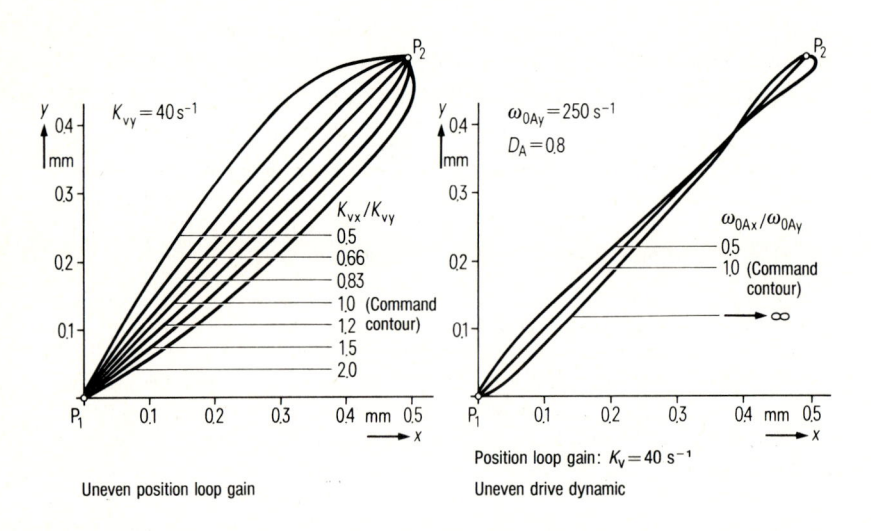

Fig. 6.7
Effects of uneven position loop gain and uneven drive dynamic, upon move P_1 to P_2

Linear signal distortions are produced by the following errors during parameter adjustment:

▷ speed control loop or position control loop not optimally adjusted
▷ uneven position loop gains of the feed axes participating in the contour generating
▷ uneven dynamic (ω_{0A}) of the feed drives.

The effect of uneven position loop gain, respectively uneven drive dynamic of the two feed axes X and Y on the production of a straight line, is shown in fig. 6.7. It is based on a position control loop with a feed drive acting as a 2nd order delay (block diagram according to fig. 1.19 without dead time).

At equal position loop gain on both feed axes, the actual position point follows the command position point on the same contour, but time-delayed. At uneven position loop gain, the actual value point follows the command contour with a parallel offset. This error can be avoided by adjusting the following error equally on both axes when driving a straight line under 45°.

An uneven drive dynamic leads to contour deviations mainly during dynamic phases. When driving a straight line, these errors are smaller than those caused by uneven position loop gains. Circles are distorted to elliptical shapes if the drive dynamic is uneven.

Non-linear signal distortions result from the false sizing of the drive or the mechanical transmission elements:

▷ activation of the current limit within the machining range
▷ reversing error within or outside the position control loop
▷ non-linear frictional behavior of the slide guides.

Contour deviations due to non-linear signal distortions occur mainly because of the reversing error. The type and size of the contour deviation are determined mainly by the location of the reversing error relative to the position control loop (inside or outside). Figures 4.12 and 4.17 show typical contour deviations caused by reversing errors occurring on a circular contour.

When driving a straight line, a parallel offset similar to that caused by an uneven position loop gain results from a reversing error outside of the position control loop (compare fig. 4.11). The flatter the slope in reference to one of the feed axes, the larger this parallel offset will be. At a reference contour incline of 45° to one of the axes, as long as the reversing errors in both axes are equal, the parallel offset will become zero (compare equation (4.9)). A frictional characteristic of the mechanical transmission elements not proportional to the speed, can lead to a ripple in the actual contour. The reason for this is not the stick-slip effect, but the oscillation response of the position control loop due to the non-linearity. This rippling is more pronounced for indirect position control than for direct position control (see figs. 4.11 and 4.16).

6.2.3.2 Measuring the Position Loop Gain

The optimal value for the position loop gain, determined in section 1.3 with equation (1.52),

$$0{,}2\,\omega_{0\,A} \leq K_v \leq 0{,}3\,\omega_{0\,A}$$

is reduced by the non-linearities in a position control loop. The effect of the dead time in the converter is demonstrated in fig. 2.12, and that of the reversing error can be estimated from fig. 4.14.

In these investigations, it is presumed that the lowest mechanical nominal angular frequency $\omega_{0\,\text{mech}}$ is at least 1.5 times higher than the nominal frequency of the drive. If circumstances are as shown, for example in fig. 4.4 on the right, the contour deviations increase. This can be deduced from fig. 4.6. The position loop gain achievable is then still further reduced.

A further limitation of the value settable for the position loop gain, can be given by the maximal admissible acceleration of the machine table $a_{i\,\text{max}}$. The maximal acceleration that can occur with pure velocity control, for a position command value without command value smoothing, according to equation (3.11), is:

$$a_{i\,\text{max}} = K_v \cdot \Delta v_B$$

(see also chapter 5)
K_v position loop gain
Δv_B maximum programmable velocity change

Since for velocity-controlled position command value generation the position loop gain is partly derived from a non-linear context of v_B and Δx, in equation (3.11) related values must be used for K_v and Δv_B.

For an acceleration-controlled command value generation process, the maximal actual acceleration will be

$$a_{i\,\text{max}} \approx \tfrac{1}{2} K_v \cdot \Delta v_B$$

if $K_v/\omega_{0\,A} = 0{,}2$, and the command value is set for an acceleration of

$$a_{f\,0} \leq \tfrac{1}{2} K_v \cdot \Delta v_B,$$

as shown in chapter 5 in equations (5.3) and (5.4).

According to equation (1.37), the position loop gain is defined as

$$K_v = \frac{v_s}{\Delta x}$$

v_s command velocity
Δx following error (lag value)

The following error can be directly displayed on some numerical controls. Deriving the position loop gain as a function of the slide velocity is then very simple:

▷ different velocities are programmed

▷ following error Δx is read off the control
▷ position loop gain K_v is computed according to equation (1.37)
▷ diagram with $K_v = f(\Delta x)$ is drawn

On numerical controls without following error display, the position loop gain can be derived with the measuring set-up in fig. 6.8.

Prerequisites for the measuring procedure are:

▷ The programmed velocity must actually be reached, i.e. the integral part of the speed control loop is active within the entire velocity range.
▷ The relationship between the voltage U at the output of the position control and the command velocity v_s must be known. (In most cases, 8 V reference voltage at the input of the speed regulator implies maximal velocity.)

The measuring procedure for the position loop gain is:

▷ A dial indicator (resolution 0.01 mm) is mounted to the machine slide.
▷ When the position control loop is closed, a drift-free DC voltage is introduced to the summing point of the speed control loop, by means of a resistor. This DC voltage acts as a disturbance which causes the machine slide to be deflected by the distance Δx. (Δx must be large as compared with the reversing error $2\varepsilon_u$.)
▷ A voltage U proportional to the command velocity v_s, is produced at the position control output, respectively the speed control input. This voltage is measured and converted into the corresponding feed rate.
▷ The position loop gain is calculated, and plotted over Δx.

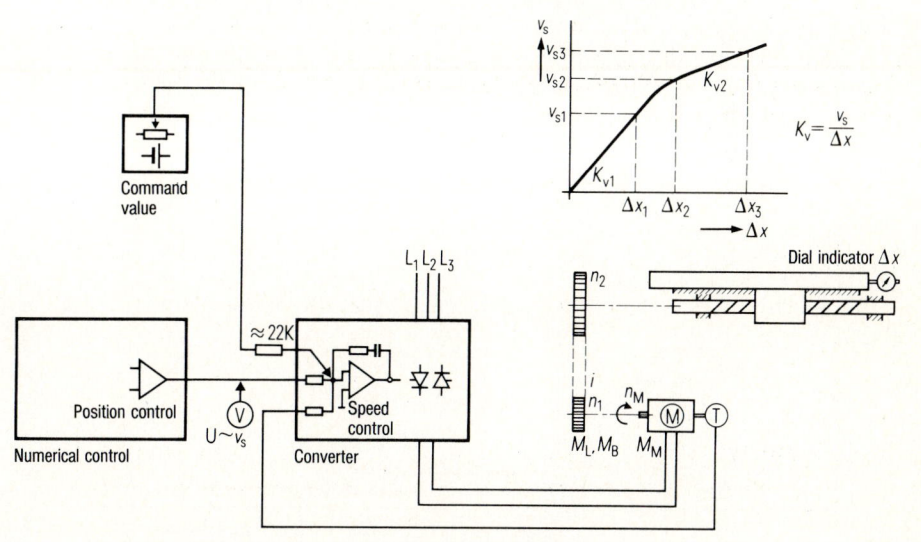

Fig. 6.8   Measuring set-up for the position loop gain

For contouring and linear path controls, the position loop gain must be constant throughout the machining range, and must be equal in all feed axes. Axis specific reduction is possible above the velocity necessary for machining. The position control gain curve for such a reduction is shown in the upper right corner of fig. 6.8. In the machining range, the position loop gain is

$$K_{v1} = \frac{v_{s1}}{\Delta x_1} \qquad (6.14)$$

In the rapid traverse range, the position loop gain is reduced to

$$K_{v2} = \frac{v_{s3} - v_{s2}}{\Delta x_3 - \Delta x_2} \qquad (6.15)$$

This process can be simplified by inhibiting the servo control with the drive enable, and incrementing the numerical control so that the position deviation will be created. The computation and the plotting are done the same way as described before, the additional reference input is not necessary.

6.2.3.3 Measurements for the Smallest Position Increments

For the uniformity of the smallest feed rates and the accuracy of the positioning, it is necessary that every position increment outputted by the control be executed by the position control, only after overcoming the reversing error between motor and position control system.

To ensure this, the following measuring process is necessary:

▷ The machine slide is moved by a few mm in the (+)direction.
▷ A dial indicator with a resolution of 0.001 mm is mounted to the slide.
▷ With the NC, the slide is moved in jogging mode with the smallest position increments, and after each command value output, its execution will be controlled with the dial indicator.
▷ The test is repeated in the (−)direction.

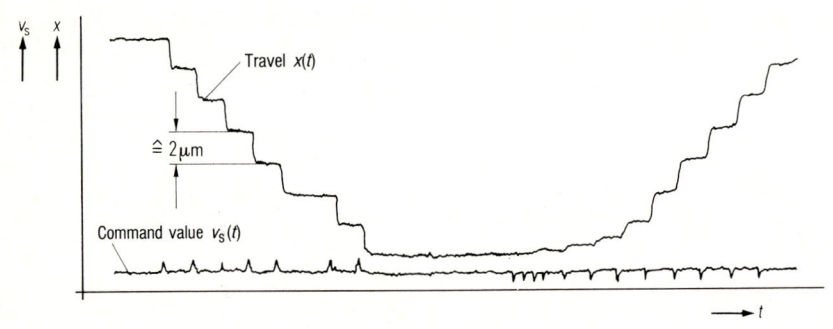

Fig. 6.9 Step-wise positioning of a lathe slide

If a record of the results of these measurements is desired, a position measurement must be conducted at the table (slide), e.g. with an inductive position transducer. Such a diagram, as recorded with a plotter, is shown in fig. 6.9. The measurements were conducted on a lathe equipped with an electrically excited DC servo motor with 3-pulse circulating-current servo controller.

After the direction change, the reversing error $2\varepsilon_{u\,1}$ outside the position control loop can be determined from the amount of the command pulses not executed. In this example it is 10 μm. After driving through the reversing error, the single position increments are recognized and executed. No "sticking" due to static friction occurs.

6.2.3.4 Positioning Response

It is required of the position control loop, that it position the machine slide at all velocities as fast as possible, and without overshoot. The position loop gain is critical for these requirements. The output voltage of the position control is recorded for testing. This velocity command value changes its sign in the case of overshoots because v_s is proportional to Δx. In the case of a position deviation, a command voltage proportional to the deviation but of opposite sign, is produced. The measurement described should be conducted on all the feed axes that are moved simultaneously during machining operations. If any axis shows an excessive overshoot, the position loop gain of that axis should be reduced accordingly. The K_v values of the other simultaneously moved axes, are always oriented according to the feed axis with the lowest nominal angular frequency $\omega_{0\,A}$. Different K_v values lead to contour deviations.

The measuring procedure is as follows:

▷ The position control output voltage $U \sim v_s$ is recorded with a fast recorder (e.g. ink jet recorder or storage oscilloscope).

▷ The machine slide is positioned with different feed rates. Voltage $U \sim v_s$ is recorded over time; the recorder is set so, that only the approach of the position is recorded. The resolution should be selected as high as possible.

In the case of a reversing error outside the position control loop, an overshoot of the velocity command value can lead to retraction of the table, only if the overshoot width is larger than the reversing error. In order to make a judgement about the movement of the table, this movement must be measured with a position encoder directly at the table.

6.3 Measuring Values of Machines in Use

The measurement results given below are mainly measured frequency response curves of different types of drives. They show, in part, the effect of dimensioning on the mechanical transmission elements. They also show the effects of optimization in the speed control loops, and of the dimensioning of electrical drives. These measurement results should provide the technician in charge of the start-up, with a comparison basis for his own measurements. They should also give him some idea of possible tests for new machines, and some insight into the ways of eliminating disturbing effects. For certain requirements, drives in practical applications can perform satisfactorily, even with lower nominal angular frequencies. By the same token, some drives will meet higher requirements by being sized accordingly, and by appropriate dimensioning of the mechanical transmission elements.

6.3.1 Evaluation of Measured Frequency Response Curves

To determine the nominal frequency from the measured frequency response curve, it is first necessary to establish from the amplitude response curve, whether we are dealing with the behavior of a 1st order, 2nd order, or an even higher order delay element. This can be accomplished with the aid of the ideal frequency response curves shown in figs. 1.13, 1.14, and 1.20.

Whether the system to which belongs a particular measured frequency response curve is of 1st, 2nd, or higher order can also be deduced from the course of the phase response curve. For an ideal 1st order delay element, at frequencies $f \gg f_{0A}$, the phase shift angle $\varphi = -90°$ electrical, and the nominal frequency is at $\varphi = -45°$ el. For an ideal 2nd order delay element at $f \gg f_{0A}$, the phase shift is $\varphi = -180°$ el, with nominal frequency f_{0A} at $\varphi = -90°$ el. For ideal 3rd order delay elements, the phase shift approaches the value of $\varphi = -270°$ el, and the cut-off frequency ω_E is at $\varphi = -135°$ el.

Depending on the type of the characteristic, we assign a straight line of slope -1 to 1st order systems, -2 to 2nd order systems, or -3 for 3rd order systems in the declining portion of the amplitude response curve, in a manner that best allows the measured values to coincide (align). The point of intersection of this line with the horizontal line passing through $|F|=1$, represents the nominal, respectively the cut-off frequency. In the double-logarithmic coordinate grid, a slope of -1 means that within a frequency decade (e.g. from 10–100 Hz) there will be an amplitude drop of one decade (e.g. from 1–0.1). Correspondingly, the amplitude drops per frequency decade for -2 and -3 slopes, are 2, respectively 3 decades.

For systems which behave like 2nd order delay elements, we can determine the damping gradient of the drive from the absolute value of the amplitude response curve, at the nominal frequency. For an ideal drive behavior like that of a 2nd order delay element, it holds that:

$$|F_A|\bigg|_{f_0} = \frac{1}{2D_A} \tag{6.16}$$

For real control loop elements, which often do not behave like the ideal 2nd order delay element, this procedure for the determination of the damping gradient D_A from the frequency response curve carries some uncertainties. For this reason, a mean value is created for D_A, from the damping gradient obtained from the overshoot width according to fig. 6.5, and the one found from the frequency response curve.

6.3.2 Measurements on a Timing Belt Gear

The significant parameters for the classification of the behavior of a timing belt gear are the reversing error, the stiffness, and the nominal frequency of the mechanical transmission elements. The measurements are shown for the feed drive of a lathe.

The technical data are:

gear ratio i	4	axis distance		259 mm
belt width	2″	ball screw	d_{Sp}	40 mm
tooth pitch	1/2″	screw lead	h_{Sp}	10 mm
diameter of pulleys		DC servo motor	M_{0M}	6 Nm
d_{w1}	55 mm		n_{maxM}	3,600 min^{-1}
d_{w2}	225 mm	3-pulse circulating current-conducting current converter		

Fig. 6.10 Shows the measured frequency response curve of the timing belt gear.

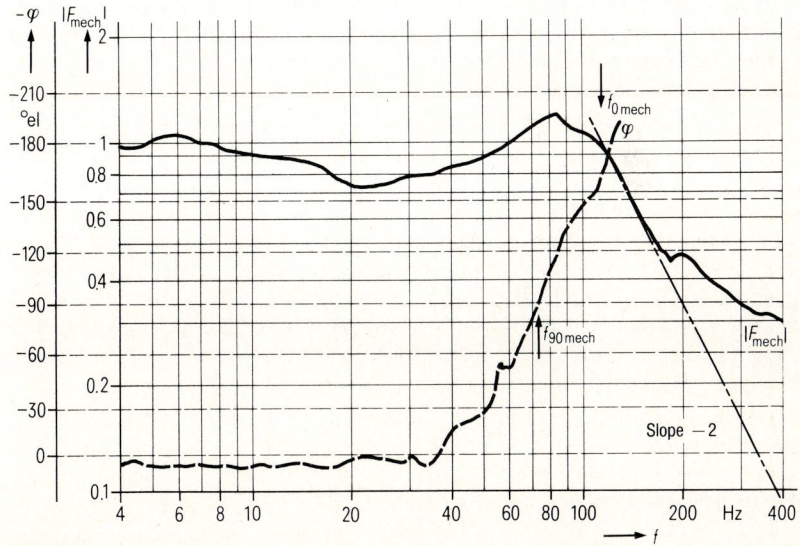

Fig. 6.10
Measured frequency response curve of a single stage timing belt gear

Table 6.1 Measurement results on a single stage timing belt gear

		with preload	without preload
Nominal frequency	$f_{0\,mech}$	115 Hz	90 Hz
90°-Frequency	$f_{90\,mech}$	70 Hz	50 Hz
Damping gradient	D_{mech}	about 0.5	about 0.35

The timing belt gear displays a behavior approximating that of a 2nd order delay element. The nominal frequency $f_{0\,mech}$ and the damping gradient D_{mech} depend on whether the timing belt is preloaded or not. Within this testing frame, the absolute value of the preload had no influence. Table 6.1 shows the results of this test study.

Frequency response measurements at different control gains in the speed control loop, show that the drop in the amplitude response curve shown in fig. 6.10 in the range of approximately 20 Hz, can be traced back to a reaction in the time behavior of the speed control loop.

The characteristic dynamic values determined basically, indicate that timing belt gears can be used as gears on numerically controlled machine tools. The timing gear nominal frequency $f_{0\,mech}$ lies far above the drive nominal frequency f_{0A}, which eliminates the possibility of negative gear effect on the frequency response curve of the total feed drive. The distance to a first mechanical nominal frequency created by a screw system, which normally would lie between 50 and 70 Hz, is also sufficiently large.

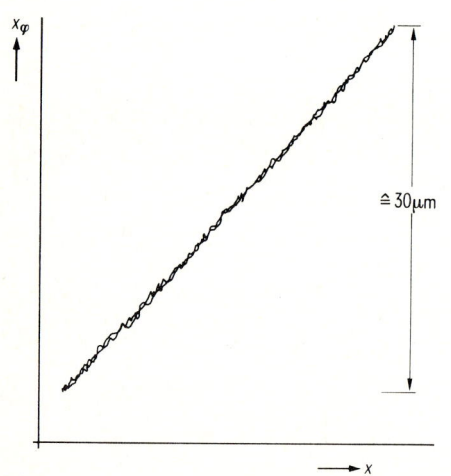

Fig. 6.11 Measured reversing error of a single stage timing belt gear

From the measurement of the reversing error, as shown in fig. 6.11, it is apparent that virtually no reversing error can be measured with the selected resolution. Recalculated for the longitudinal motion, the reversing error of this timing belt gear lies under 2 µm.

In order to test the mechanical nominal frequency $\omega_{0\,mech}$, the stiffness of the timing belt was measured, and when it was recalculated for longitudinal distance, it resulted in the deforming diagram shown in fig. 6.12. From this diagram, it can be interpreted that the stiffness k_{Ges} amounts to approximately 330 N/µm, in which case the nominal angular frequency $\omega_{0\,mech}$ for a table mass of around 600 kg will be, according to equation (6.7):

$$\omega_{0\,mech} = \sqrt{\frac{330 \cdot 10^6}{600}}\ s^{-1} = 740\ s^{-1}$$

or

$$f_{0\,mech} = \frac{740\ s^{-1}}{2\pi} = 119\ Hz$$

We thus can see, that there is close agreement between the measured frequency response and the value calculated here.

The conclusion which can be drawn from these measurements, is that this particular timing belt drive is suitable as a feed drive gear for contour controlled machine tools. The stiffness, nominal angular frequency $\omega_{0\,A}$, and the damping gradient D_{mech} are sufficiently high, and the reversing error is negligeably low.

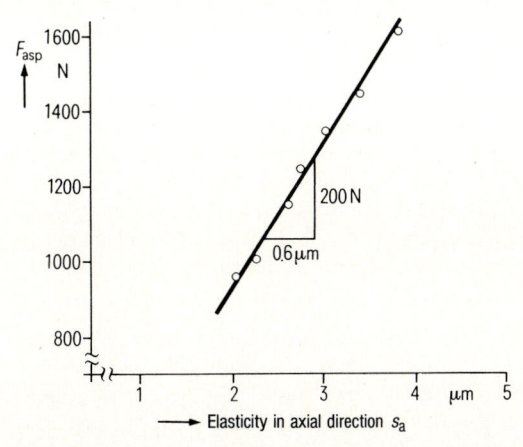

Fig. 6.12
Measured force/shift distance diagram of the timing belt gear

6.3.3 Effect of the Moment of Inertia

The effect of the external moment of inertia on the nominal angular frequency derived in section 2.2.2.6, can also be demonstrated through a practical test. For this purpose, the external moment of inertia of a DC servo motor 1HU3078-0AC01 equipped with a transistor chopper, will be changed between 0% and 120% of the motor internal moment of inertia. The measured frequency response curves of the feed drive are shown in fig. 6.13.

The optimization of the speed control loop is not changed for these measurements. If the gain were to be adjusted, the motor with the higher moment of inertia would reach a nominal frequency f_{0A} which would show slight improvement. At 120% external moment of inertia, the nominal frequency is only about 70% of the nominal frequency of the idle motor.

6.3.4 Motors in Short Version

Compared to the regular construction form, short versions of permanent magnet-excited DC servo motors have approximately double the mechanical time constant. This results from the fact that in reference to the torque, their armature diameter is larger. Figure 6.14 shows a frequency response curve measured on such a motor.

On a machine, the motor is connected directly to the ball screw. The external moment of inertia is approximately 60% of the motor internal moment of inertia. The current converter unit is a transistor chopper. The step function response shown indicates a damped transient behavior. The response time is approximately 8 ms.

The interpretation of the characteristical behavior is difficult in this case. Neither straight lines of slope -1, or -2 characterize the behavior unambiguously. We must remember however, that the characteristic frequency of a system approaching 2nd order (f_{0A2}), is about twice as high as that of a system approaching 1st order. This is the reason why, at the optimization of a speed control loop, one must always try to make the behavior of the drive to approximate as closely as possible the behavior of a 2nd order delay element. In this particular case, the frequency response curve could have been improved with a higher gain in the speed regulator.

Under the same conditions, a regular version of the motor, with equal torque, has a measured nominal frequency of $f_{0A2} \approx 40$ Hz. In comparison to the nominal frequency $f_{0A2} \approx 33$ Hz which can be read out of fig. 6.14, this second nominal frequency is not much higher. This means, that short motors also have good potential for position control, and that sufficiently high position loop gains can be reached within their position control loops.

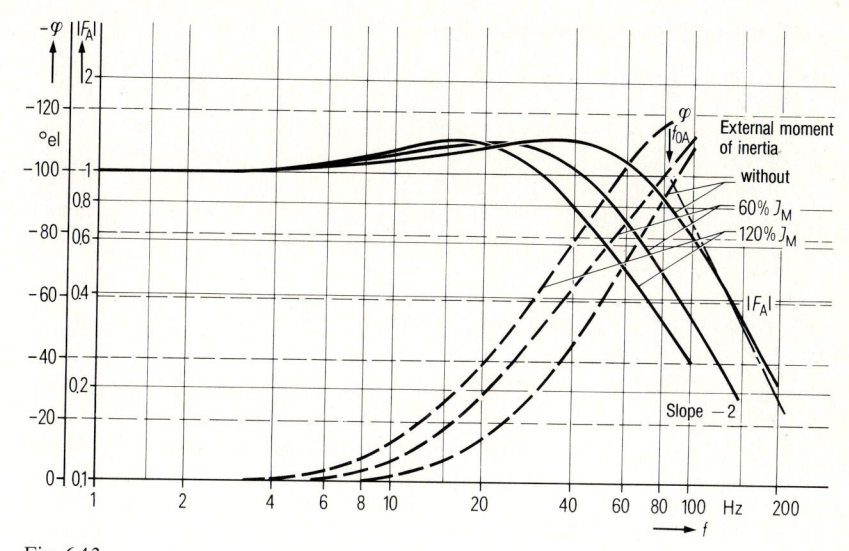

Fig. 6.13
Frequency response curves of a DC servo motor 1HU3078-0AC01 with transistor chopper. Effects of the external moment of inertia

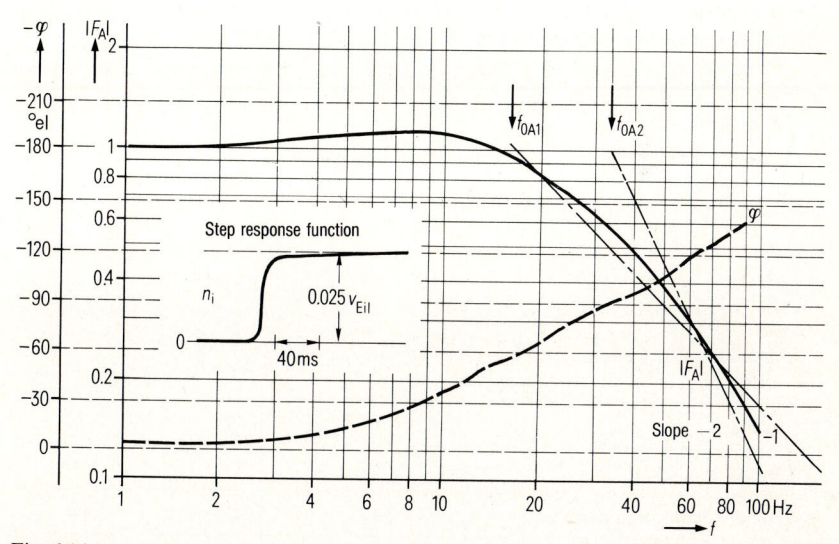

Fig. 6.14
Frequency response curve of a DC servo motor 1HU3100-0AC01 (short version)

315

6.3.5 Effects of Optimization of the Speed Control

The nominal angular frequency ω_{0A} possible for a drive, can be calculated from the data of the drive shown under 2.2.2.5. With a suitably selected gain, the speed regulator circuit should behave following the pattern of a 2nd order delay with a damping gradient $D_A \approx 0.5$. Choosing the gain too high, results in oscillations within the speed regulator circuit, and contour deviations during contour changes. Choosing the gain too low, reduces the angular frequency ω_{0A}, and thus also leads to contour deviations and to overshooting the actual position.

The influence of gain adjustment in the speed regulator is shown in fig. 6.15.

The measurements were taken on a DC servo motor 1HU3078-0AC01 with transistor chopper and without machine tool. The external inertia represented 60% of the motor inertia. The gain factor of the speed regulator was changed from 33.5, to 39 and then to 44. The step responses are shown in the figure; the response times are approximately 10 ms, 9 ms, and 8 ms. The nominal frequencies f_{0A} reached, are listed in table 6.2. The damping gradient was calculated as a mean of both values, according to equation (6.16) and to fig. 6.5.

6.3.6 Frequency Response Curve of a Shell Magnet Motor

In contrast with the frequency response curves of DC servo motors presented before, which are constructed according to the flux concentration principle, fig. 6.16 shows the frequency response of a shell magnet motor.

The test was conducted with a transistor chopper, without machine. The external moment of inertia amounts to about 30% of the motor moment of inertia.

The nominal frequency f_{0A} reached is approximately 55 Hz, the damping gradient D_A is around 0.5. The step response function, shown left in fig. 6.16, indicates approximately 20% overshoot.

Table 6.2
Nominal frequency and nominal angular frequency for different gains of the speed regulator

Gain K_{gn}		33.5	39	44
Nominal frequency	Hz	60	72	80
Nominal angular frequency ω_{0A}	s^{-1}	375	450	500
Damping gradient D_A		0.65	0.55	0.5

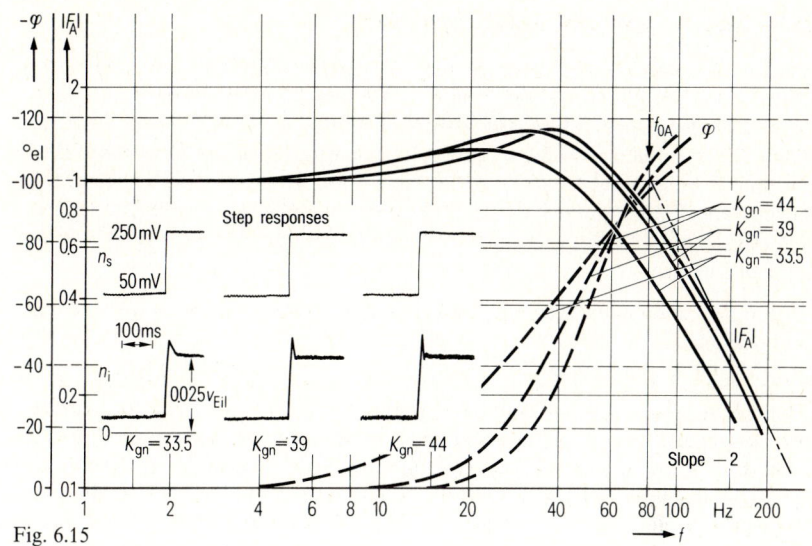

Fig. 6.15
Effects of optimization of the speed regulator on the frequency response curve (DC servo motor 1HU3078-0AC01)

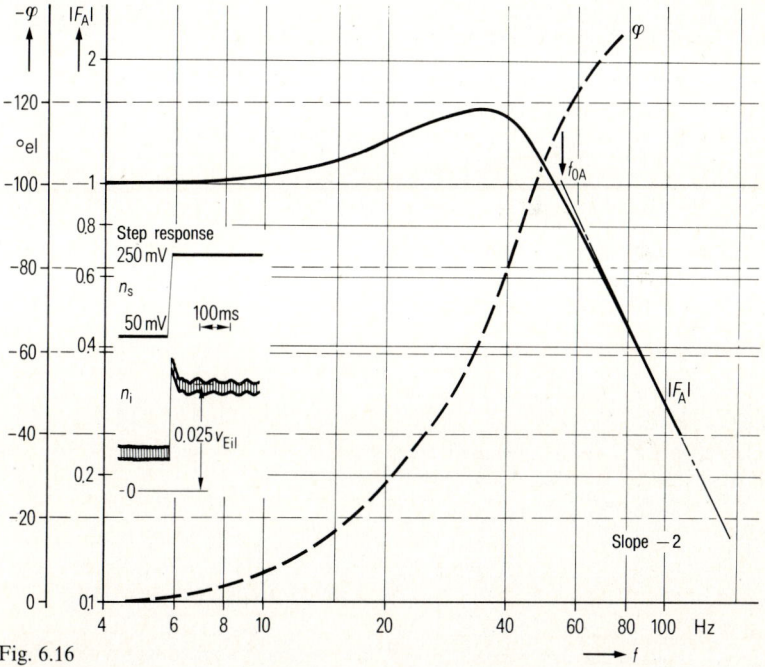

Fig. 6.16
Frequency response curve of a permanent magnet excited shell magnet motor 1HU5040-0AC01

6.3.7 Influence of the Converter Circuit

The following frequency response curves were recorded on a feed drive, for the table of a milling machine. The first mechanical nominal frequency $f_{0\,mech}$ of this feed screw drive was at about 30–35 Hz, thus relatively low. It had a significant influence on the frequency response curve of the drive. The reason for this low nominal frequency was the low total stiffnes, which amounted to only about 20 N/μm. The sum of the frictional moments was approximately 28% of the motor rated torque, and the external moment of inertia was around 25% of the motor internal moment of inertia.

Figure 6.17 shows the frequency response curve of a feed drive, with no gear and with feed screw, the permanent magnet-excited DC servo motor 1HU3104-0AD01, and 3-pulse circulating current-conducting current converter.

The set DC portion of the command velocity was, for this test, approximately 5% of the rapid traverse feed rate. From the diagram, we can deduce a nominal frequency $f_{0\,A}$ of about 32 Hz. The amplitude drop below this frequency can be traced back to the reaction of the mechanical transmission elements, and so can the steep slope of the phase response curve at about 25 Hz. Also because of this reaction, no exact statement can be made concerning the damping gradient D_A. It will be about 0.55.

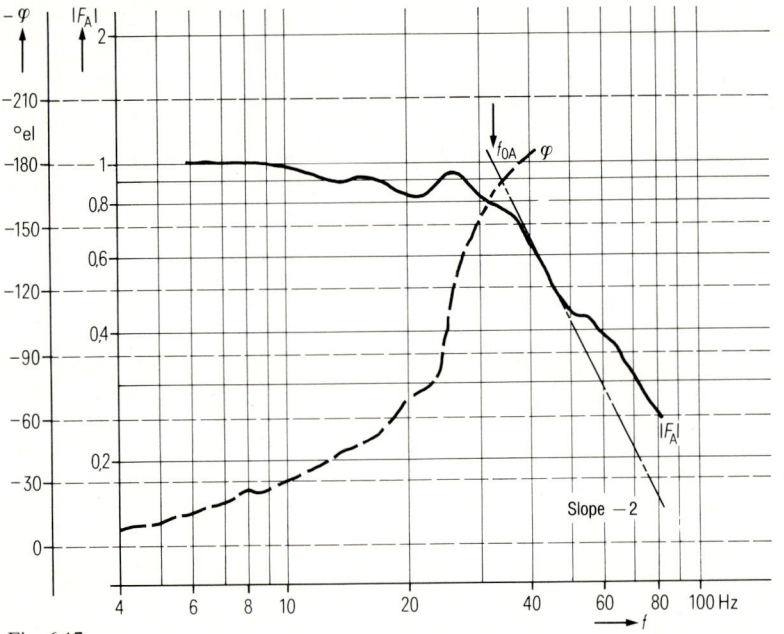

Fig. 6.17
Frequency response curve of a feed drive with the permanent magnet-excited DC servo motor 1HU3104-0AD01, and 3-pulse circulating current-conducting converter

The same motor was operated with a 6-pulse circulating current-free current converter circuit. The previously given values also apply in this case. Figure 6.18 shows the frequency response curve.

The test results in a nominal frequency f_{0A} of about 26 Hz. The mechanical transmission elements reaction between approximately 30 and 35 Hz can be seen, from the declining portion of the amplitude response curve, as an insignificant influence. The mechanical nominal frequency $f_{0\,mech}$ now lies above the nominal frequency of the drive. The step transient response for 10% of the rapid traverse speed, which is shown on the left in fig. 6.18, displays approximately 36% overshoot. The response time is about 20 ms, and the damping gradient can be derived to be $D_A \approx 0.4$.

When comparing figures 6.17 and 6.18, we can observe that the 6-pulse circulating current-free current converter causes a nominal frequency f_{0A}, that is only slightly lower than that obtained with a 3-pulse circulating current-conducting converter. We can conclude from this, that this converter circuit is suitable for many applications, and that the position loop gain it allows in the position control loop, is sufficiently high.

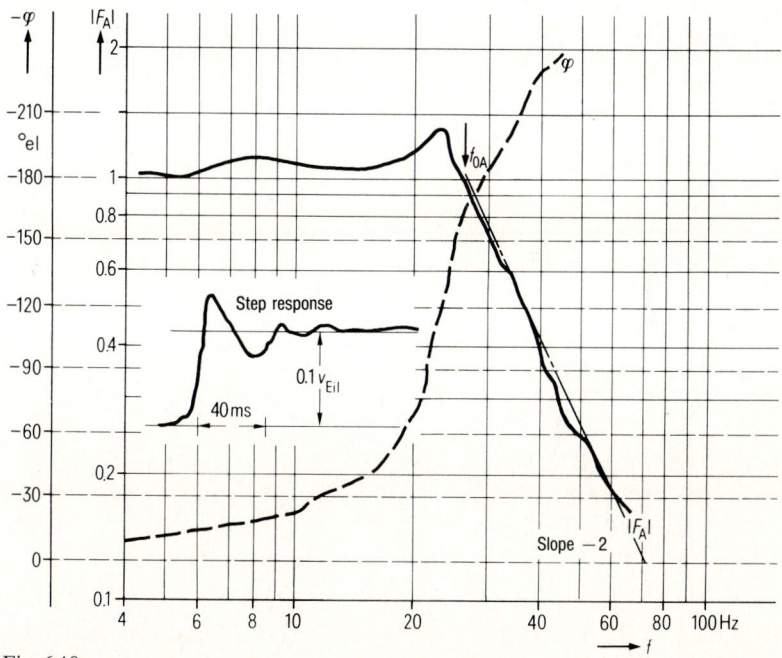

Fig. 6.18
Frequency response curve of a feed drive with the permanent magnet-excited DC servo motor 1HU3104-0AD01, and 6-pulse circulating current-free current converter circuit

6.3.8 Frequency Response Curve of an Electrically Excited DC Servo Motor

On the feed table used for the tests in section 6.3.7, we mount an electrically excited DC servo motor over a tooth wheel gear with $i=4$. The friction torque reflected on this motor is about 30% of the motor rated torque, and the external moment of inertia amounts to about 60% of the motor moment of inertia. Figure 6.19 shows the frequency response curve at a creep feed rate of 2% of the rapid traverse speed.

The nominal frequency f_{0A} lies at about 64 Hz, the damping gradient D_A at 0.45. In the lower frequency ranges, the frequency response curve is affected strongly by the low nominal frequency of the mechanical transmission elements. This reaction causes the noticeable amplitude drop.

By comparison with the figures 6.17 and 6.18, the nominal frequency of this drive lies considerably higher. This is a consequence of the low electrical time constant of the electrically excited DC servo motor, as compared to its mechanical time constant.

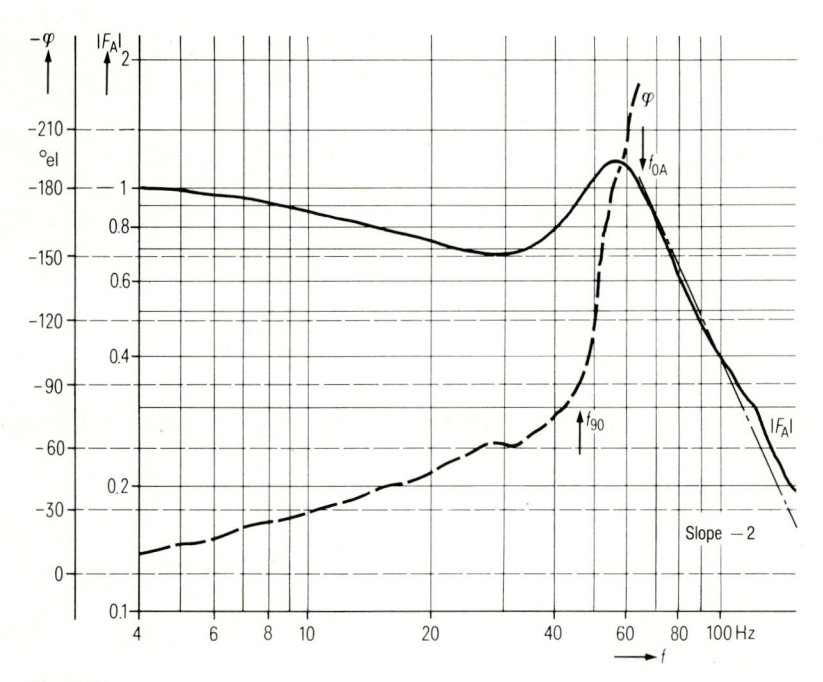

Fig. 6.19
Frequency response curve of a feed drive with the electrically excited DC servo motor 1GS3107-5SW41, and 3-pulse circulating current-conducting converter circuit

6.3.9 Conclusions

The measured frequency response curves, shown in figures 6.13 through 6.19, show in part better results than those calculated with equations (2.21) and (2.22). This is understandable in view of the fact that the presumptions stated in chapter 2, concerning the ideal behavior of control loop elements, do not hold absolutely in the presence of the non-linearities existing in reality. The influence of the dependent current regulation is neglected when calculating the nominal angular frequency of the drive. This means higher nominal angular frequencies than calculated for the transistor chopper. However, since the pre-computation of the optimal position loop gain is also based on the behavior of ideal control elements, a complete calculation of the position loop gain to be set results in an entirely realistical value.

It is not possible to do a partial measurement, for instance of the nominal angular frequency ω_{0A}, and then to base the calculation of the position loop gain K_V on the results obtained from such measurement, because the resulting position loop gain would be too high. The values given in table 3.1 (page 173), are results that can be transfered to practice, and meet all the requirements.

7 Technical Data

The requirements presented in the preceding chapters for servo motors, as well as for current converter units to be used on position controlled feed drives, show that the motors must have special contruction, and the current converter must be especially developed for this task.

The drives produced by Siemens meet all the practical requirements, due on one side to extensive research and development work, and on the other side, to the constant cooperation established with the manufacturers and users of such machine tools. The DC servo motors and converter units can be offered in multiple combinations, depending on the technology of the machine tool and on the requirements imposed by the user. Tables 7.1 through 7.12 present the standard program of DC servo motors and current converters for feed drives on machine tools. Detailed technical descriptions and quotations for these products can be obtained from any Siemens company.

Special development and adaptations for special technology are planned and executed upon request by the experienced engineers of our company.

7.1 DC Servo Motors

7.1.1 General Data

DC servo motors come in permanent magnet-excitation, as well as electrical external excitation versions. They conform to both, DIN norms and VDE requirements, especially the VDE 0530/DIN 57530 rules concerning rotating electrical machines. The rated torques are valid up to a limit overtemperature of the armature winding of 100 K, corresponding to insulation materials of class F. The torques given in the tables are admissible for continuous operation. The permissible values are higher for intermittent operation. The significances are:

S1 continuous operation with rated torque
S3 intermittent operation with a relative switch-on duration of 25% or 40%, for a duty cycle of 10 minutes

The torque/speed diagrams pertaining to the motors can be derived from the technical descriptions.

The self-cooling motors eliminate the heat from losses, through radiation and natural convection processes. A sufficient heat elimination must be ensured by means of a suitable mounting on the machine tool. Forced cooled motors have a blower mounted on the side, with attached dry filter. It is mounted though an intermediary flange.

All motors are equipped with roller bearings. The drive side bearing is built as a fixed bearing, and the bearing on the commutator side can shift axially and is fixed through a spring disk. This way, the armature heat expansion is contained. When tooth wheel gears are used, it is best to utilize straight gear pinions only, since otherwise significant axial forces will be transmitted to the roller bearings. Helical gear pinions lead to axial movements and to oscillations of the armature of the motor and tacho generator, within the elasticity of the bearing.

Motors in standard construction version are delivered with normal cylindrical shaft ends according to DIN 748, and keys and key ways, according to DIN 6885. The shaft end on the commutator side holds the armature of the tacho generator through a conical friction coupling.

With timing belt drives, the shafts and bearings are under stronger radial stress. The shear forces allowed depend on the shaft-shoulder distance of the load attack points, and on the average operating speed. The admissible shear forces can be determined from the appropriate shear force diagrams. The connection between motor shaft and pinion is, under the load, in a multi-axis tension state, which results from torsional, radial and axial forces, as well as from a bending torque.

Shaft connections with key or tapered key couplings are *positive* connections, and under continuous load, change with alternating torques. Rotational assymetrical shifts reduce the smooth quality of the run. Increased deformation can lead the shaft end to break. With *friction couplings,* torque transmission is accomplished exclusively through surface pressure. This type of coupling allows safe force transmission, and can be manufactured either with shrink seat or with clamp elements. The thickness of the hub must be found from the manufacturer's data. The key way present in the motor shaft is of no consequence.

The values presented in tables 7.1 through 7.6 are defined as follows:

M_{0M} Rated torque of the DC servo motor:
Is the torque available as continuous torque at small speeds, thus virtually within the feed range. At longer stillstand of the motor in the same angular position (>5 minutes), is has to be reduced to about 50%.

I_{0M} Rated current of the DC servo motor:
Is the current necessary for the rated torque.

U_A Armature voltage:
Is the rated voltage of the DC servo motor at the rated speed.

n_M Motor speed:
Is reached by the DC servo motor at the rated voltage.

J_M Moment of inertia of the DC servo motor.

$T_{mech\,M}$ Mechanical time constant of the DC servo motor:
           Is calculated according to equation (2.13), and describes the mechanical motor time behavior, without additional external moment of inertia.

T_{el} Electrical time constant of the DC servo motor:
           Is calculated according to equation (2.10), and describes the electrical time behavior of the armature circuit alone, without additional resistances or inductivities.

T_{th} Thermal time constant of the DC servo motor:
           Definition hereby disregards any heat removal through the mounting flange, and refers only to normal heat removal through the surface.

K_T Torque constant of the DC servo motor.

K_E Voltage constant of the DC servo motor.

R_{A+B} Armature resistance of the DC servo motor.

L_M Armature inductivity of the DC servo motor.

Values which have not been listed above, can be derived from the given technical descriptions.

7.1.2 Design Series 1HU

DC servo motors in the construction series 1HU are permanent magnet-excited, high-quality dynamic motors for feed drives on machine tools. They excel in the following properties:

- ▷ constant torque over a wide speed range (feed range)
- ▷ good, smooth run even at low speeds
- ▷ high overload capability
- ▷ high speed stiffness
- ▷ low level run noise and oscillation
- ▷ integrated DC tacho generator
- ▷ possibility of integrating a measuring gear with resolver, or an incremental pulse coder (position encoder)
- ▷ possibility of mounting a permanent magnet holding brake
- ▷ permanent magnet-excitation, and therefore lack of additional heating due to excitation coils.

Because of the permanent magnet-excitation, these motors have the behavior of a shunt motor. The magnet material used is highly coercive ferrite material, which is suitable for armature voltage control down to the lowest speeds. The motors can be operated with a limit overtemperature of 130 K, without damaging the insulation, and can thus deliver higher rated torques.

For all self-cooling types, the noise level of the motors is within the type-determined speed range, under 58 dB (A). In the case of forced-cooled versions, the machine noise levels are determined entirely by the blower; it is, independently of motor speed, at 65 dB (A).

Values of oscillation severity range N (DIN 45665) are not reached. Shock loads are admissible up to 6 g. The functioning of the motor is not disturbed either during or after the shock.

According to their applications, the DC servo motors of series 1HU are manufactured for the highest quality of run smoothness. The slot frequent torque ripple lies under 1.5%.

The most important technical data are presented in tables 7.1 through 7.4.

Table 7.1
Permanent magnet-excited DC servo motors of series 1HU
Technical data of the self-cooling motors 1HU5, in protection type IP 54

Rated torque M_{0M}	Type (order no.)	Weight approx. kg	Rated current at M_{0M} I_{0M} A	Armature voltage U_A V	Speed n_M min^{-1}	Moment of inertia J_M kg m^2	Time constants Mechanical T_{mechM} ms	Electrical T_{el} ms	Thermal T_{th} min	Torque const. K_T Nm/A	Voltage const. K_E V/1000 min^{-1}	Armature resistance with brushes at 20 °C R_{A+B} Ω	Armature inductivity L_M mH
Nm													
1.2	1HU5040-0AC01	5.9	1.9	180	2,000	0.00068	16.4	2.5	50	0.691	72.3	11.50	28.6
	1HU5040-0AF01		2.8	174	3,000		17.3	2.4	50	0.477	49.9	5.76	13.6
1.75	1HU5042-0AC01	7.4	2.7	172	2,000	0.00093	12.9	2.9	60	0.684	71.6	6.49	19.1
	1HU5042-0AF01		4.0	166	3,000		13.9	2.7	60	0.466	48.8	3.24	8.9
2.5	1HU5044-0AC01	8.8	3.6	178	2,000	0.0012	11.1	2.9	70	0.719	75.2	4.77	13.8
	1HU5044-0AF01		5.3	176	3,000		11.9	2.7	70	0.498	52.1	2.45	6.6

Table 7.2
Permanent magnet-excited DC servo motors of series 1HU
Technical data of the self-cooling short-version motor 1HU3, in protection type IP 54

Rated torque M_{0M} Nm	Type (order no.)	Weight approx. kg	Rated current at M_{0M} I_{0M} A	Armature voltage U_A V	Speed n_M min^{-1}	Moment of inertia J_M kg m^2	Time constants Mechanical T_{mechM} ms	Electrical T_{el} ms	Thermal T_{th} min	Torque const. K_T Nm/A	Voltage const. K_E V/1000 min^{-1}	Armature resistance with brushes at 20 °C R_{A+B} Ω	Armature inductivity L_M mH
3.2	1HU3070-0AC01 1HU3070-0AF01	12	4.5 6.4	187 190	2,000 3,000	0.0022	11.0 12.0	5.1 4.7	68	0.786 0.548	82 57	3.100 1.650	15.9 7.7
5	1HU3071-0AC01 1HU3071-0AF01	14	7.1 10.7	176 180	2,000 3,000	0.0028	7.6 8.4	6.4 5.8	90	0.772 0.536	81 56	1.620 0.860	10.3 5.0
7	1HU3073-0AC01 1HU3073-0AF01	21	10.0 13.8	163 173	2,000 3,000	0.0042	7.2 7.6	7.2 6.8	90	0.719 0.523	75 55	0.889 0.497	6.4 3.4
7	1HU3100-0AC01 1HU3100-0AF01	17	9.5 14.2	171 168	2,000 3,000	0.0086	15.4 16.1	11.4 10.9	100	0.765 0.510	80 53	1.044 0.486	11.9 5.3
10.0	1HU3101-0AC01 1HU3101-0AF01	25	13.4 20.5	167 163	2,000 3,000	0.0120	10.6 12.1	13.5 11.8	120	0.766 0.501	80 53	0.515 0.253	6.9 3.0
12.5	1HU3103-0AC01 1HU3103-0AF01	30	15.7 23.5	190 190	2,000 3,000	0.0160	10.5 11.8	17.3 15.4	140	0.823 0.549	86 57	0.443 0.222	7.7 3.4

Table 7.3
Permanent magnet-excited DC servo motors of series 1HU
Technical data of the self-cooling motor 1HU3, in protection type IP 54

Rated torque M_{0M} Nm	Type (order no.)	Weight approx. kg	Rated current at M_{0M} I_{0M} A	Armature voltage U_A V	Speed n_M min^{-1}	Moment of inertia J_M kg m^2	Time constants Mechanical T_{mechM} ms	Time constants Electrical T_{el} ms	Thermal T_{th} min	Torque const. K_T Nm/A	Voltage const. K_E V/1000 min^{-1}	Armature resistance with brushes at 20°C R_{A+B} Ω	Armature inductivity L_M mH
2.2	1HU3054-0AC01 1HU3054-0AF01	8.7	3.3 4.7	174 170	2,000 3,000	0.0012	14.9 16.0	18.7 17.5	75	0.688 0.475	74 51	6.08 3.09	113.7 54.1
4.5	1HU3056-0AC01 1HU3056-0AF01	15.5	6.7 10.0	176 163	2,000 3,000	0.0022	10.7 10.9	23.5 23.2	90	0.72 0.458	75 50	2.42 1.09	56.7 25.2
6	1HU3058-0AC01 1HU3058-0AF01	22	8.8 12.7	171 169	2,000 3,000	0.0033	10.0 10.7	26.3 25.6	105	0.691 0.560	72 51	1.39 0.727	36.5 186
7	1HU3074-0AC01 1HU3074-0AF01	23	9.7 13.5	167 177	2,000 3,000	0.0048	7.1 7.6	7.6 7.1	90	0.735 0.535	77 56	0.804 0.452	6.1 3.2
10	1HU3076-0AC01 1HU3076-0AF01	31.5	12.5 20.0	175 163	2,000 3,000	0.0065	5.8 7.1	8.9 7.3	120	0.802 0.501	84 53	0.575 0.273	5.1 2.0
14	1HU3078-0AC01 1HU3078-0AF01	40	17 25	183 183	2,000 3,000	0.0085	5.6 6.3	9.1 8.1	150	0.837 0.558	88 58	0.458 0.23	4.2 1.9
18	1HU3102-0AD01 1HU3102-0AH01	42	15 24	173 170	1,200 2,000	0.0200	8.9 10.3	16 14	120	1.280 0.770	136 81	0.74 0.309	12 4.5
25	1HU3104-0AD01 1HU3104-0AH01	58	19.5 31	175 179	1,200 2,000	0.0280	7.4 8.0	20 18	120	1.300 0.820	138 86	0.450 0.193	8.9 3.5

32	1HU3106-0AD01	73	24	179	1,200	0.0370	6.3	31	120	1.350	143	0.312	9.6
	1HU3106-0AH01		42	172	2,000		7.3	27		0.788	83	0.123	3.3
38	1HU3108-0AD01	88	31	168	1,200	0.0450	6.0	34	120	1.270	134	0.216	7.3
	1HU3108-0AH01		46	184	2,000		6.9	29		0.850	89	0.111	3.2
47	1HU3132-0AC01	114	31	167	1,000	0.1100	9.9	52	120	1.540	163	0.214	11.0
	1HU3132-0AF01		45	171	1,500		11.2	45		1.050	111	0.114	5.2
65	1HU3134-0AC01	143	43	167	1,000	0.1500	8.2	50	120	1.540	162	0.13	6.5
	1HU3134-0AF01		62	172	1,500		10.1	42		1.060	113	0.073	3.1
90	1HU3136-0AC01	180	59	167	1,000	0.2300	8.0	47	120	1.550	164	0.084	4.0
	1HU3136-0AF01		83	178	1,500		9.4	40		1.100	117	0.05	2.0
115	1HU3138-0AC01	224	80	158	1,000	0.2900	7.7	44	120	1.460	155	0.057	2.5
	1HU3138-0AF01		134	170	1,800		10.1	33		0.877	93	0.027	0.9

Table 7.4
Permanent magnet-excited DC servo motors of the series 1HU
Technical data of the forced-cooled motors 1HU3, in protection type IP 21 for construction form B5, respectively IP 20 for construction form V1

Rated torque M_{0M} Nm	Type (order no.)	Weight approx. kg	Rated current at M_{0M} I_{0M} A	Armature voltage U_A V	Speed n_M min^{-1}	Moment of inertia J_M kg m^2	Time constants Mechanical $T_{mech\,M}$ ms	Electrical T_{el} ms	Thermal T_{th} min	Torque const. K_T Nm/A	Voltage const. K_E V/1000 min^{-1}	Armature resistance with brushes at 20 °C R_{A+B} Ω	Armature inductivity L_M mH
29	1HU3102-0SD01 1HU3102-0SH01	46	23 39	173 170	1,200 2,000	0.0200	8.3 9.3	18 16	90	1.270 0.762	136 81	0.69 0.278	12 4.5
40	1HU3104-0SD01 1HU3104-0SH01	62	31 50	175 179	1,200 2,000	0.0280	6.8 7.1	22 21	90	1.290 0.810	138 86	0.413 0.169	8.9 3.5
50	1HU3106-0SD01 1HU3106-0SH01	77	38 65	179 172	1,200 2,000	0.0370	5.7 6.3	34 31	90	1.337 0.780	143 83	0.282 0.105	9.6 3.3
60	1HU3108-0SD01 1HU3108-0SH01	92	48 72	168 184	1,200 2,000	0.0450	5.4 6.0	38 34	90	1.260 0.840	134 89	0.191 0.095	7.3 3.2
72	1HU3132-0SC01 1HU3132-0SF01	119	48 70	167 171	1,000 1,500	0.1100	8.9 9.8	58 52	90	1.520 1.040	163 111	0.188 0.099	11.0 5.2
100	1HU3134-0SC01	147	66	167	1,000	0.1500	7.2	57	90	1.520	162	0.113	6.5
135	1HU3136-0SC01 1HU3136-0SF01	184	89 125	167 178	1,000 1,500	0.2300	7.0 7.9	55 48	90	1.530 1.100	164 117	0.072 0.042	4.0 2.0
165	1HU3138-0SC01	228	116	158	1,000	0.2900	6.7	51	90	1.450	155	0.049	2.5

7.1.3 Design Series 1GS3

The DC servo motors of the design series 1GS3 are electrically excited, high-quality dynamic motors for feed drives on machine tools. They excel in the following properties:

- ▷ low mechanical time constant, due to low armature moment of inertia
- ▷ low electrical time constant, due to the compensation windings
- ▷ the linearity between armature current and torque is up to 8, respectively 10 times the motor rated current
- ▷ high overload capability, and smoothness of run at low speeds
- ▷ low levels run noise and oscillation
- ▷ integrated DC tacho generator
- ▷ friction couplings between motor and tacho generator
- ▷ speed-independent forced cooling through the mounted blower.

The motors are operated in the armature control range up to a maximal speed of 6,000 min^{-1}. Continuous operation (S1), according to the torque characteristic curves, is allowed with constant rated torque, in the lower speed range, up to about 1,000 min^{-1}. At constant power, field weakening operation is allowed for each curve point of the diagram $U_A = f(n_M)$, respectively $M_{0M} = f(n_M)$, up to 1.5 times the starting speed. During field weakening operations, the ramp-up time increases inversely proportional to the torque reduction.

Within the type-dependent speed range, the noise levels of self-cooling motors of the 1GS3 series lies under 66 dB (A).

The noise levels of forced-cooled versions depends on the type of motor. For DC servo motors 1GS3107, at a speed of 6,000 min^{-1}, it lies at 65 dB (A), while it is at 78 dB (A) for 1GS3168 motors at the same speed. These figures do not include gear noises. The oscillating behavior corresponds to oscillation severity range R of DIN 45665. According to their applications, motors of the 1GS3 series are manufactured for the highest quality of run smoothness. The slot frequent torque ripple lies under 3%.

The most important technical data are shown in tables 7.5 and 7.6.

Table 7.5
Electrically excited DC servo motors
Technical data of the self-cooling motors 1GS3, in protection in type IP 44

Rated torque M_{0M}	Type (order no.)	Weight approx.	Rated current at M_{0M}	Armature voltage	Speed	Moment of inertia	Time constants — Mechanical	Time constants — Electrical	Thermal	Torque const.	Voltage const.	Armature resistance with brushes at 20 °C	Armature inductivity
Nm		kg	I_{0M} A	U_A V	n_M min^{-1}	J_M kg m^2	$T_{mech M}$ ms	T_{el} ms	T_{th} min	K_T Nm/A	K_E V/1000 min^{-1}	R_{A+B} Ω	L_M mH
2.5	1GS3107-5UV91	37	12.4	137	6,000	0.002	28.30	1.17	90	0.207	21.7	0.61	0.71
	1GS3107-5UW91		7.8	220	6,000		25.23	1.32		0.331	34.7	1.39	1.82
	1GS3107-5UT91		3.9	400	5,400		23.03	1.44		0.663	69.4	5.07	7.30
6.0	1GS3137-5UV91	63	28.4	130	6,000	0.005	18.88	1.84	90	0.215	22.5	0.17	0.32
	1GS3137-5UW91		14.2	283	6,000		14.17	2.45		0.430	45.0	0.52	1.28
	1GS3137-5UT91		8.1	400	5,100		12.51	2.77		0.752	78.8	1.42	3.92
	1GS3137-5US91		21.3	167	6,000		16.86	2.06		0.286	30.0	0.28	0.57
11.0	1GS3167-5UV91	122	55.2	130	6,000	0.014	32.33	3.44	90	0.203	21.3	0.09	0.32
	1GS3167-5UW91		27.6	263	6,000		20.59	5.40		0.408	42.7	0.24	1.28
	1GS3167-5UT91		15.1	400	5,000		17.67	6.29		0.747	78.2	0.68	4.31
	1GS3167-5US91		41.4	170	6,000		26.98	4.12		0.271	28.4	0.14	0.57
18.0	1GS3168-5UW91	152	10.8	280	6,000	0.035	25.11	2.41	90	0.450	47.1	0.15	0.35
	1GS3168-5UT91		22.7	400	4,700		20.48	2.96		0.809	84.7	0.38	1.13

Table 7.6
Electrically excited DC servo motors
Technical data of the forced-cooled motors 1GS3, in protection type IP 21 for version B5, respectively in protection type IP 20 for version V1

Rated torque M_{0M} Nm	Type (order no.)	Weight approx. kg	Rated current at M_{0M} I_{0M} A	Armature voltage U_A V	Speed n_M min^{-1}	Moment of inertia J_M kg m^2	Time constants Mechanical T_{mechM} ms	Time constants Electrical T_{el} ms	Thermal T_{th} min	Torque const. K_T Nm/A	Voltage const. K_E V/1000 min^{-1}	Armature resistance with brushes at 20 °C R_{A+B} Ω	Armature inductivity L_M mH
6.8	1GS3107-5SV41	41	23.3	190	6,000	0.002	12.28	1.34	60	0.294	30.8	0.53	0.71
	1GS3107-5SW41		14.6	320	6,000		11.41	1.44		0.472	49.4	1.27	1.82
	1GS3107-5ST41		7.3	400	3,600		10.86	1.51		0.942	98.7	4.82	7.29
15.6	1GS3137-5SV41	68	58.4	168	6,000	0.005	9.57	2.32	60	0.268	28.1	0.14	0.32
	1GS3137-5SW41		29.2	350	6,000		7.82	2.84		0.538	56.3	0.45	1.28
	1GS3137-5ST41		16.7	400	3,800		7.29	3.04		0.941	98.5	1.29	3.92
	1GS3137-5SS41		43.8	230	6,000		8.91	2.49		0.358	37.5	0.23	0.57
32.0	1GS3167-5SV41	127	107.7	180	6,000	0.014	12.31	4.23	60	0.297	31.1	0.08	0.32
	1GS3167-5SW41		53.9	378	6,000		8.32	6.34		0.595	62.3	0.20	1.28
	1GS3167-5ST41		29.4	400	3,400		7.52	6.94		1.090	114.1	0.62	4.31
	1GS3167-5SS41		80.8	250	6,000		10.51	4.96		0.396	41.5	0.11	0.57
43.0	1GS3168-5SW41	157	76.4	370	6,000	0.035	12.31	2.86	60	0.589	61.7	0.12	0.35
	1GS3168-5ST41		42.4	400	3,600		10.64	3.31		1.061	111.1	0.34	1.13

7.2 Current Converter Units

7.2.1 Transistor DC Choppers

Transistor DC choppers regulate the speed of the drive through a pulse-width modulated output DC voltage. They function in 4-quadrant operation, and meet the highest requirements for dynamic control behavior. Complete units are offered for feed drives with up to four feed axes, with a maximal output DC voltage of 200 V.

The design series 6 RB 26 is in modular construction form. These modules can be plugged into a base plate, and they are interconnected with pluggable signal ribbon cables for the electronic portion, and with bus bars in the power section. The connection to external signals is made directly to the PC boards through soldered-in terminals, and the connection to the power circuits is made with flat connectors. The control principle is like that of a speed regulator with detaching current limit control. The pulse frequency of the power section is approximately 9 kHz.

Design series 6 RB 20 is equipped with PC boards and a board rack. These are connected with each other and to the terminal through pluggable ribbon cables for the electronic signals, and the power connections are also made through the terminals. The speed control with PI controller has a subordinated current control. The pulse frequency of the power section is 2.5 kHz.

Both construction series have monitor circuits with LED displays. These monitor circuits prevent inadmissible operating states for the converter as well as for the DC servo motor, and give to the outside a ready-operation signal.

Tables 7.7 and 7.8 give the following values:

▷ Rated line voltage:
  Is the rated voltage on the 3-phase AC side. The frequency is 50/60 Hz, and the admissible voltage tolerance $\pm 10\%$.
▷ Rated DC voltage:
  Is achieved on the output side at rated load and rated line voltage.
▷ Rated DC current:
  Enables the converter to be continuously loaded, by exploiting the short term limit current. For multi-axis choppers, a reduction, which depends on the simultaneously occurring load, is necessary.
▷ Short-term limit current:
  Is the maximal admissible current which can be delivered by the converter for acceleration and deceleration processes. The rated DC current is assumed as preload. For a load period of 10 s, the load duration is maximum 200 ms.

The technical descriptions contain more detailed data.

Table 7.7
Transistor DC choppers
Technical data of series 6 RB 26

Feed axes no.	Rated line voltage (50/60 Hz) V	Rated DC voltage V	Rated DC current A	Short-term limit current (200 ms) A	Type (order no.)	Weight approx. kg
1	3~165	1 × 200	1 × 7	1 × 20	6 RB 26 07-1 MA 00	9.5
2	3~165	2 × 200	2 × 7	2 × 20	6 RB 26 07-2 MA 00	13
3	3~165	3 × 200	3 × 7	3 × 20	6 RB 26 07-3 MA 00	18
4	3~165	4 × 200	4 × 7	4 × 20	6 RB 26 07-4 MA 00	21.5

Table 7.8
Transistor DC choppers
Technical data of series 6 RB 20

Feed axes no.	Rated line voltage (50/60 Hz) V	Rated DC voltage V	Rated DC current A	Short-term limit current (200 ms) A	Type (order no.)	Weight approx. kg
1	3~165	1 × 200	1 × 12	1 × 24	6 RB 20 12-1 BA 00	15,5
	3~165	1 × 200	1 × 25	1 × 50	6 RB 20 25-1 BA 00	16
	3~165	1 × 200	1 × 30	1 × 75	6 RB 20 30-1 BA 00	16
2	3~165	2 × 200	2 × 12	2 × 24	6 RB 20 12-2 BA 00	17
	3~165	2 × 200	2 × 25	2 × 50	6 RB 20 25-2 BA 00	17,5
	3~165	2 × 200	2 × 30	2 × 75	6 RB 20 30-2 BA 00	17,5
3	3~165	3 × 200	3 × 12	3 × 24	6 RB 20 12-3 BA 00	29.5
	3~165	3 × 200	3 × 25	3 × 50	6 RB 20 25-3 BA 00	30.5
	3~165	3 × 200	3 × 30	3 × 75	6 RB 20 30-3 BA 00	31.5
4	3~165	4 × 200	4 × 12	4 × 24	6 RB 20 12-4 BA 00	31
	3~165	4 × 200	4 × 25	4 × 50	6 RB 20 25-4 BA 00	32
	3~165	4 × 200	4 × 30	4 × 75	6 RB 20 30-4 BA 00	33

7.2.2 Line-synchronized Thyristor Controllers

Thyristor current converter units 6 RA 26 are used in combination with DC servo motors of series 1HU and 1GS, as drives for numerically controlled feed axes. They function in 4-quadrant operation and are available in the following circuit forms:

▷ Circulating current-conducting cross-connection of two 6-pulse bridge connections (B6C) X(B6C), as drive of one feed axis

▷ Circulating current-conducting anti-parallel connection of two 3-pulse middle-point connections (M3C) A(M3C), as drive of one feed axis

▷ Circulating current-conducting anti-parallel connection of two times two 3-pulse middle-point connections $2 \times$ (M3C) A(M3C), as drive for two feed axes

▷ Circulating current-free anti-parallel connection of two 6-pulse bridge connections (B6C) A(B6C), as drive for one feed axis

▷ Circulating current-conducting anti-parallel connection of two 2-pulse middle-point circuits (M2C) A(M2C), as drive for one feed axis

A speed regulator with detaching current limitation control is used on all thyristor converter units.

The power section and the control section of a thyristor converter unit are encased in a compact unit. The PC boards of the control section are spaced behind each other in front of the power section, and are interconnected with ribbon cables. The motor and the current converter are protected by different monitor circuits. Ready and error states are indicated and signaled to the outside.

The data given in tables 7.9 through 7.12 offer an overview of the available spectrum of products. The definitions of the variables described are according to the explanations given in section 7.2.1, with the exception of the output voltage tolerance which here is fixed with $+10\%$, -5%. Further details can be obtained from the technical descriptions.

The switch gears necessary for the proper functioning of a drive must have their controls, interlocks, and further possible measuring and monitoring devices selected according to the detailed data from the technical descriptions.

Table 7.9
Thyristor current converter units 6 RA 26
Technical data of the 6-pulse circulating current-conducting circuit

Rated line voltage (50/60 Hz) V	Rated DC voltage V	Rated DC current A	Short-term limit current (200 ms) A	Type (order no.)	Power denomination acc. to DIN 41 752	Weight approx. kg
3 ~ 190	200	20	80	6 RA 26 15-6 MW 30	D200/20	14
		26	105	6 RA 26 17-6 MW 30	D200/26	14
		40	160	6 RA 26 21-6 MW 30	D200/40	14
		60	240	6 RA 26 25-6 MW 30	D200/60	14
		85	340	6 RA 26 27-6 MW 30	D200/85	16
		125	500	6 RA 26 71-6 MW 30	D200/125	28
		175	700	6 RA 26 74-6 MW 30	D200/175	28
3 ~ 380	400	40	160	6 RA 26 21-6 DW 30	D400/40	14
		60	240	6 RA 26 25-6 DW 30	D400/60	14
		85	340	6 RA 26 27-6 DW 30	D400/85	16
		125	500	6 RA 26 71-6 DW 30	D400/125	28
		175	700	6 RA 26 74-6 DW 30	D400/175	28

Table 7.10
Thyristor current converter circuits 6 RA 26
Technical data of the 3-pulse circulating current-conducting circuit. (For two-axes versions, instead of the letter U the order designation contains the letter X)

Rated line voltage (50/60 Hz) V	Rated DC voltage V	Rated DC current A	Short-term limit current (200 ms) A	Type (order no.)	Power denomination, acc. to DIN 41 752	Weight approx. kg
3 ~ 380	200	11	45	6 RA 26 10-6 DU 30	D200/11	12
		22	90	6 RA 26 16-6 DU 30	D200/22	13
		30	120	6 RA 26 18-6 DU 30	D200/30	13
		44	175	6 RA 26 22-6 DU 30	D200/44	13
		65	260	6 RA 26 25-6 DU 30	D200/65	13
		95	380	6 RA 26 28-6 DU 30	D200/95	15
		135	540	6 RA 26 72-6 DU 30	D200/135	27
		175	700	6 RA 26 74-6 DU 30	D200/175	27

Table 7.11
Thyristor current converter circuits 6 RA 26
Technical data of the 6-pulse circulating current-free conection

Rated line voltage (50/60 Hz) V	Rated DC voltage V	Rated DC current A	Short-term limit current (200 ms) A	Type (order no.)	Power denomination acc. to DIN 41752	Weight approx. kg
3~190	200	11	45	6 RA 26 10-6 MV 30	D200/11	12
		20	80	6 RA 26 15-6 MV 30	D200/20	13
		26	105	6 RA 26 17-6 MV 30	D200/26	13
		40	160	6 RA 26 21-6 MV 30	D 200/40	13
		60	240	6 RA 26 25-6 MV 30	D200/60	13
		85	340	6 RA 26 27-6 MV 30	D200/85	15
		125	500	6 RA 26 71-6 MV 30	D200/125	27
		175	700	6 RA 26 74-6 MV 30	D200/175	27
3~380	400	40	160	6 RA 26 21-6 DV 30	D400/40	13
		60	240	6 RA 26 25-6 DV 30	D400/60	13
		85	340	6 RA 26 27-6 DV 30	D400/85	15
		125	500	6 RA 26 71-6 DV 30	D400/125	27
		175	700	6 RA 26 74-6 DV 30	D400/175	27

Table 7.12
Thyristor current converter units 6 RA 26
Technical data of the 2-pulse circulating current-conducting circuit

Rated line voltage (50/60 Hz) V	Rated DC voltage V	Rated DC current A	Short-term limit current (200 ms) A	Type (order no.)	Power denomination acc. to DIN 41752	Weight approx. kg
2~275	200	11	44	6 RA 26 10-6 RM 30	E200/11	2.5
2~275		15	60	6 RA 26 13-6 RM 30	E200/15	2.5
2~275		20	80	6 RA 26 15-6 RM 30	E200/20	2.6

8 Technical Appendix

8.1 Symbols and Units Used

A	crossectional area	m^2
a_f	command variable of acceleration	m/s^2
a_{f0}	limit value for the command variable of acceleration	m/s^2
a_{fx}	command variable of acceleration for the X-axis	m/s^2
a_{fy}	command variable of acceleration for the Y-axis	m/s^2
$a_{i\,Grenz}$	limit value of the actual acceleration	m/s^2
$a_{i\,max}$	maximal value of actual acceleration	m/s^2
a_w	linear acceleration of the work piece	m/s^2
b_1	length factor for toothed belt	—
b_2	loading factor for toothed belt	—
b_3	gear ratio factor for toothed belt	—
b_{Kn}	correction factor for calculating the buckling strength	—
b_{Krit}	correction factor for calculating the critical speed	—
b_{St}	factor for calculating the ball screw nut complex	—
C	capacitance	μF
c_E	excitation constant	V/min^{-1}
c_M	motor constant	Vs/rad, Nm/A
c_{WM}	specific heat capacity of the motor	J/kg
c_v	speed-dependent damping constant	$N\,m^{-1}\,s$
D	damping gradient	—
D_A	damping gradient of the drive	—
D_A^*	damping gradient of uncontrolled drive	—
D_L	damping gradient of position control loop	—
D_{mech}	damping gradient of the mechanical transmission elements	—
$D_{mech\,L}$	damping gradient of feed screw bearing	—
d	diameter	m
d_G	thread diameter	m
d_{Gt}	gear wheel diameter	m
d_{KSp}	core diameter of feed screw	m
d_{Sp}	diameter of feed screw	m
d_W	active diameter of toothed belt	m
d_{mL}	mean diameter of feed screw bearing	m

E	elasticity modulus	N/m²
E_M	stationary induced counter voltage in motor (electro motive force, EMF of motor)	V
E_{0M}	nominal value of induced counter voltage in motor	V
E_{maxM}	maximal induced counter voltage in motor	V
e	base of natural logarithms	(e = 2.718...)
e	final deviation	mm
e_E	corner deviation	mm
e_{maxE}	maximal corner deviation	mm
$e_{max\,zul\,E}$	maximal admissible corner deviation	mm
e_{maxKr}	maximal circular deviation	mm
e_{statKr}	static circular deviation	mm
e_x	final deviation from position in the X-direction, respectively tolerance band at threading	mm
e_M	time-variable induced countervoltage in motor (motor EMF)	V
F	frequency response curve	—
F_A	frequency response curve of the feed drive	—
F_I	frequency response curve of the integral element	—
F_{LR}	frequency response curve of the position control loop	—
F_R	frequency response curve of the control device	—
F_{RI}	frequency response curve of the current regulator	—
F_{Rn}	frequency response curve of the speed regulator	—
F_S	frequency response curve of the control system	—
F_T	frequency response curve of the dead time element	—
F_a, F_b, F_c	frequency response curve of transfer elements a, b, c	—
F_{mech}	frequency response curve of the mechanical transmission elements	—
F_o	frequency response curve of the open control loop	—
F_r	frequency response curve of feedback	—
F_{res}	resultant frequency response curve	—
F_w	command frequency response curve	—
F_{wL}	command frequency response curve of the position control loop	—
F_z	disturbance frequency response curve	—
F	force	N
F_{KnSp}	buckling force of the feed screw	N
F_V	feed force	N
F_{VB}	acceleration force	N
F_{VL}	machining force	N

F_{VT}	component of cutting force perpendicular to the table	N
F_{ZV}	preload force of timing belt	N
F_{aL}	bearing load in axial direction of the feed screw	N
F_{aSp}	force in axial direction of the feed screw	N
F_{aVL}	preload force of the feed screw bearing	N
F_{aVM}	preload force of the reciprocating ball screw nut	N
$F_{max\,a\,Sp}$	maximal force in the axial direction of the feed screw	N
F_i	form factor of the current	—
f	frequency	Hz
f_{0A}	nominal frequency of the feed drive	Hz
$f_{0\,mech}$	nominal frequency of the mechanical transmission elements	Hz
$f_{0\,To}$	torsional resonance frequency	Hz
f_1	frequency of the fundamental oscillation on the DC side	Hz
f_N	line or pulse frequency of the current converter	Hz
G	shearing modulus	N/m²
g	gravitational acceleration (standard = 9.81 m/s²)	m/s²
h_G	thread lead	mm
h_{Sp}	feed screw lead	mm
I	current	A
I_{0M}	rated current of the DC servo motor at rated torque	A
I_A	stationary armature current	A
I_E	excitation current	A
I_{Grenz}	set value of current limitation	A
I_{Kr}	circulating current	A
I_M	motor current	A
I_{MR}	motor current for friction torque	A
I_d	arithmetical mean of the converter output current	A
I_{eff}	rms value of converter output current	A
$I_{eff\,M}$	rms value of motor current	A
I_i	actual current value	A
$I_{max\,A}$	maximal armature current, stationary	A
i_A	time-variable armature current	A
i_{Gr}	time-variable current in rectifier	A
i_Z	time-variable current in DC bus	A
i_C	time-variable charge current of capacitor	A
i	gear ratio	—
i_n	gear ratio of a gear section	—
i_{opt}	gear ratio for maximal acceleration	—

J	moment of inertia	kgm²
J_1	moment of inertia reflected on shaft 1	kgm²
J_2	moment of inertia reflected on shaft 2, without gear wheel 2	kgm²
J_{ext}	external moment of inertia reflected on motor shaft	kgm²
J_{Ges}	total moment of inertia reflected on motor shaft	kgm²
J_{Getr}	total gear moment of inertia reflected on motor shaft	kgm²
$J_{Gt\,1}$	moment of inertia of gear wheel 1	kgm²
$J_{Gt\,2}$	moment of inertia of gear wheel 2	kgm²
J_M	motor moment of inertia	kgm²
J_{Ri}	pinion moment of inertia	kgm²
J_{Sp}	feed screw moment of inertia	kgm²
J_{T+W}	moment of inertia of linearly moved masses, reflected on shaft 2	kgm²
j	imaginary unit	—
$j\omega$	imaginary angular frequency	s⁻¹
K	gain	—
K_E	voltage constant of DC servo motor	V/1,000 min⁻¹
K_N	normalized gain	—
K_S	control system gain	—
K_T	torque constant of DC servo motor	Nm/A
K_{gI}	current regulator gain	—
K_{gn}	speed regulator gain	—
K_v	position loop gain	s⁻¹
$K_{v\,err}$	achievable position loop gain	s⁻¹
k	order numbers	—
k	spring constant	N/μm
k_B	bending strength of shafts	N/μm
k_G	gear spring constant	N/μm
k_{Ges}	total stiffness of the mechanical transmission elements	N/μm
k_M	ball screw nut spring constant	N/μm
k_{Sp}	feed screw spring constant	N/μm
k_{TM}	spring constant of the ball screw nut mounting to the table	N/μm
k_{To}	torsion spring constant	Nm
$k_{To\,Sp}$	feed screw torsion spring constant	Nm
k_Z	tooth stiffness of gear wheel pairs	N/μm
k_{ZRi}	tooth stiffness of the rack and pinion pair	N/μm
k_{aL}	bearing spring constant in axial direction	N/μm
$k_{ers\,B}$	bending strength reflected on the table movement	N/μm
$k_{ers\,To}$	torsion spring constant reflected on the table movement	N/μm

$k_{\text{ers res rL}}$	resultant spring constant of the bearing in radial direction, reflected on the table movement	N/µm
$k_{\text{ers Z}}$	tooth stiffness of gear wheel pairs reflected on table motion	N/µm
k_{rL}	bearing spring constant in radial direction	N/µm
$k_{\text{res rL}}$	bearing resultant spring constant in radial direction	N/µm
L_A	total armature inductivity of feed drive	mH
L_D	reactor inductivity	mH
L_E	excitation inductivity of generator	mH
L_M	motor armature inductivity	mH
l	length	m
l_A	shaft distance for timing belts	mm
l_{Sp}	feed screw length	m
M	torque	Nm
M_{0M}	rated torque of DC servo motor	Nm
M_B	acceleration torque	Nm
$M_{\text{Grenz M}}$	motor torque at current limitation	Nm
M_L	load torque reflected on motor shaft	Nm
M_M	motor torque	Nm
ΔM_M	torque difference in motor characteristic curves	Nm
M_R	torque for friction and losses reflected on motor shaft	Nm
ΣM_R	sum of torques for friction and losses reflected on motor shaft	Nm
M_{RF}	torque for friction of the table guide, reflected on feed screw or pinion	Nm
M_{RSL}	torque for friction of the feed screw bearing reflected on the feed screw	Nm
M_{Sp}	feed screw torque	Nm
$M_{\text{St A}}$	short-circuit torque of the feed drive	Nm
$M_{\text{St M}}$	short-circuit torque of the motor	Nm
M_V	torque for the machining force reflected on the motor shaft	Nm
$M_{\text{eff M}}$	effective value of the motor torque	Nm
$M_{\text{zul M}}$	admissible motor torque	Nm
m	mass	kg
m_G	mass of the rotating gear section	kg
m_{Sp}	mass of feed screw	kg
m_T	feed table mass	kg
m_W	work piece mass	kg
n	number sequence	–

n	speed	min^{-1}
n_0	rated speed of generator	min^{-1}
n_1	speed of gear shaft 1	min^{-1}
n_2	speed of gear shaft 2	min^{-1}
n_{Krit}	critical speed of feed screw	min^{-1}
n_M	motor speed	min^{-1}
Δn_M	speed difference in motor characteristic curves	min^{-1}
n_{Sp}	feed screw speed	min^{-1}
n_i	actual speed value	—
n_{mM}	arithmetic mean value of motor speeds	min^{-1}
n_{maxM}	maximal motor speed	min^{-1}
n_{maxSp}	maximal feed screw speed	min^{-1}
n_{minM}	minimal motor speed	min^{-1}
n_s	speed command value	—
n_{maxs}	maximal speed command value	—
P	power	kW
P_B	calculated power for timing belt drives	kW
P_M	power stated for the motor	kW
P_{RSL}	power loss in the bearing of the feed screw	kW
P_V	power loss	kW
P_{VSR}	power loss in the current converter	kW
P_{ab}	real power of the drive	kW
p	complex angular frequency	s^{-1}
p_{SR}	pulse number of the converter	—
Q	quotient, ratio	—
q_M	active winding crossectional area in the armature circuit of the DC servo motor	mm^2
R	resistance	Ω
R_1	input resistance	Ω
R_2	feedback resistance	Ω
R_A	total armature resistance in the feed drive	Ω
R_{A+B}	motor armature resistance	Ω
R_E	excitation circuit resistance	Ω
R_L	resistance external to the motor in the armature circuit of the drive	Ω
r	radius	mm
r_{Ri}	pinion diameter (pitch circle)	mm
r_i	actual radius	—
s	complex angular frequency	s^{-1}

s	distance, position	m
Δs	positioning increment	μm
s_0	initial distance	m
s_H	auxiliary distance for referencing the coordinate motions	m
Δs_{SpM}	shift between feed screw and ball screw nut	μm
Δs_{ZSp}	position change due to pull or thrust on the feed screw	μm
Δs_{ToSp}	position change due to torsion of the feed screw	μm
s_a	distance in axial direction	μm
Δs_a	position change in axial direction due to elasticity	μm
Δs_{aL}	position change in axial direction due to bearing elasticity	μm
T	time constant	ms
$T_{0\,mech}$	reference mechanical time constant	ms
T_{An}	response time	ms
T_{AnI}	armature circuit ramp time	ms
T_{Aus}	settling time	ms
T_E	excitation time constant	ms
T_{Last}	load cycle	s
T_T	dead time	ms
T_{el}	electrical time constant	ms
T_{elA}	electrical time constant of the feed drive	ms
T_{mechA}	mechanical time constant of the feed drive	ms
T^*_{mechA}	mechanical time constant of the drive without external moment of inertia	ms
$T_{mechA(mG)}$	mechanical time constant of the drive with gears	ms
$T_{mechA(oG)}$	mechanical time constant of the drive without gears	ms
T_{mechM}	mechanical time constant of the DC servo motor	ms
T_{nI}	integral action time of the current regulator	ms
T_{nn}	integral action time of the speed regulator	ms
T_{th}	thermal time constant	ms
T_u	delay time	ms
t	time	s
dt	time differential	—
Δt	time difference	ms
t_A	switch-off time	—
t_{Br}	braking time	ms
t_E	switch-on time	—
t_H	ramp time	ms
t_T	dead time	ms
t_V	dwell time	ms
U	voltage	V
U_{0E}	rated excitation voltage of the generator	V
U_{0G}	generator rated voltage	V
U_A	stationary armature voltage	V

Symbol	Description	Unit
U_C	stationary capacitor voltage	V
U_E	excitation voltage of the generator	V
ΔU_E	excitation potential difference on the generator	V
U_G	generator voltage	V
ΔU_G	potential difference on the generator	V
U_{HE}	auxiliary voltage for the excitation circuit	V
ΔU_{HE}	auxiliary potential difference in the excitation circuit	V
U_{di}	ideal idle voltage of the converter	V
$U_{max A}$	maximal value of the stationary armature voltage	V
U_φ	rotational angle-dependent voltage at the measuring potentiometer	V
u	input variable of a transfer element	—
$\hat{u}$	amplitude of an input variable of a transfer element	—
u_s	input step for a transfer element	—
u	time-variable voltage	V
u_A	time-variable armature voltage	V
u_s	time-variable phase voltage in the 3-phase AC system	V
$\ddot{u}_a$	overshoot width	%
$\ddot{u}_{ax}$	position overshoot width in the X-axis	mm
v	output variable of a transfer element	—
$\hat{v}$	amplitude of an output variable of a transfer element	—
v	velocity	m/s
v_B	contour velocity	m/min
Δv_B	programmable velocity change	m/min
v_{Eil}	rapid traverse feed rate	m/min
v_{Sch}	cutting feed rate	m/min
v_V	feed rate	m/min
v_f	command-variable for the contour velocity	—
v_{fx}	command value for contour velocity in the X-axis	—
v_{fy}	command value for contour velocity in the Y-axis	—
v_i	actual velocity value	—
v_i^*	actual velocity value after processing the measuring value	—
v_s	velocity command value	—
W	energy	J
W_C	energy content of capacitor	J
W_{Kin}	energy of moving masses	J
W_{LA}	energy content of the inductivity in the armature circuit	J
w	command variable in the control loop	—

X	real portion of the frequency response curve	—
X	feed axis X	—
x	control variable in the control loop	—
x	position in the X-axis	mm
Δx	position control deviation (following error) in the X-axis	mm
x_{an}	ramp distance at thread cutting	mm
x_i	actual position value in the X-axis	—
x_i^*	actual position value in the X-axis after processing the measuring value	—
$x_{i\,dir}$	actual position value in the X-axis with direct position measuring system	—
$x_{i\,in}$	actual position value in the X-axis with indirect measuring system	—
x_s	position command value for the X-axis	—
x_z	position deviation in the X-axis due to a disturbance variable	—
$x_{max\,z}$	maximal position deviation in the X-axis due to disturbance variables	—
x_φ	position in the X-axis corresponding to rotational angle φ	—
Y	imaginary portion of the frequency response curve	—
Y	feed axis Y	—
y	correcting variable in the control loop	—
y	position in the Y-axis	mm
y_s	position command value for the Y-axis	—
Z	teeth number of a timing belt drive	—
z	disturbance variable in the control loop	—
α	angular acceleration	rad/s^2
α	control angle for thyristor converters	°el
α_G	rectifier limit in the current converter	°el
α_W	inverter limit in the current converter	°el
α_0	pressure angle	°
β	angle	°

$2\varepsilon_u$	total reversing error in the mechanical transmission elements	µm
$2\varepsilon_{u1}$	reversing error between transducer and table	µm
$2\varepsilon_{u2}$	reversing error in the position control system	µm
$2\varepsilon_{u3}$	reversing error between motor shaft and transducer	µm
$2\varepsilon_{u4}$	range of insensitivity in the position controller	—
$2\varepsilon_{uR}$	reversing error due to friction and elasticity	µm
$2\varepsilon_{uS}$	back lash	µm
$2\varepsilon_{ux}$	resultant reversing error in the X-axis	µm
$2\varepsilon_{uy}$	resultant reversing error in the Y-axis	µm
η_A	drive efficiency	—
η_G	gear efficiency	—
η_{SM}	ball screw nut efficiency	—
η_{SR}	current converter efficiency	—
ϑ	temperature	°C
ϑ_A	initial temperature	°C
ϑ_E	final temperature	°C
$\Delta\vartheta_{adi}/\Delta t$	adiabatic temperature change per unit time	K/s
$\varkappa$	specific conductivity	$\dfrac{m}{mm^2 \cdot \Omega}$
μ_F	friction factor of table guides	—
μ_{SL}	friction factor of feed screw bearing	—
ν	harmonic order numbers	—
ρ	density	kg/m³
φ	phase shift	°el
φ	rotational angle at the motor shaft	°
ω	angular velocity	rad/s⁻¹
ω	angular frequency	s⁻¹
ω_0	nominal angular frequency	s⁻¹
ω_{0A}	drive nominal angular frequency	s⁻¹
ω_{0A}^*	nominal angular frequency of the uncontrolled drive	s⁻¹
$\omega_{0A(mG)}$	nominal angular frequency of the drive with gear	s⁻¹
$\omega_{0A(oG)}$	nominal angular frequency of the drive without gear	s⁻¹
ω_{0L}	nominal angular frequency of the position control loop	s⁻¹
$\omega_{0\,GrenzA}$	limit value of the nominal angular frequency of a feed drive	s⁻¹
$\omega_{0\,maxA}$	maximal nominal angular frequency of the feed drive	s⁻¹

$\omega^*_{0\,\text{maxA}}$	maximal nominal angular frequency of the drive without external moment of inertia	s^{-1}
$\omega_{0\,\text{mech}}$	nominal angular frequency of the mechanical transmission elements	s^{-1}
$\omega_{0\,\text{mech}\,1}$	nominal angular frequency of the table/screw system	s^{-1}
$\omega_{0\,\text{mech}\,3}$	nominal angular frequency of the gear/coupling system	s^{-1}
ω_{90}	90°-angular frequency	s^{-1}
ω_E	cut-off angular frequency	s^{-1}
ω_{EL}	cut-off angular frequency of the position control loop	s^{-1}
ω_G	limit angular frequency	s^{-1}
ω_R	resonance angular frequency	s^{-1}
ω_M	motor angular velocity	rad/s^2
$\omega_{\text{max}\,M}$	maximal motor angular velocity	rad/s^2
$\omega_{\text{max}\,2}$	maximal angular velocity of shaft 2	rad/s^2

8.2 SI Units

According to the law for units in the measurement system, starting January 1st, 1978, only units of the SI System (Systeme International d'Unités) are to be used (DIN 1301). The units and corresponding basic dimensions of this system are listed in table 8.1.

Related units can be derived through products and/or ratios of these units. Some of these derived SI units have special names and especially designated symbols. The units and dimensions important for electric drives are listed in table 8.2.

Table 8.1. Basic dimensions of the SI system

Basic dimension	SI basic unit	
	Name	Symbol
Length	Meter	m
Mass	Kilogram	kg
Time	Second	s
Electrical current intensity	Ampere	A
Thermodynamic temperature	Kelvin	K
Matter quantity	Mole	mol
Illumination	Candle	cd

Table 8.2 Some derived SI units for the electrical drive engineering

Dimension	SI Unit Name	Symbol	Conceptualization
Plane angle	Radian	rad	
Speed, Frequency of rotation	Reciprocal second	s^{-1}	
Angular velocity	Radian per second	rad/s	
Velocity	Meter per second	m/s	
Acceleration	Meter per second squared	m/s^2	
Angular acceleration, rotary acceleration	Radian per second squared	rad/s^2	
Force	Newton	N	$1\,N = 1\,kg \cdot m/s^2$
Torque	Newtonmeter	Nm	$1\,Nm = 1\,J = 1\,Ws$
Inertia	Kilogram meter squared	$kg\,m^2$	$1\,kg\,m^2 = 1\,Nm\,s^2$
Pressure, mechanical tension	Pascal	Pa	$1\,Pa = 1\,N/m^2$
Energy, work, heat	Joule	J	$1\,J = 1\,N \cdot m = 1\,W \cdot s$
Power, heat flow	Watt	W	$1\,W = 1\,J/s$
Angular frequency	Reciprocal second	s^{-1}	
Frequency, period frequency	Hertz	Hz	$1\,Hz = 1\,s^{-1}$
Electric potential, electric voltage	Volt	V	$1\,V = 1\,W/A = 1\,A \cdot \Omega$
Capacitance	Farad	F	$1\,F = 1\,H/\Omega^2 = 1\,s/\Omega$
Resistance	Ohm	Ω	$1\,\Omega = 1\,V/A$
Conductivity	Siemens	S	$1\,S = 1\,\Omega^{-1}$
Magnetic flux	Weber	Wb	$1\,Wb = 1\,V \cdot s$
Magnetic flux density, magnetic induction	Tesla	T	$1\,T = 1\,Wb/m^2$
Inductivity	Henry	H	$1\,H = 1\,Wb/A = 1\,\Omega \cdot s$
Celsius temperature	Celsius degree	°C	$1\,°C = 1\,K$

The fractions and multiples of SI units derived by multiplication by a factor of $10^{\pm 1}$, $10^{\pm 2}$, $10^{\pm 3k}$ (where k 1,2...6), are given special names and symbols. The most common of these prefixes are presented in table 8.3.

There are other units, outside the SI, which may also be used. Of these, the dimensions and units used in drive engineering are shown in table 8.4. Note that these last units cannot be used with the prefixes of table 8.3.

Table 8.3 Prefix symbols for SI units

10^{-12}	pico	p
10^{-9}	nano	n
10^{-6}	micro	µ
10^{-3}	milli	m
10^{-2}	centi	c
10^{-1}	deci	d
10^{1}	deca	da
10^{2}	hecto	h
10^{3}	kilo	k
10^{6}	mega	M

Table 8.4 Some units permitted in drive engineering

Dimension	Unit		Conceptualization
	Name	Symbol	
Plane angle	Degree	°	$1° = (\pi/180)$ rad
Time	Minute	min	1 min = 60 s
Speed, frequency of rotation	Reciprocal minute	$\min^{-1}$	
Velocity	Meter per minute	m/min	1 m/min = 0.01667 m/s
Mass	Ton	t	$1 \text{ t} = 10^3$ kg

8.3 Conversion Tables

The following tables show conversions of units of the earlier technical measurement system or of American/English units, into units of the SI.

Table 8.5   Conversion of length units

	µm	mm	m	mil	in	ft
µm	1	10^{-3}	10^{-6}	0.03937	$3.937 \cdot 10^{-5}$	$3.281 \cdot 10^{-6}$
mm	10^3	1	10^{-3}	39.37	$3.937 \cdot 10^{-2}$	$3.281 \cdot 10^{-3}$
m	10^6	10^3	1	$3.937 \cdot 10^4$	39.31	3.281
mil	25.4	$2.54 \cdot 10^{-2}$	$2.54 \cdot 10^{-5}$	1	0.001	$8.33 \cdot 10^{-5}$
in	$2.54 \cdot 10^4$	25.4	$2.54 \cdot 10^{-2}$	1,000	1	$8.33 \cdot 10^{-2}$
ft	$3.048 \cdot 10^5$	$3.048 \cdot 10^2$	0.3048	$1.2 \cdot 10^4$	12	1

Table 8.6   Conversion of mass units

	g	kg	t	oz	lb	British (long) ton	US (short) ton
g	1	10^{-3}	10^{-6}	$3.53 \cdot 10^{-2}$	$2.205 \cdot 10^{-3}$	$0.984 \cdot 10^{-6}$	$1.102 \cdot 10^{-6}$
kg	10^3	1	10^{-3}	35.3	2.205	$0.984 \cdot 10^{-3}$	$1.102 \cdot 10^{-3}$
t	10^6	10^3	1	$3.53 \cdot 10^4$	$2.205 \cdot 10^3$	0.984	1.102
oz	28.35	$2.84 \cdot 10^{-2}$	$2.84 \cdot 10^{-5}$	1	$6.25 \cdot 10^{-2}$	$2.79 \cdot 10^{-5}$	$3.125 \cdot 10^{-5}$
lb	$4.54 \cdot 10^2$	0.454	$4.54 \cdot 10^{-4}$	16	1	$4.46 \cdot 10^{-4}$	$5 \cdot 10^{-4}$
British (long) ton	$1.016 \cdot 10^6$	$1.016 \cdot 10^3$	1.016	$3.58 \cdot 10^4$	2240	1	1.12
US (short) ton	$9.07 \cdot 10^5$	907	0.907	$3.2 \cdot 10^4$	2000	0.893	1

Table 8.7 Conversion of force units

	N	kN	p	kp	ozf	lbf	ltonf
N	1	10^{-3}	$1.02 \cdot 10^2$	0.102	3.597	0.2248	$1.004 \cdot 10^{-4}$
kN	10^3	1	$1.02 \cdot 10^5$	$1.02 \cdot 10^2$	$3.597 \cdot 10^3$	$2.248 \cdot 10^2$	0.10036
p	$9.807 \cdot 10^{-3}$	$9.807 \cdot 10^{-6}$	1	10^{-3}	$3.53 \cdot 10^{-2}$	$2.205 \cdot 10^{-3}$	$9.842 \cdot 10^{-7}$
kp	9.807	$9.807 \cdot 10^{-3}$	10^3	1	35.274	2.2046	$9.842 \cdot 10^{-4}$
ozf	0.278	$2.78 \cdot 10^{-4}$	28.35	$2.835 \cdot 10^{-2}$	1	$6.25 \cdot 10^{-2}$	$2.79 \cdot 10^{-5}$
lbf	4.448	$4.45 \cdot 10^{-3}$	$4.53 \cdot 10^2$	0.4536	16	1	$4.46 \cdot 10^{-4}$
ltonf	$9.964 \cdot 10^3$	9.964	$1.016 \cdot 10^6$	1016	$3.58 \cdot 10^4$	2240	1

Table 8.8 Conversion of torque units

	N cm	Nm	kp m	oz·in	lbf·in	lbf·ft
N cm	1	10^{-2}	$1.02 \cdot 10^{-3}$	1.42	$8.85 \cdot 10^{-2}$	$7.38 \cdot 10^{-3}$
Nm	10^2	1	0.102	142	8.85	0.738
kp m	981	9.81	1	$1.389 \cdot 10^3$	86.8	7.23
oz·in	0.706	$7.06 \cdot 10^{-3}$	$7.2 \cdot 10^{-4}$	1	$6.25 \cdot 10^{-2}$	$5.2 \cdot 10^{-3}$
lbf·in	11.3	0.113	$1.15 \cdot 10^{-2}$	16	1	$8.33 \cdot 10^{-2}$
lbf·ft	135.6	1.356	0.138	192	12	1

When converting inertia units into other units according to table 8.9, it should be noted that the dimensions have been differently defined:

▷ moment of inertia = mass × inertia radius^2,
▷ flywheel effect = force × inertia diameter^2,
▷ flywheel effect = force × inertia radius^2.

Physical dimensions, formula symbols, and units are interrelated, and cannot be interchanged for other dimensions and units. For this reason, when calculating with these dimensions, their corresponding formulas must be used. In the formulas given in chapters 1 through 6, the calculations use exclusively the moment of inertia.

Table 8.9
Conversion of moments of inertia, and relation to other dimensions (see remarks in the text, page 354)

	Moment of inertia		Polar moment of inertia			Flywheel effect GD^2		Flywheel effect	
	$kg\,m^2\,(Nm\,s^2)$	$kp\,ms^2$	$lbf\cdot ft\cdot s^2$	$lbf\cdot in\cdot s^2$	$ozf\cdot in\cdot s^2$	Nm^2	$kp\,m^2$	$lbf\cdot ft^2$	$lbf\cdot in^2$
$kg\,m^2\,(Nm\,s^2)$	1	0.102	0.738	8.85	141.6	39.2	4	23.73	$3.42\cdot 10^3$
$kp\,ms^2$	9.81	1	7.23	86.8	1389	385	39.2	232	$3.35\cdot 10^4$
$lbf\cdot ft\cdot s^2$	1.355	0.138	1	12	192	53	5.43	32.15	$4.63\cdot 10^3$
$lbf\cdot in\cdot s^2$	0.113	0.015	0.0833	1	16	4.43	0.452	2.68	386
$ozf\cdot in\cdot s^2$	$7.06\cdot 10^{-3}$	$7.2\cdot 10^{-4}$	$5.21\cdot 10^{-3}$	$6.25\cdot 10^{-2}$	1	0.278	$2.82\cdot 10^{-2}$	$1.68\cdot 10^{-1}$	24.13
Nm^2	$2.55\cdot 10^{-2}$	$2.6\cdot 10^{-3}$	$1.88\cdot 10^{-2}$	$2.03\cdot 10^{-1}$	3.52	1	0.102	0.605	87.2
$kp\,m^2$	0.25	$2.55\cdot 10^{-2}$	0.184	2.21	35.4	9.81	1	5.94	854
$lbf\cdot ft^2$	$4.21\cdot 10^{-2}$	$4.3\cdot 10^{-3}$	$3.1\cdot 10^{-2}$	0.363	5.97	1.65	0.168	1	144
$lbf\cdot in^2$	$2.93\cdot 10^{-4}$	$2.98\cdot 10^{-5}$	$2.16\cdot 10^{-4}$	$2.6\cdot 10^{-3}$	$4.14\cdot 10^{-2}$	$1.15\cdot 10^{-2}$	$1.19\cdot 10^{-3}$	$6.94\cdot 10^{-3}$	1

8.4 Equations

For the presentation of the physical relationships of chapters 1 through 6, the following equations have been used:

> dimensional equations and
> adjusted dimensional equations.

Numerical value equations have not been used (see DIN 1313). The dimensional symbols for values are presented in italics, e.g.:

> length l, velocity v.

Regular letters are used for the symbols of the units, e.g.:

> m, s, N.

The dimensional symbol represents a value in dimensional equations:

> value = numerical value × unit

Numerical values and units are treated as independent factors, and in calculations they must be taken into account up to the point where a result has been obtained. Units can be multiplied, reduced, or substituted with other units. Dimensional equations are thus independent of the units used in the presentation of the given values.

Example:

Equation (2.1) for the determination of the angular velocity, states:

$$\omega = 2\pi \cdot n$$

If we give the speed n in the unit $\min^{-1}$, we obtain for $n = 1{,}000\ \min^{-1}$:

$$\omega = 2\pi\ \text{rad} \cdot 1000\ \min^{-1} = 2\pi\ \text{rad} \cdot 1000\ \min^{-1} \cdot \frac{1\ \min}{60\ \text{s}}$$

$$= 2\pi\ \text{rad} \cdot \frac{1000}{60}\ \text{s}^{-1} = 104{,}7\ \text{rad/s}$$

The unit rad is equivalent to 1.

According to equation (2.41), the angular acceleration is:

$$\alpha = \frac{d\omega}{dt}$$

For $\omega = 104.7$ rad/s, and $dt = 100$ ms, we obtain:

$$\alpha = \frac{104.7\ \text{rad/s}}{100\ \text{ms}} = \frac{104.7\ \text{rad/s}}{100\ \text{ms}} \cdot \frac{1\ \text{ms}}{10^{-3}\ \text{s}} = 1{,}047\ \text{rad/s}^2 = 1{,}047\ \text{s}^{-2}$$

In adjusted dimensional equations, every value is divided by the unit valid for that particular equation. These ratios thus produce numerical values which must be used for the calculation of the equation given. In this manner, the adjusted dimensional equations can be manipulated as would numerical equations, but become difficult to grasp when the calculations are extensive.

Example:

Equation (2.35) expresses the ramp time as

$$\frac{t_H}{s} = \frac{\dfrac{J_{Ges}}{kg\,m^2} \cdot \dfrac{\Delta n_M}{min^{-1}}}{9.55 \cdot \dfrac{M_B}{Nm}}$$

In order to keep the ramp time in units of s, we must substitute:

the moment of inertia in $kg\,m^2$,
the speed difference in min^{-1},
the acceleration torque in Nm.

Index

Absolute value, frequency response curve 26, 29
acceleration 108, 119
– controlled command value generation 272
– force 219
–, maximal 109, 164, 276
– process 109
– torque 77, 104
actual acceleration 276
– position value 36
adaptive speed control 169
adiabatic temperature change slope 117
aerostatic guide 239, 244
amplitude drop 32
– ratio 29
– response curve 29
analog position control 38
angular frequency 29
–, complex 27
–, imaginary 26
–, 90°- 46
angular velocity 76
–, motor 77
anti-parallel circuit 146
–, 2-pulse circulating current-conducting 159
–, 3-pulse circulating current-conducting 153
–, 6-pulse circulating current-free 157
armature circuit inductivity 76, 77
– circuit resistance 76, 77
auxiliary acceleration 279
– coordinates system 16
– distance 276
axial-radial bearing, combined 223

Ball circuit 220, 221
– screw nut 220
basic dimension 350
– transfer element 18
– units 350
bearing, mounting parts 226

belt wrap, timing belt 257
bending strength, gear wheels 233
–, shafts 231
binary signal 14
block diagram 13
– symbol 23
Bode diagram 29
break point 271
break away torque 153, 192
brush life time 74
buckling stiffness, feed screw 218

Characteristic curve, linear 15
–, non-linear 15
–, stationary 14
chatter marks 188
– oscillation 188, 242
circulating current 150
– – conducting converter 135, 150
– – suppression 135, 162
– – -free converter 135, 149
coefficient of thermal expansion 139
command frequency response curve 35
–, drive 85
command stage 149
command value 13
–, acceleration 276
– modification 271
–, position control loop 45, 52
command response, uncontrolled drive 83
commutating limit 73
continuous duty S1 74
contour deviation 51, 304
–, circle 63
contour distortion 273
control 13
– angle 147
– deviation 13
– device 13
control characteristic curve, converter 147
control loop 13
– element 13

–, structure   163
– system   13
– variable   13
control range, converter   147
conversion tables   353
coordinate system   16
corner angle   61
– deviation   61, 275
correcting variable   13
counterbalance   238
countervoltage, induced   76
couplings   266
cross connection   147
–, 6-pulse circulating current conducting   151
current converter   134
–, characteristics   172
–, selection   174
current limit   102, 164, 301
current ripple   137, 302
current take-over   166
cut-off angular frequency   45
– position control loop   47
cutting feed rate   59
– force   100, 123

Damping correction value   186
damping gradient   31
–, drive   44, 86, 299, 310
–, mechanical   185, 245
–, position control loop   45, 49
DC bus   145
DC shunt motor   71
DC servo motor, computation   96
–, electrically excited   74, 331
–, permanent magnet excited   71, 325
dead band   189
deadtime, drive system   95, 136
– element   19, 22
–, influence (effects)   94
–, medium   95, 136
–, statistical   136
delay time   24
D-element, differential element   19, 21
demagnetization   72, 114
differential equation, drive   76
differential element   19, 21
– equation   18, 19
digital position control   38
digital signal   14

dimensional equation   356
–, adjusted   356
direct position control   40, 198
disturbance frequency response curve   35
–, position control loop   67
disturbance response behavior   66
drive, command frequency response curve   85
– comparison   121
–, damping gradient   44, 86, 299, 310
–, electrical time constant   78, 84
–, frequency response curve   42
–, limit nominal angular frequency   87
–, maximal nominal angular frequency   86
–, mechanical time constant   79, 84, 120, 259
–, nominal angular frequency   44, 86, 300
–, nominal frequency   303
–, non-stationary operation   70, 102
–, stationary characteristic   80
–, stationary operation   70, 97
–, with feed screw   97
–, with rack   99
drive system, dead time   95, 136
duty cycle   115
dwell time   276
dynamic behavior, controlled drive   84
–, uncontrolled drive   82
dynamic load, drive   102

Effective torque   111, 132
efficiency, drive   139
–, feed screw   215
elasticity modulus   209
electrical time constant, drive   78, 84
electro-motive force   77
energy content   118
equations   356
equivalent bending strength   231
external inertia   76, 127
–, effect of   92, 314

Feed force   38
feed rate   123, 126
– screw bearing   223
– screw drive   97, 123, 179, 205
– screw lead   216
– table mass   123
feedback circuit, transmission elements   33

359

ferrite material 73
final deviation 24
firing angle 147
firing angle slope 167
first order delay element, PT_1 19, 20, 28
flux concentration 72
following error 41
form factor 137, 302
frequency response curve 26
–, absolute value 26
–, command 35
–, disturbance 35
–, drive 42, 302, 310
–, equation 26
–, measured 310
–, open control loop 35
–, position control 41
frequency range 27
friction 190, 296
–, bearing 227, 228
–, dry 192
friction factor 98, 123
friction guide 240
friction reversing error 192
function, steady 14
–, unsteady 15
fundamental oscillation 138

Gain 15
–, normalized 16
–, operating point dependent 15
gear 247
–, for maximum acceleration 119
gear range 123, 259
gib 240
guide, aerostatic 239, 244
–, friction 239, 240
–, hydrostatic 239, 243
–, roller 239, 241

Harmonic oscillation current 138
heat behavior 111
hydrostatic guide 239, 243
hysteresis element 94

Idle speed 82
idle voltage, ideal 148
idler pulleys 256
I-element, integral element 19, 21, 43
imaginary part 26

indirect position control 39, 195
induced countervoltage 76
inertia, at gear ratio 205, 251
–, calculation 202
–, cylinder 202
–, feed screw 213
–, linearly moved mass 127, 204
–, timing belt drive 258
–, total 76
informations 13
input value 13
integral element, I-element 19, 21, 43
intermittent duty S3 74
inverter limit 149
inverter mode 147

K_v-factor 41

Lead accuracy, feed screw 216
limit value 24
limit acceleration 284
limit angular frequency 45
limit nominal angular frequency, drive 87
line frequency 136
line synchronized thyristor controller 146, 336
linear acceleration 119
linear characteristic 15
linearization 15
load cycle 111
load period 111
load torque 77, 97, 126
losses, gear 99
lubrication, bearing 227

Machine table (slide) 238
machining force 100, 123
mass force 98, 238
mass oscillator second order 50
maximum acceleration 109, 164, 276
maximum nominal angular frequency, drive 86
measurements, mechanical transmission elements 289
–, position control loop 304
–, speed control loop 298
measuring procedure, cyclic-absolute 39
measuring system, analog 38
–, digital 38

mechanical nominal angular frequency 180, 295
–, transmission elements 38, 177
mechanical time constant, drive 78, 84, 120, 259
mode of operation 115
modified dimensional equation 356
modulus of elasticity 209
modulus of shearing 209
monitor 168
motor constant 77
motor torque 76

Negative feedback (reaction), transfer elements 33
nominal angular frequency 28, 31
–, drive 44, 86, 300
–, mechanical 180, 295
–, position control loop 45
nominal frequency, mechanical 184
non-linear characteristic curve 15
non-linearity, insensitivity range 50
–, position control loop 50
–, reversing error 50
non-pulsating mode 149
non-stationary operation, drive 70, 102
–, uncontrolled drive 82
normal form, complex number 26

Oil film 243
open loop control 13
operating point 15
optimization criteria 118, 186
–, speed regulator 316
oscillating element 31
oscillation phenomenon 40, 201
output value 13
overload behavior 112
overshoot width 24, 47, 61, 275, 299
overtemperature 116

Parallel circuit, transfer elements 33
parallel offset, actual contour 197
P-element, proportional element 18, 19
phase angle 26, 29
phase response curve 29
phase shift 27
pitch error, thread cutting 57
pole shoe 72
position command value 36
– frequency response curve 45, 52
– smoothing 271

position control 36
– control deviation 41
–, measurements 304
– measuring system 38
position control loop 36
– controller 37
–, cut-off angular frequency 47
–, damping gradient 45, 49
–, direct 198, 40
–, disturbance frequency response curve 67
–, frequency response curve 41
–, gain 41, 44, 306
–, indirect 39, 195
–, linear 44
– measurement 38
–, nominal angular frequency 45
–, non-linear 50
–, requirements 67
position command generation, acceleration controlled 272
–, velocity controlled 272
position increments, smallest 308
position measuring procedure absolute 39
–, incremental 39
positioning 53
– process 109
– response 47, 309
positive feedback, transfer elements 33
power 140
–, transmission (timing belt) 257
–, theoretical (timing belt) 257
preload 231, 241
preload, bearing 224
–, electrical 237
– factor 113
–, feed screw 208
– force 221, 225
–, gear wheels 254
–, pinion 236
–, threaded nut 221
pressure angle 231
proportional element, P-element 18, 19
–, with first order delay, PT_1-element 19, 20, 28
–, with second order delay, PT_2-element 19, 20, 30
pulsating current range 149
– mode 149
pulse frequency 136
– width control 41

361

2-pulse circulating current-conducting anti-parallel circuit 159
3-pulse circulating current anti-parallel connection 153
6-pulse, circulating current-conducting cross connection 151
– circulating current-free anti-parallel circuit 157

4-Quadrant drive 134

Rack and pinion drive 99, 123, 179, 229, 99, 123, 179, 229
radial stiffness, shaft bearing 231
ramp distance 57
ramp-up time 104, 127
rapid traverse 123, 126
rare earth elements 73
rated torque 73
reaction of the mechanical transmission elements 184
reaction time 146, 159
real component 26
reciprocal channel system 221
reciprocal tube system 220
reciprocal system, axial 221
rectifier limit 149
rectifier mode 147
referenced values 18
referenced time constant, mechanical 259
referenced control area, squared 186
resonance angular frequency 46
resonance phenomenon 138
– point 181
– ratio 49
response time 24, 102, 299
reversing error 189, 254
reversing error, total 189, 194, 290
roller guide 241

Second order delay element, PT_2 19, 20, 30
serial connection, transmission elements 32
settling time 24
shear modulus 209
shell magnet 72
– magnet principle 72, 316
shift, axial (ball nut) 222
short-circuit torque, drive 78, 81
–, motor 25

SI units 350
signal 13
–, analog 14
–, binary 14
–, digital 14
single mass oscillator 180
speed, arithmetic mean 112
– control 84
–, critical 217
–, motor 77
– setting range 70
– stiffness 70, 80
speed control, adaptive 169
speed control loop, measurements 298
spring constant 184
squared reference control area 186
squeeze film effect 244
stationary, characteristic curve 14
– systems behavior 14
stationary load, drive 97
–, uncontrolled drive 80
stationary operation, drive 70, 97
–, uncontrolled drive 80
status monitor 168
steady function 14
step function 23, 299
step response 23
stick-slip effect 153, 156, 192, 244
stiffness, bearing 224
–, feed screw 209
structure, control loops 163
substitute time constant 163
substitute torsional stiffness 230
subordinated current control 165
summation point 33
symbol 339
symmetrical optimum 163

Table (slide) guide 239
take-over current limit 166
tension rollers 255
thermal time constant 116
thread cutting 56
thyristor controller, line synchronized 146, 336
time range 27
time constant, drive, mechanical 79, 84, 120, 259
–, thermal 116
toothed belt gear 255, 311

362

toothed belt, wrap angle   257
–, transmission power   257
tooth wheel gear   250
torque constant   77
torque, effective   111
–, for friction, losses   97, 126
– for machining force   101, 126
–, slide guide   98, 99
–, spindle bearing   98
torsional resonance frequency   211
torsional stiffness   209
–, shafts   230
total inertia   127
total reversing error   189, 194, 290
total stiffness   184, 206, 229, 295
tractor roller bearing   241
transfer function   27
– response   13
transient response   22
– response function   23, 299
– response tolerance   23

transistor DC chopper   141, 334
transmission elements, measurements   289
–, mechanical   38, 177
transmission power, timing belt   257
two way converter   147

Unit   351
unsteady function   15

Value   356
–, referenced   18
velocity controlled position command value generation   272
voltage constant   77

Waviness (rippling), actual contour   197
work piece mass   123
working range   15
worm gear   179